AUTOMATION OF WASTEWATER TREATMENT FACILITIES

Prepared by Automation of Wastewater Treatment Facilities Task Force of the Water Environment Federation

Dr. Bob Hill, P.E., *Chair*	Salil M. Kharkar, P.E.
Richard G. Atoulikian	Phil Kiser
Charles K. Aycock	Ryan Kowalski, P.E.
Edward Baltutis	Robert Lagrange, Ph.D.
Peter Brady, B.E.	David Olson, P.E.
James Conboy, P.E., BCEE	Tony M. Palmer
Peter R. Craan, P.E., CAP	Albert C. Petrasek, Jr., Ph.D., P.E.
Alex Ekster, Ph.D., P.E., DEE	William Phillips, P.E.
Richard V. Frykman, P.E.	Ignasi Rodriguez-Roda, Ph.D.
Robert A. Gillette, P.E., DEE	Maureen C. Ross
John D. Haase	Bob Rutemiller
Gene Heyer, P.E., PMP	Andrew R. Shaw, P.E.
Sam Irrinki, P.E.	David A. Vaccari, Ph.D., P.E., DEE
D. Scott Joslyn, P.E.	Troy D. Vassos, Ph.D., P.Eng.
John C. Kabouris, Ph.D., P.E.	John Weber

Under the Direction of the Municipal Subcommittee of the Technical Practice Committee

2006

Water Environment Federation
601 Wythe Street
Alexandria, VA 22314-1994 USA
www.wef.org

AUTOMATION OF WASTEWATER TREATMENT FACILITIES

WEF Manual of Practice No. 21
Third Edition

Library Commons
Georgian College
One Georgian Drive
Barrie, ON
L4M 3X9

WEF Press

Water Environment Federation Alexandria, Virginia

McGraw-Hill

New York Chicago San Francisco Lisbon London Madrid
Mexico City Milan New Delhi San Juan Seoul
Singapore Sydney Toronto

TD741 .A87 2007

0134110 300 776

Automation of wastewater treatment facilities c2007.

2007 10 17

The McGraw-Hill Companies

Copyright © 2007 by the Water Environment Federation. All rights reserved. Printed in the United States of America. Except as permitted under the United States Copyright Act of 1976, no part of this publication may be reproduced or distributed in any form or by any means, or stored in a data base or retrieval system, without the prior written permission of the publisher.

Water Environment Research, WEF, and WEFTEC are registered trademarks of the Water Environment Federation.

1 2 3 4 5 6 7 8 9 0 DOC/DOC 0 1 2 1 0 9 8 7 6

ISBN-13: 978-0-07-147937-0
ISBN-10: 0-07-147937-6

The sponsoring editor for this book was Larry S. Hager and the production supervisor was Pamela A. Pelton. It was set in Palatino by Lone Wolf Enterprises, Ltd. The art director for the cover was Anthony Landi.

Printed and bound by RR Donnelley.

This book is printed on acid-free paper.

McGraw-Hill books are available at special quantity discounts to use as premiums and sales promotions, or for use in corporate training programs. For more information, please write to the Director of Special Sales, McGraw-Hill Professional, Two Penn Plaza, New York, NY 10121-2298. Or contact your local bookstore.

IMPORTANT NOTICE

The material presented in this publication has been prepared in accordance with generally recognized engineering principles and practices and is for general information only. This information should not be used without first securing competent advice with respect to its suitability for any general or specific application.

The contents of this publication are not intended to be a standard of the Water Environment Federation (WEF) and are not intended for use as a reference in purchase specifications, contracts, regulations, statutes, or any other legal document.

No reference made in this publication to any specific method, product, process, or service constitutes or implies an endorsement, recommendation, or warranty thereof by WEF.

WEF makes no representation or warranty of any kind, whether expressed or implied, concerning the accuracy, product, or process discussed in this publication and assumes no liability.

Anyone using this information assumes all liability arising from such use, including but not limited to infringement of any patent or patents.

Water Environment Federation

Improving Water Quality for 75 Years

Founded in 1928, the Water Environment Federation (WEF) is a not-for-profit technical and educational organization with members from varied disciplines who work toward the WEF vision of preservation and enhancement of the global water environment. The WEF network includes water quality professionals from 79 Member Associations in more than 30 countries.

For information on membership, publications, and conferences, contact

Water Environment Federation
601 Wythe Street
Alexandria, VA 22314-1994 USA
(703) 684-2400
www.wef.org

Manuals of Practice of the Water Environment Federation

The WEF Technical Practice Committee (formerly the Committee on Sewage and Industrial Wastes Practice of the Federation of Sewage and Industrial Wastes Associations) was created by the Federation Board of Control on October 11, 1941. The primary function of the Committee is to originate and produce, through appropriate subcommittees, special publications dealing with technical aspects of the broad interests of the Federation. These publications are intended to provide background information through a review of technical practices and detailed procedures that research and experience have shown to be functional and practical.

Water Environment Federation Technical Practice Committee Control Group

B. G. Jones, *Chair*
A. B. Pincince, *Vice Chair*

S. Biesterfeld
R. Fernandez
L. Ford
S. S. Jeyanayagam
Z. Li
M. D. Moore
M. D. Nelson
S. Rangarajan
J. D. Reece
E. P. Rothstein
A. T. Sandy
J. Witherspoon

Contents

Preface xxix

Chapter 1 Introduction
What Is Complete Automation Design? 1
Motivation for Updating This Manual 2
Technology Now a Strategy, Not a Limitation 3
What This Manual Does and Does Not Include 4
Intended Audience 4
Manual of Practice Chapter Contents 4
References 6

Chapter 2 The Business Case for Automation
The Benefits of Automation 8
    Meet Regulatory Requirements 8
    Improve Process Performance and Reliability 8
    Record Data and Create Reports 9
    Save Chemicals and Energy 9
    Save Labor 9
    Reduce Risks 10
    Ensure a Good Night's Sleep 10
The Cost of Automation 10
Return on Investment—Why We Do Projects 10
    Payback Period 11
    Cost–Benefit Ratio 11
    Rate of Return 13
Ten Keys to a Successful Project 13
    System Advocate 13
    Management Support 14
    Owner Involvement 15

Operator Involvement 15
Stakeholder Involvement 16
Maintenance Plan 16
Full System Testing 16
Contingency Plans 17
Training 17
Goals 18
References 18

Chapter 3 What Is a Complete Automation Design?

The Design Process 20
   Project Structure 20
   The Predesign Phase 23
   The Detailed Design Outline Phase 24
   The Final Design Phase 24
   Design Documents 25
   Process and Instrumentation Diagrams 25
   Elementary Wiring Diagrams 25
   Process-Control Narratives 25
      Tag Numbers 28
      Monitored Variables 28
      Valve-Position Switches and Indicators 29
      Alarms 29
      Reset Function 30
      Remote Indicators 31
      Local Pushbutton Stations 31
      Uninterruptible Power Supplies 31
      Vendor-Supplied Packaged-Equipment Panels and Controls 31
   Process-Control-System Configuration and Communication
   Schematics 32
   Control-Panel Designs 32
   Instrumentation- and Panel-Location Drawings 32
   Instrument Loop Diagrams 32
   Instrumentation Installation Details 33
   Instrument and Input/Output Lists 33
   Specifications and Bid Documents 33
Sources of Technical Design Information 33
   Technical Societies 33
   Vendors 34
      Selecting Vendors and Technologies 35
      Incorporating Selected Technologies into Design 35
   Training 35

    Web Sites 36
    Books and Periodicals 36
    Previous Design Experience 36
Post-Design Phase 36
    Design Services during Construction 36
        Reviewing Shop Drawings 36
        Processing Change Orders................................ 37
        Responding to Requests for Information 37
        Meeting with Contractors and Suppliers 37
        Making Regular Site Visits 37
            Witnessing Factory Acceptance Tests 37
            System Performance 38
            Hardware Testing 38
            Software Testing 38
            Simulation Software 39
            Operator Interface Terminals 40
    Construction Management 41
        Field Tests 41
            Loop Testing 41
            Integrated System Testing 41
        Final Acceptance 42
References 42
Suggested Readings 44

Chapter 4 Process and Instrumentation Diagrams

Introduction 48
How To Create Process and Instrumentation Diagrams 49
    Process-Flow Diagrams 50
    Symbols 51
    Computer-Aided Design Software 53
    Interactions with Other Design Documents 54
        Instrument List 54
        Input/Output List 54
        Panel List 55
        Cable and Conduit List 55
        Process-Control Narratives 55
        Panel and Field-Instrument Specifications 55
        Electrical Schematics 56
        Instrument Loop Diagrams 56
        Equipment-Location Drawings 56
        Control-System Architecture Diagram 56
Process and Instrumentation Diagrams as Communication Tools 57

    Contracts 57
    User and Owner Feedback 57
    Operator Training 58
    Operator-Interface Graphics 58
    As-Built Documentation 59
New Developments 59
    Intelligent Process and Instrumentation Diagrams 59
        Streamlined Diagram Development 60
            Object-Based Schematics 60
            Modular System 61
            Integration with External Data Sources 61
            Interoperability 61
            Symbol-Management Tools 61
            Review and Navigation Tools 61
            Component-Management Tools 61
        Automatically Generated Lists. 62
        Design History for As-Built Drawings. 62
        Construction, Startup, Operation, Maintenance, and Asset Management Uses. . . 62
    Process and Instrumentation Diagrams for HVAC Systems 62
References 63
Suggested Reading 64

Chapter 5 General Characteristics
Background 66
Instrument Classifications 67
    Signal Types 67
    Sampling Method 68
        Point Sensors 68
        Nonpoint Sensors 68
    Measurement Method 69
Instrument Characteristics 69
    Properties 69
        Range 69
        Span 70
        Rangeability. 70
        Overrange 70
        Zero Suppression, Elevation, and Offset 70
        Damping 70
    Static Characteristics 71
        Zero error 72
        Span error 72
        Hysteresis or hysteretic error. 73

Dead band	73
Linearity	74
Drift	74
Repeatability	75
Sensitivity	75
Dynamic Characteristics	77
Value Conversions	77
Linear Value Conversion	78
Nonlinear Value Conversion	79
Nonlinear, Non-zero-based Conversion	80
Environmental Concerns and Design Considerations	82
Temperature	82
Moisture	83
Short-circuits	83
Corrosion	83
Swelling	83
Corrosion	84
Responses to Environmental Concerns	84
Enclosures and Electronic Standards	84
Enclosed Spaces	86
Temperature Controls	87
Fire Ants	87
Testing and Quality Assurance	87
Need for Instrumentation Testing	87
Instrumentation Test Reports	88
Standards	91
Sources and Types	92
Calibration Standards	93
Primary Standards	93
Traceable Standards	93
Portable Standards	93
Internal Electronics	93
Terminal Strip	94
Power Supply	94
Printed Circuit Cards	94
Connectors	94
Calibration Points	94
Display	94
Electrical Concerns	94
Voltage Spikes	94
Transient-Voltage Surge Suppressors	95
Grounding	96

Shielding	96
Smart Process Instrumentation	97
Documentation	98
Suggested Readings	98

Chapter 6 Sensors

Sensor Characteristics	105
Wastewater Characteristics	105
Plastic Products	106
Hair and Fibers	106
Solids	106
Grease	106
Color	107
Sulfides and Hydrogen Sulfide	107
Diurnal Changes	107
Flow Meters	107
Magnetic	108
Operating Principles	108
Construction Materials	110
Accuracy and Precision	110
Installation	112
Maintenance Requirements	114
Ultrasonic	115
Operating Principles	116
Construction Materials	116
Accuracy and Precision	116
Installation	119
Maintenance Requirements	121
Weirs and Flumes	121
Operating Principles	121
Construction Materials	127
Accuracy and Precision	127
Installation	127
Maintenance Requirements	128
Differential-Pressure	129
Operating Principles	129
Construction Materials	130
Accuracy and Precision	130
Installation	131
Maintenance Requirements	131
Mechanical	132
Rotary-Element	132

  Operating Principles 132
  Construction Materials 134
  Accuracy and Precision 134
  Installation 134
  Maintenance Requirements 135
 Variable-Area 136
  Operating Principles 136
  Construction Materials 136
  Accuracy and Precision 137
  Installation 137
  Maintenance Requirements 137
Mass 137
 Coriolis 138
  Operating Principles 138
  Construction Materials 139
  Accuracy and Precision 139
  Installation 140
  Maintenance Requirements 140
 Thermal-Dispersion 140
  Operating Principles 140
  Construction Materials 140
  Accuracy and Precision 141
  Installation 141
  Maintenance Requirements 141
Level-Measurement Devices 141
 Continuous 141
  Bubbler 142
   Operating Principles 143
   Construction Materials 144
   Accuracy and Precision 144
   Installation 145
   Maintenance Requirements 146
  Capacitance and Impedance 146
   Operating Principles 146
   Construction Materials 149
   Accuracy and Precision 149
   Installation 149
   Maintenance Requirements 150
  Differential-Pressure 150
   Operating Principles 150
   Construction Materials 153
   Accuracy and Precision 153

Contents

    Installation 153
    Maintenance Requirements 155
   Sonic and Ultrasonic 156
    Operating Principles 156
    Construction Materials 158
    Accuracy and Precision 158
    Installation 158
    Maintenance Requirements 159
   Microwave (Radar) 160
    Operating Principles 160
    Construction Materials 161
    Accuracy and Precision 161
    Installation 161
    Maintenance Requirements 162
  Point 164
   Operating Principles 164
   Construction Materials 165
   Installation 165
   Maintenance Requirements 165
Pressure 166
 Operating Principles 166
 Construction Materials 167
 Accuracy and Precision 167
 Installation 167
 Maintenance Requirements 168
Temperature 168
 Thermocouple 168
 Resistance 171
 Thermistor 172
 Thermal Bulb 172
 Accuracy and Precision 172
 Installation 173
On-line Analyzers 174
 Dissolved Oxygen 175
  Typical Membrane Sensor Design 175
   Conventional Galvanic Measuring Cell 176
   Clark Polarographic Measuring Cell 177
   Ross Polarographic Measuring Cell 179
  Zullig Sensor 180
  Luminescent Dissolved Oxygen Sensor 180
  Accuracy and Precision 181
  Installation 181

Contents

Maintenance Requirements	181
Chlorine Residual	182
DPD Colorimetric Method Analyzer	183
Amperometric Bare-Electrode Analyzer	186
Amperometric Membrane-Covered-Electrode Analyzer	186
Oxidation–Reduction Potential Probes	187
Accuracy and Precision	188
Installation	188
Maintenance Requirements	189
Polymer	189
Operating Principles	190
Construction Materials	191
Installation	192
Maintenance Requirements	192
Nitrogen	192
Ammonia and Ammonium	192
Operating Principles	193
Construction Materials	194
Accuracy and Precision	194
Installation	194
Maintenance Requirements	194
Nitrate and Nitrite	195
Operating Principles	195
Construction Materials	196
Accuracy and Precision	196
Installation	196
Maintenance Requirements	196
pH	196
Operating Principles	197
Typical pH Sensor	197
Ion-Sensitive Field Effect Transistor	199
Accuracy and Precision	202
Installation	202
Phosphorus and Orthophosphate	204
Orthophosphate	204
Operating Principles	204
Construction Materials	204
Accuracy and Precision	204
Installation	204
Maintenance Requirements	204
Total Phosphorus	205
Operating Principles	205

    Construction Materials 205
    Accuracy and Precision 205
    Installation 205
    Maintenance Requirements 205
  On-Line Solids Measurement 205
    Operating Principles 207
      Near-Infrared Analyzers 207
      Microwave Analyzers 207
    Construction Materials 210
    Accuracy and Precision 210
    Installation 210
    Maintenance Requirements 210
  Interface/Sludge Blanket Level (ISBL) Analyzers 211
    Operating Principles 212
      Ultrasonic ISBL Analyzers 212
      Optical ISBL Analyzers 213
    Applications 213
      Variable Blanket Height 213
      Variable Blanket Height and Suspended Solids Profile 213
      Blanket Height Alarm 214
    Construction Materials 215
    Accuracy and Precision 216
    Installation 216
      In-Line with Skimmers or Flights 216
      Off-Line with Skimmers or Flights 217
      Fixed Submerged Location 217
    Maintenance Requirements 217
  Vibration 218
    Operating Principles 218
      Velocity Transducers 218
      Acceleration Transducers 219
      Displacement Transducers 220
    Construction Materials 220
    Accuracy and Precision 221
    Installation 222
      Flush-Mount 222
      Nonflush-Mount 223
    Maintenance Requirements 224
 References 224

Chapter 7 Final Control Elements
Control Valves 227

Characteristics	228
Types	228
Globe	229
Solenoid	230
Ball	231
Diaphragm	231
Plug	231
Butterfly	231
Gate	231
Other	231
Standards	232
Selection Process	232
Sizing	233
Actuators	234
Solenoid	234
Electric Motor	235
Pneumatic	237
Diaphragm	237
Piston	238
Hydraulic	238
Positioners	239
Pumps, Aerators, and Mixers	243
Pumps	243
Displacement Pumps	243
Nondisplacement Pumps	245
Fans, Blowers, and Compressors	245
Aerators and Mixers	245
Electric Motors	248
Starters	250
Drives	250
Variable-Frequency Drives	250
Wire-Wound Motor Control	251
Variable-Torque Speed Controls	252
Control Signal Interface and Smart Pumps and Drives	253
Other Final Control Devices	254

Chapter 8 Control Panels or Stations

Panel Devices	256
Analog Displays	257
Digital Displays	257
Operator Interfaces	258

Contents

        Graphic Terminals 258
        PC-Based Terminals 258
        Portable Terminals 259
    Annunciators 259
        Horns 259
        Strobes 260
    Pilot Lights 260
    Pushbuttons and Switches 261
    Miscellaneous Panel Hardware 261
    Programmable Logic Controllers 261
  Human–Machine Interface Software 261
  Types of Control Panels and Stations 262
    Local Control Panels and Stations 262
    Vendor-Supplied Panels 262
    Area Control Panels, Stations, and Consoles 262
    Main Control Panel, Station, or Console 263
    Graphic Panels 263
    Alarm Stations 263
    Termination Panels 263
    Fiber-Optic-Patch Panels 264
  Control-Panel Specifications 264
    CSI Format 264
    Standards, Codes, and Regulations 265
    Submittal Requirements 265
  General Construction Requirements 266
    Identification 267
    Environmental Requirements 267
        NEMA Rating 267
        Thermal Management 268
    Electrical Systems 269
        Power Source 269
        Safety Concerns 269
        Motor Control Centers 270
        Wiring 271
        Surge Protection 272
    Factory Testing 272
Installation 274
    Panel List 274
    Special Requirements 274
        UL 508: The Underwriters Laboratory Standard for Industrial Control
          Equipment 274
        Options for Panels in Hazardous or Corrosive Areas 275

  Intrinsically Safe Systems 275
  Purged Enclosures 275
 New Developments or Trends 277
  Operator Interfaces 277
  Touchscreens 278
   Resistive 278
   Capacitive 278
   Infrared 278
   Acoustic Wave 279
   Near Field Imaging 281
   Dispersive Signal Technology 282
  Flat-Panel Displays 283
   PC-Based Displays 284
   Monitors 285
    Plasma 285
    Digital Light Processing 285
  Wireless Operator Interface 287
  Digital Recorders 289
  Data Loggers 289
  Selection Considerations for Recorders or Data Loggers 289
  Remote Annunciation 290
References 291
Suggested Readings 293

Chapter 9 Connectivity Options for Process Control Systems

Basic Communications-System Terms 297
 Client 297
 Driver 298
 I/O Server 298
 Master 298
 Media 299
 Peer-to-Peer 299
 Protocol 299
 RS-232 299
 RS-422/485 299
 Serial 299
 Server 299
 Slave 299
 Transport 300
Basic Communications-Network Components 300
 Physical Transport Media 300

Contents

- Network Transport Protocol 300
- Application Protocols 301
- Communication System Networks 301
  - Serial 301
    - Modbus RTU and Modbus ASCII 301
    - DF1 302
    - Other Proprietary Protocols 302
  - Proprietary 303
    - Data Highway, Data Highway Plus, Data Highway 485 304
    - Modbus Plus 304
    - Genius 304
  - Fieldbus 305
    - Speed of Operation 306
    - Plant Layout 306
    - User Friendliness 306
  - Ethernet 306
- Ethernet in Depth 307
  - Copper Twisted-Pair Wires 307
  - Fiber Optic Cable 307
  - Fiber Optic ≠ Ethernet 308
  - Wireless Media 309
  - Bridging the Gap to Non-Ethernet Devices 309
    - Ethernet Encapsulation 309
    - Ethernet–Proprietary Network Bridges 312
  - Common Ethernet Application Protocols 313
  - Wireless Connectivity 313
    - In-Plant Applications 314
    - Remote Sites 314
  - Wireless Security 315
  - Wireless Reliability 316
- Wireless Solutions and Communications Software 317
  - Communications and Connectivity Software 318
    - Proprietary Drivers 318
    - ActiveX Drivers 319
    - DDE Drivers 321
    - OPC Standards-Based Drivers 321
  - Communications Redundancy 323
    - Instrument or Controller 323
    - Network Interfaces 324
    - Communications Software 325
    - HMI or SCADA Software 326
  - Web Connectivity 326

Chapter 10 Automatic Process Control

Process Control Objectives 330
Benefits of Automatic Process Control 332
Control Theory 332
    Process Identification and Modeling 333
    Control Strategies 336
        Sequential Process Control 336
        Regulatory Process Control 336
        Controllability Problem 337
Control Options 337
    Feedback 338
        PID 338
        PID Tuning 340
    Feedforward 342
    Combined Feedback–Feedforward 343
    Advanced Options 343
Role of Microprocessors and Computers 345
Process Control Systems 346
    Single-Loop Controllers 346
    Multiple-Loop Controllers 347
    Supervisory Setpoint Control Systems 347
    Distributed Networked Control Systems 347
Control System Configuration 347
References 348

Chapter 11 Human–Machine Interfaces

Real-Time Display and Control Devices 354
    Types of Control Panels 354
        Wall-Mount Vertical Panels 354
        Free-Standing Vertical Panels 354
        Console Panels 354
        Breakfront Panels 354
    Device Arrangement 354
        Put Devices within the Operator's Line of Sight 356
        Group Displays and Associated Controls by Equipment Function....357
        Put Frequently Used Devices in the Center 357
        Put Displays above Related Controls 357
        Make Labels Easy to Read 357
        Avoid Abbreviations 357
        Provide Test Functions 357
    Visual and Aural Stimuli 357
        Annunciators 357

   Status-Indicating Lights 357
   Switches and Pushbuttons 359
   Strobe Lights and Horns 359
   Bar Graphs and Loop-Controller Displays 360
  Control-Panel Distribution 361
   Local Control Panels 361
   Vendor-Supplied Control Panels 361
   Area, Master, and Main Control Panels 362
  Graphic Panels 364
   Sandwich Displays 364
   Mosaic Tile Displays 364
  Computerized Displays 365
   Text Displays 365
   Proprietary Graphic Displays 365
   PC-Based Displays 367
   Video Projectors 368
   Large, Multisegment Flat-Panel Displays 368
   Browser-Based HMIs 368
   Wireless HMIs 369
  HMI Software 370
   Configuration Standards 370
   Object-Oriented Programming 371
   Screen Navigation 371
   Screen Layout 375
    System Navigation Bar 376
    Graphics Area 376
    Alarm Summary 376
    Colors and Shapes 377
    Graphics and Bitmaps 377
   Screen Refresh Rate 378
    Limit the Number of Active Tags on a Screen 378
    Avoid Bitmap Graphics 378
    Stagger Process-Loop Updates 378
   Calculations 378
Alarm and Event Communications 378
  Software Alarms 379
   Alarm Triggered 379
   Alarm Acknowledged, but Alarm Condition Still Present 379
   Alarm Acknowledged, and Condition Has Returned to Normal 379
  Printers 380
  Audible Alarms 380
   Autodialers 380

Contents   xxiii

       Control-Room Alarms 380
       Pagers 381
   Real-Time and Historical Trends 381
      Recorders 381
      Displays 381
       Local HMI Stations 382
       Historian 384
      Update Times and Sampling Intervals 384
       Deadband 384
       Smoothing 385
   Control Rooms and Human Engineering 385
      Ergonomics 390
      Equipment Layout 390
      Physical Security 390
      Fire Protection 390
      HVAC 391
      Lighting and Electricity 391
      Consoles and Furniture 391
       Console Subsystem 392
       Console Surface and Appurtenances 392
       Chairs 393
Security 393
   Security Vulnerabilities 393
   Network/PCS Security 395
      Separate the SCADA System from Other Networks 395
      Separate SCADA and Security Functions 397
      Control Access to SCADA Equipment 397
      Secure Remote Connections 397
      Implement Firewall Protection 397
      Leverage All Logging Features 398
      Install Virus-Protection Software 398
      Implement the Highest Operating-System Security 398
      Implement Backup and Recovery Procedures 398
   HMI Access 398
      Level I: View Only 399
      Level II: Shift Supervisor/Control Room Operator 399
      Level III: Process Engineer 399
      Level IV: Software Engineer 399
      Level V: Administrator 399
      Level IIA: Secondary Systems Operator 399
      Level IIB: Solids Handling Building Operator 399
   Security Policy 400

References 400

Chapter 12 Process Controllers

Applications 404
    Collection Systems 404
    Wastewater Treatment Plants 407
Types of Controllers 407
    Programmable Logic Controllers 407
        Components 408
            Input Relays (Coils) 408
            Internal Utility Relays (Contacts) 408
            Counters 409
            Timers 409
            Output Relays (Contacts) 409
            Data Storage 409
            Analog I/O, Monitoring, and Control Capability 409
        Operations 410
            Step 1: Check Input Status 410
            Step 2: Execute Program 410
            Step 3: Update Output Status 411
        Programming 411
        Soft PLCs 412
    Remote Terminal Units 412
        Hardware 412
        Software 412
        Diagnostics 415
        Basic Operations 415
        Size 415
        Standards 416
        Recommended Specifications 416
        PLCs versus RTUs 417
    Distributed Control Units 419
    Single-Loop Controllers 420
    Embedded Controllers 420
    Input/Output Modules 420
        Adapter 420
        Discrete 421
        Analog 421
        Specialty 421
        Communications 421
        Ethernet 421
        Wireless 422

Design Guidelines 422
  Market Position and Reputation 422
  Maintenance Costs and Local Support 423
  I/O Capacity and Program Memory 423
  Environmental Constraints 424
  Programming Languages and Program Maintenance Costs 424
  Security Features 425
References 425

Chapter 13 Process Control Narratives

Purpose of Process Control Narratives 428
Typical Process Control Narrative Components 428
  Equipment Tag Numbers 428
  Process and Instrumentation Diagrams 437
  Process and Equipment Descriptions 437
  Local Controls 437
  Motor-Control-Center Functions 437
  PLC Functions 437
  Remote Manual Controls 437
  Remote Automatic Controls 437
  Alarms 438
  System Components 438
Other Methods of Conveying Process Descriptions 438
  Animations 438
  Static Process Models 439
  Interactive Process Models 439

Chapter 14 Advanced Applications for Wastewater Treatment

Energy Management 442
  Conservation 442
  Real-Time Power Monitoring 442
  Reducing On-Peak Energy Demand 443
  Capital Improvements 443
Artificial Intelligence 444
  Expert Systems 446
    The Knowledge Base 446
    The Inference Engine 446
  Case-Based Reasoning Systems 449
  Fuzzy Logic 451
  Artificial Neural Networks 452
  Intelligent Decision-Support Systems 453

        Data Gathering 453
        Diagnosis 454
        Decision Support 454
    A Case Study 455
Biological Nutrient Removal 457
    Why Is BNR Control Needed? 457
    Selecting Setpoints for a BNR Plant 460
    Ammonia Control 460
    Denitrification Control 462
    Respirometry 463
    Intermittent Aeration 464
    Sequencing Batch Reactors 464
    IWA/COST Benchmark Simulation Environment for Control
    Strategy Comparison 465
References 466

Chapter 15 Instrumentation and Control System Specifications

The Documentation Process 474
    Process and Instrumentation Diagrams 474
    Control-System Block Diagrams 474
    Physical Drawings 475
    Special Mounting Details 475
    Signal I/O and Instrument Lists 475
    Loop Drawings 475
The Specifications Process 475
    Related Documents 477
    Organization 481
        General 481
        Products 483
        Control System 483
        Instrumentation 483
        Execution 484
Suggested Readings 486

Chapter 16 Instrumentation Maintenance

Maintenance Management System Components 488
Computerized Maintenance Management Systems 489
Verification of Instrument Performance 498
    Quality Control Procedures 498
    Control Charts 499
    Example 500

Implementation Methods	502
Textbook	502
Modular	505
Build as You Go	506
Turnkey	506
Reference	507
Suggested Readings	507

Chapter 17 Instrumentation Troubleshooting

The Importance of Details	510
Useful References	511
Troubleshooting Strategy	511
Define the Problem	512
Observe the System Closely	512
Isolate the Problem Area	512
Develop a Theory	513
Test the Theory	513
Implement Changes	513
Prepare a Report	513
Test Cases	513
Example 1	514
Define the Problem	514
Observe the System	514
Isolate the Problem Area	514
Develop a Theory	516
Test the Theory	516
Implement Changes	516
Prepare a Report	516
Example 2	516
Define the Problem	517
Observe the System	517
Isolate the Problem Area	518
Develop a Theory	518
Test the Theory	519
Implement Changes	521
Prepare a Report	521
Troubleshooting Equipment	521
Safety	523
Suggested Readings	523

Chapter 18 Instrumentation Training

| Training Is a Necessity | 526 |

Training Program Elements 526
Training-Course Classifications 527
    Operations 527
    Maintenance 527
    Configuration 527
    Programming 527
    Management Overview 528
Class Types 528
    On the Job 528
    Home Study 528
    Onsite Classrooms 528
    Offsite Classrooms 528
    Conferences and Web Sites 529
Sources of Onsite Instructors 529
Sources of Onsite Training Materials 530
Training Materials 530
Training Program Implementation 531
Training Continuity 531
Example 532
Suggested Readings 535

Index 537

Preface

This manual focuses on the elements of a complete automation design. It primarily focuses on the automation system designer and the design process. However, this manual also will show wastewater treatment facility owners, managers, and operators what types of design documents to expect when undertaking an automation project and the standards used to evaluate them.

Many chapters of the 1993 manual were simply updated to current standards, while others were greatly expanded to reflect technological changes. New chapters also were added to discuss modern control-system issues (e.g., energy management).

This third edition of this manual was produced under the direction of Dr. Bob Hill, P.E., *Chair*.

Principal authors of the publication are

Charles K. Aycock	(6)
Edward Baltutis	(7)
Duncan Browne	(6)
James Conboy, P.E., BCEE	(11)
Peter R. Craan, P.E., CAP	(3, 4, 8)
Alex Ekster, Ph.D., P.E., DEE	(14)
Richard V. Frykman, P.E.	(15)
Robert A. Gillette, P.E., DEE	(6)
Dr. Bob Hill, P.E.	(1, 2, 3, 4, 14)
Sam Irrinki, P.E.	(12)
D. Scott Joslyn, P.E.	(6)
Salil Kharkar, P.E.	(6)
Phil Kiser	(6)
Ryan Kowalski, P.E.	(11)
David Olson, P.E.	(16, 17, 18)
Albert C. Petrasek, Jr., Ph.D., P.E.	(13)
William Phillips, P.E.	(12)
Ignasi Rodriguez-Roda, Ph.D.	(14)
Bob Rutemiller	(9)

Andrew R. Shaw, P.E.	(14)
Mark Snyder, P.E.	(6)
David A. Vaccari, Ph.D., P.E., DEE	(10)
Troy D. Vassos, Ph.D., P.Eng.	(5)
John Weber	(9)

Authors' and reviewers' efforts were supported by the following organizations:

Alpine Technology, Inc., Austin, Texas
Automation Consulting & Education, Inc., Tampa Florida
Black & Veatch, Kansas City, Missouri
Carollo Engineers, Sacramento, California; Sarasota, Florida
Carter & Burgess, Inc., Dallas, Texas
Cedar Rapids Water Pollution Control Facility, Cedar Rapids, Iowa
CH2M Hill, Inc., El Paso, Texas; Santa Ana, California
City of Roseville, California
D.C. Water and Sewer Authority, Washington, D.C.
Diversified Technical Services, Stamford, Connecticut
Donohue and Associates, Inc., Sheboygan, Wisconsin
Ekster and Associates, Fremont, California
EMA, Inc., Conroe, Texas
Endress and Hauser Instruments, Greenwood, Indiana
GE Energy-Bently Nevada, L.L.C.
Hach Company, Loveland, Colorado
Hazen and Sawyer, PC, New York, New York
Instrumentation Testing Association, Henderson, Nevada
Keppel Seghers, Inc., Houston, Texas
Malcolm Pirnie, Inc., White Plains, New York
MWH, Cleveland Ohio
Ocean County Utilities Authority, Bayville, New Jersey
SISLtech, Girona, Spain
Software Toolbox, Inc., Charlotte, North Carolina
Stephens Institute of Technology, Hoboken, New Jersey
Weston Engineering, Inc., Chicago, Illinois

Chapter 1

Introduction

What Is Complete Automation Design?	1	What This Manual Does and Does Not Include	4
Motivation for Updating This Manual	2	Intended Audience	4
Technology Now a Strategy, Not a Limitation	3	Manual of Practice Chapter Contents	4
		References	6

WHAT IS COMPLETE AUTOMATION DESIGN?

This relatively simple question requires this entire manual of practice (MOP) to answer fully. Basically, this manual specifies the information required to make the decisions needed to build an effective automation system. It also introduces the types of documents—both text and drawings—that constitute a complete design and provides guidance on what data these documents should contain.

Automation system design requires tremendous attention to detail—seemingly more so than other engineering disciplines. The design documents discussed in this manual reflect that attention to detail. Does every project need every document discussed in this manual? No, but each project typically will require decisions on every detail before the automation system can be installed. *Someone* will decide what to install, where to install it, where to put the wires, and how to program it. Whether that person is a design engineer, a general contractor, an electrical subcontractor, a systems integrator, a programmer, a computer-aided design (CAD) specialist, an installing electrician, or a trench digger depends on how complete the design documents are, which in turn depend on the owner's design budget and the designer's technical competence.

MOTIVATION FOR UPDATING THIS MANUAL

This is the third edition of *Instrumentation in Wastewater Treatment Facilities* (MOP No. 21), and the title was changed to *Automation of Wastewater Treatment Facilities*. The first edition was published in 1978, and the second was published in 1993. Although the automation field has matured considerably, it is still changing rapidly. Since the last edition was published,

- Researchers and practitioners have improved our understanding of physical, chemical, and biological treatment processes and the best strategies to control them;
- Manufacturers have begun creating field instruments specifically for the wastewater treatment industry (rather than adapting ones made for other industries);
- Field instruments typically have become less expensive, more accurate, easier to calibrate, and sometimes self-diagnosing because of digital electronics, better human interfaces, and other new technologies;
- Networking and wireless (radio) communications have become less expensive, provide more capabilities, and are widely available (e.g., fiber-optic cabling is now common);
- Programmable logic controllers (PLCs) and distributed control systems (DCSs) are smaller, perform better, have more capabilities, and are less expensive;
- "Standardized" PLC-programming tools based on the 1993 International Electrotechnical Commission (IEC) 1131-3 standards are now used extensively;
- Personal computers (PCs) operate even faster, have more capacity, are less expensive, and—along with their related software—dominate the automation market;

Introduction

- Shrink-wrapped, PC-based, supervisory control and data acquisition (SCADA) software is now readily available and relatively inexpensive (but requires extensive configuration);
- SCADA systems now typically include the monitoring, control, trend, and data features that most users need, and many also include historical data archiving, reporting, fuzzy control (a formal methodology for representing, manipulating, and implementing a human's heuristic knowledge about how to control a system), artificial neural networks, and model predictive control (a class of controllers that use a model of the process to compute a sequence of manipulated variable adjustments to optimize the future behavior of the process);
- The Internet now plays an important role in dispersing information and is evolving into a platform for wide-area control systems;
- Physical and cyber security have become important issues.

Twenty-eight years ago, the authors of the first edition noted that because automation is a rapidly changing field, the manual would need frequent updates. This is still true today.

TECHNOLOGY NOW A STRATEGY, NOT A LIMITATION

Automation technology has largely matured. Today, instruments, control elements, and strategies can handle most processes and problems, and wastewater treatment facility owners, managers, and operators simply expect every part of an automation system to work and work well. Now, most of the keys to a successful project are organizational and management issues rather than technical ones (Chapter 2).

Admittedly, some challenges, such as automation of solids treatment and dewatering processes, remain. However, researchers, manufacturers, and practitioners have recently made tremendous progress in addressing them. For example, a 2001 Water Environment Research Foundation project, *Thickening and Dewatering Processes: How to Evaluate and Implement an Automation Package,* describes how to evaluate and implement an automation system for thickening and dewatering processes. Further research is being conducted in Europe.

WHAT THIS MANUAL DOES AND DOES NOT INCLUDE

When the Water Environment Federation's (WEF's) Automation of Wastewater Treatment Facilities Task Force first defined the scope of this update, they estimated that the manual would be about 500 pages—considerably longer than the 1978 edition (108 pages) and 1993 edition (332 pages)—to accommodate new developments in the field.

After considerable discussion on what to keep, add, and leave out, the task force decided to update much of the material in the previous editions, except for obsolete technologies, such as pneumatic transmission systems. The task force also incorporated much of the content of the 1984 edition of WEF's MOP No. OM-6, *Process Instrumentation and Control Systems*. However, the task force decided not to duplicate WEF's 1997 Special Publication, *Automated Process Control Strategies*, which is currently the latest on specific process-control strategies for wastewater treatment systems. The task force also decided not to debate the relative merits of traditional design-bid-build and design-build contracts or address construction management issues. Nor did the task force address collection system controls in this edition, because that topic deserves its own publication.

INTENDED AUDIENCE

This manual primarily focuses on the automation system designer and the design process. However, this manual also will show wastewater treatment facility owners, managers, and operators what types of design documents to expect when undertaking an automation project and the standards used to evaluate them.

MANUAL OF PRACTICE CHAPTER CONTENTS

Each chapter is written to stand alone, so the manual does not need to be read from cover to cover. Readers looking for specific information should simply turn to the appropriate chapter. For more information on a particular topic, see the references or general bibliography included in each chapter.

In Chapter 2, a business case for automation is made, stressing that control systems can help utilities be more effective and efficient. Investments in automation can and should reduce labor, chemical, and power needs while improving performance and reliability. Ten "Keys to Success" are introduced.

In Chapter 3, the elements of a complete automation design project are introduced. Each type of document is discussed briefly, and standards and references are delineated.

In Chapter 4, the process and instrumentation diagram (P&ID)—typically one of the first design documents developed for a project—is introduced, its elements are described, and its interactions with other design documents are discussed. Recent developments in "smart" P&IDs also are introduced.

In Chapter 5, general instrumentation characteristics, including properties and measures of accuracy, are introduced. Design considerations, such as temperature, moisture, corrosion, and grounding, are discussed.

In Chapter 6, major types of instrumentation used at wastewater treatment plants are discussed. This section has been revised to account for recent improvements, as well as new instruments specifically developed for solids treatment.

In Chapter 7, final control elements are discussed. Topics include valves, valve actuators, pumps, pumping characteristics, blowers, blower characteristics, motors, and variable frequency drives.

In Chapter 8, the characteristics of local control panels are discussed. Special consideration is given to environmental requirements, thermal management, and panel instrumentation.

In Chapter 9, signal transmission and data communications are discussed. This section has been revised to include material on networking and new wireless communication technologies.

Chapter 10 is a tutorial on process-control basics—from feedback controllers to advanced model-based controls.

In Chapter 11, process-control system functionality is discussed. The human-machine interface (HMI) is addressed.

In Chapter 12, the design of control system hardware, including programmable logic controllers (PLCs) and distributed control systems (DCSs), is discussed.

In Chapter 13, the process-control narrative—a text-based method of describing a process-control strategy—is discussed. Several sample narratives for common processes are developed.

In Chapter 14, advanced applications and tools, including energy management, decision support, modeling and simulation, artificial intelligence, and control strategies for biological nutrient removal (BNR) plants, are introduced.

Chapter 15 is an introduction to writing specifications. It describes the Construction Specifications Institute (CSI) format and provides an example of a specification, as well as a list of potential specifications for a "typical" automation project.

Chapter 16 is an introduction to instrumentation maintenance. The difference between instrumentation and rotating-equipment maintenance is discussed.

Chapter 17 is a brief introduction to troubleshooting instrumentation systems.

Finally, in Chapter 18, instrumentation documentation and training are discussed.

REFERENCES

Water Pollution Control Federation (1984) *Process Instrumentation and Control Systems*, Manual of Practice No. OM-6; Water Pollution Control Federation: Washington, D.C.

Water Environment Federation (1993) *Instrumentation in Wastewater Treatment Plants*, 2nd Ed.; Manual of Practice No. 21, Water Environment Federation: Alexandria, Virginia.

Water Environment Federation (1997) *Automated Process Control Strategies*, Special Publication; Water Environment Federation: Alexandria, Virginia.

Water Environment Research Foundation (2001) *Thickening and Dewatering Processes: How to Evaluate and Implement an Automation Package*, Project 03-REM-3; Water Environment Research Foundation: Alexandria, Virginia.

Chapter 2

The Business Case for Automation

The Benefits of Automation	8	Cost–Benefit Ratio	11
Meet Regulatory Requirements	8	Rate of Return	13
Improve Process Performance and Reliability	8	Ten Keys to a Successful Project	13
		System Advocate	13
Record Data and Create Reports	9	Management Support	14
		Owner Involvement	15
Save Chemicals and Energy	9	Operator Involvement	15
Save Labor	9	Stakeholder Involvement	16
Reduce Risks	10	Maintenance Plan	16
Ensure a Good Night's Sleep	10	Full System Testing	16
The Cost of Automation	10	Contingency Plans	17
Return on Investment—Why We Do Projects	10	Training	17
		Goals	18
Payback Period	11	References	18

THE BENEFITS OF AUTOMATION

There are several good reasons to automate a wastewater treatment facility:

- Meet regulatory requirements;
- Improve process performance and reliability;
- Record data and create reports;
- Save chemicals, energy, and labor;
- Reduce risks, and
- Ensure a good night's sleep.

MEET REGULATORY REQUIREMENTS. Many instruments are required under water-quality regulations. State and federal discharge permits, for example, require wastewater treatment facilities to record their daily flowrates and report average monthly and maximum daily flows. Most treatment plants also must collect flow-weighted composite samples for analysis. The samples may be collected by an operator and composited manually based on the flowrate during sampling or collected by an automatic sampler paced by a flow meter's signal. In Texas, treatment plants must measure the underflow rate of each activated sludge clarifier and keep it between 0.34 and 0.68 $m^3/m^2 \cdot d$ (200 and 400 gpd/sq ft) (Texas Administrative Code, 1990).

Design engineers and wastewater treatment plant staff should carefully study design criteria and permit requirements and make sure that all required instruments are properly installed and maintained.

IMPROVE PROCESS PERFORMANCE AND RELIABILITY. Proper instrumentation and control are essential for good process performance. Biological nutrient removal systems, for example, depend on effective dissolved oxygen control in every basin. Nitrification requires sufficient dissolved oxygen concentrations, while denitrification requires the dissolved oxygen level to be near zero.

Virtually every wastewater management conference now includes presentations documenting how instrumentation and automation improved process performance in every area of the treatment plant. Several publications, including *Sensing and Control Systems: A Review of Municipal and Industrial Experiences* (Water Environment Research Foundation, 2002) and *Efficient Redundancy Design Practices* (Water Environment Research Foundation, 2003), also note instrumentation's key role in reliable treatment plant operations.

RECORD DATA AND CREATE REPORTS. All wastewater utilities must collect various data and use them to produce reports for regulatory and management purposes. For many utilities, however, these efforts are inefficient and time-consuming. Automation systems can streamline these processes by collecting some of the data, making it readily available, and even generating related reports.

A utility's automation goals should include collecting data automatically whenever feasible, entering manually collected data only once, making it available electronically to all who need it, and storing it in only one database.

SAVE CHEMICALS AND ENERGY. Many treatment plants could save chemicals and energy by implementing closed-loop control of chemical dosing and other processes. Accurate estimates of these savings, however, are needed to make good design decisions for automation. Typically, closed-loop control of chemical addition saves 10 to 20%.

Saving 1.0 mg/L of chlorine at a 375 000-m^3/d (100-mgd) treatment plant, for example, could justify the cost of several chlorine analyzers and a part-time technician, with substantial funds left over. Saving 1.0 mg/L of chlorine at a 375-m^3/d (100 000-gpd) treatment plant, however, may not justify the installation of any automation equipment.

In the chlorination process, changes in flow and effluent quality result in a varying chlorine demand. If the automation system can match this demand, substantial savings in chlorine are possible. Control of chlorination therefore requires accurate flow and chlorine residual measurements. In the case of a large facility, Hill and Martin (1994) reported a chlorine savings of $200,000 per year (33%) while only incurring $12,500 of additional maintenance labor.

There are often many opportunities to save energy in wastewater treatment (Chapter 14). Controlling dissolved oxygen, for example, saves 15 to 20% of electrical costs.

SAVE LABOR. Automation saves operator labor by

- Consolidating data so staff can observe it all at one location rather than having to walk the plant for it,
- Eliminating the need for "rounds" to check that equipment is operating correctly,
- Eliminating such repetitive tasks as filling day tanks and draining sumps,

- Eliminating the need to check and adjust chemical flows, and
- Eliminating unnecessary visits to lift stations and other offsite facilities.

However, automation can increase maintenance staff's requirements. A well-designed project should show a net labor savings for comparative performance.

REDUCE RISKS. Today, the penalties for improper risk management include fines; incarceration; wasted resources; and public outrage at plant odors, unsightly receiving waters, and discharged toxics. Instrumentation and controls can reduce the risk of permit violations and keep utility finances under control.

ENSURE A GOOD NIGHT'S SLEEP. As one anonymous manager said, the best benefit of automation may be getting a good night's sleep knowing that the automation system will oversee operations; make prompt, economical control decisions; and contact staff if conditions arise that it cannot handle well.

THE COST OF AUTOMATION

Automation equipment, including sensing instrumentation, control elements, controllers, software, and programming, typically adds 4 to 12% or more to the total cost of building a treatment facility. Project costs are site-specific and depend on the treatment processes involved and managers' decisions about the tradeoffs between automation and labor.

Once installed, an automation system then has ongoing maintenance costs. Such costs should be considered early in the design phase, because an automation system that is not maintained properly will eventually fail and fall into disuse.

RETURN ON INVESTMENT—WHY WE DO PROJECTS

The wastewater treatment plants built or expanded 15 to 20 years ago often included instruments and control systems just because utilities needed something to operate the plant. The related cost justifications typically were based on estimates with little supporting data. It did not matter then, however, because automation systems typically were a small percentage of the total construction budget.

Cost justification is more important today. Accurate estimates of automation costs and benefits are required to justify replacing or upgrading current systems.

Common methods for measuring return on investment include the payback period, cost–benefit ratio, or rate of return. All three will be demonstrated on the following dissolved oxygen control example, in which the following cost estimates were made:
- Initial investment = $500,000;
- Power savings = $200,000 per year;
- Additional maintenance requirements = $50,000 per year;
- Equipment life = 10 years; and
- Interest rate = 5% per year.

PAYBACK PERIOD. The payback period is a relatively simple calculation in which the initial investment is divided by the net annual savings. The underlying assumption is that the interest rate is 0%.

$$\text{Payback period} = \frac{\text{Initial Investment}}{\text{Net Savings}} = \frac{500,000}{200,000 - 50,000} = 3.33 \text{ years} \quad (2.1)$$

Payback periods of less than 5 years typically are considered good investments. If the payback period is more than 6 or 7 years, however, the project task force should thoroughly evaluate the accuracy of all estimates before making a strictly economic decision.

COST–BENEFIT RATIO. The cost–benefit ratio is a calculation in which the initial investment (cost) is compared to the overall net savings (benefit). First, however, the designer must calculate the present worth of such savings—the amount that would have to be deposited in an interest-bearing account today to yield the savings total at a future date.

Most books on engineering economics will have various formulae for calculating present worth and related parameters. The formula relating future values to present worth is

$$\frac{P}{F} = \left[\frac{1}{(1+i)**n} \right] \quad (2.2)$$

Where

P = present worth of money,

F = future payment or savings,
i = interest rate per interest period, and
n = number of interest periods.

In this case, the present worth is $1,158,260.24 (Table 2.1), so the cost–benefit ratio is 2.32 to 1.

cost–benefit ratio = $1,158,260.24/$500,000 = 2.32 to 1

The ratio must be at least 1.2 or 1.3 or more for a project to be considered a good investment.

TABLE 2.1 Calculation of present worth.

Year	Investment ($)	Net savings ($)	P/F factor	Present worth ($)
0	500,000		1	
1		150,000	0.952381	142,857.14
2		150,000	0.907029	136,054.42
3		150,000	0.863838	129,575.64
4		150,000	0.822702	123,405.37
5		150,000	0.783526	117,528.92
6		150,000	0.746215	111,932.31
7		150,000	0.710681	106,602.20
8		150,000	0.676839	101,525.90
9		150,000	0.644609	96,691.34
10		150,000	0.6139113	92,086.99
				Sum = 1,158,260.24

P/F factor = the ratio of present value of money divided by future value as defined in eq (2.2).

RATE OF RETURN. The *rate of return* is the interest rate (*i*) needed so the equivalent present worth of the future annual net savings equals the initial investment. The following equation describes the relationship between annual savings (*A*) and initial investment (*P*):

$$\frac{A}{P} = \left[\frac{i(1+i)^n}{(1+i)^{n-1}}\right] \quad (2.3)$$

Where

A = end-of-period payment or receipt in a uniform series continuing for n periods, the entire series equivalent to P at interest rate i.

In our example, where A = $150,000 and P = $500,000, eq 2.3 can be solved iteratively to find that i = 0.273, or 27.3%. If i is substantially greater than the lending rate, the project is considered a good investment.

TEN KEYS TO A SUCCESSFUL PROJECT

During the 2000 Water Environment Federation Technical Exhibition and Conference (WEFTEC), a Water Environment Research Foundation (WERF) project team asked leading utility managers, consultants, industrial specialists, instrumentation suppliers, and academics to identify factors that they thought contributed to control-system success (WERF, 2002). The team then prioritized the success factors, made a list of the top 10, and asked attendees of the WEFTEC 2001 workshop, State-of-the-Art WWTP Sensing and Control Systems, to validate it. Thirty-nine attendees (93%) agreed with the list, and three (7%) disagreed.

Interestingly, most of the success factors on this list are organizational and managerial issues rather than technical ones. (Successful projects, however, still require complete specifications, drawings, and other documents.) The following sections describe these "keys to success," which should be implemented throughout the system's lifetime (Table 2.2).

SYSTEM ADVOCATE. The success or failure of any automation project depends on the people leading it. Successful projects typically had an advocate, someone who was technically competent, really excited about the project, and respected throughout the organization.

TABLE 2.2 The keys to success and their importance during each phase of a project.

Key	Planning	Design	Implementation	Operations
System advocate	X	X	X	X
Management support	X	X	X	X
Owner/operator involvement	X	X	X	X
Trust in operators	X	X	X	X
What's in it for me	X	X	X	X
Plan for maintenance	X	X	X	X
Full system testing			X	X
Preparation for failure	X	X	X	X
Training			X	X
Setting goals	X			X

If the staff turnover rate is high, the role of system advocate should be assumed by a project task force of several people. This task force should be responsible for the project from planning through operations and should develop a common, consensual understanding of system requirements, capabilities, and features. This will maintain continuity if one member leaves the organization.

The task force should be led by a key manager and include representatives from the operations, maintenance, engineering, and information technology (IT) groups. The task force should not be led by IT staff unless that person also has a thorough understanding of wastewater treatment processes and real-time control. One task force member should be a project manager responsible for task force logistics.

MANAGEMENT SUPPORT. Successful projects must have the utility managers' full backing. A high-level statement committing the organization to the project and its task force is essential.

Managers also need to provide sufficient resources for the project (both procurement and maintenance) and support its integration into the organization. If the system will result in staff reassignments or reduction, then managers need to develop and communicate the specific plan for reorganization (e.g., staff reductions only through attrition).

OWNER INVOLVEMENT. Large control-system projects typically are planned, designed, and implemented by consulting engineers because the related risks and level of effort are suitable for contract work. However, the results may not meet the owner's needs and intentions unless these are well-defined before the contract is signed.

During initial project definition, owners and designers need to define their needs and expectations, which should be the basis of project objectives. It is then the owner's responsibility to stay involved in the project and confirm that the system will meet these objectives.

During the design phase, consulting engineers are responsible for the design's details and accuracy. The owners, however, should be involved in establishing the overall project goals and should review design documents for operations- and maintenance-related issues.

During the implementation phase, owners should be involved in system testing, both in the factory and in the field.

OPERATOR INVOLVEMENT. Ultimately, a system's success depends on whether it is used for its intended purpose. The best way to get operators to use a new control system is by involving them in all phases of the project and creating a system that meets their needs. Also, operators can provide valuable information that will help consulting engineers design and build a better control system.

During the planning phase, operators should be represented on the project task force and should evaluate the proposed control system's overall operability (e.g., ease of use and functionality). They can provide a practical, "hands on" perspective that is frequently overlooked or understated by engineers and managers.

During the design phase, operators can provide valuable information needed to define control strategies, which is especially important during a plant retrofit. (The task force should verify and validate this information.) Plant retrofits may require revising instrumentation and controls to change the way operators run a process—either because it previously was run incorrectly or because the process itself changed.

During the implementation phase, operators must be integrally involved in system testing. This is a critical opportunity to introduce operators to the system and give them a chance to use it without worrying about degrading the treatment process.

STAKEHOLDER INVOLVEMENT. Stakeholders' reasons for wanting the control system will vary, and the project objectives need to reflect each stakeholder's specific needs. Stakeholders also should be represented on the project task force. Ideally, these representatives' visions should be broad enough to reflect the entire organization's goals.

Ongoing stakeholder involvement throughout the project will help ensure that the resulting system will meet everyone's needs. It also increases the likelihood that the system will be accepted and considered a success.

MAINTENANCE PLAN. should determine how reliable the control system must be and what steps must be taken to maintain that level of reliability. One of these steps is maintaining equipment in accordance with their manufacturers' requirements.

During the design phase, the designers should address system-reliability issues, including redundancy, power source and backup, surge and lightning protection, response to control-system equipment failure, and equipment location and accessibility, in the design documents.

During the construction phase, construction inspectors should review the equipment for ease of access, maintenance, and replacement. Provisions also should be made for maintaining equipment until it is turned over to the plant staff so no warranty is voided. Afterward, ongoing equipment maintenance should be the responsibility of plant staff or an appropriate vendor.

FULL SYSTEM TESTING. Complete control-system testing and approval is critical; any part of a new control system that does not work contributes to a negative perception of the entire system. Also, the owner should witness all system testing because unwitnessed testing often is compromised when budgets or schedules get tight.

During the planning phase, the owner and project task force should allocate sufficient money, resources, and time so the necessary tools and documents are available for timely testing. Plant staff also must be authorized to participate in all testing activities.

During the design phase, the contract and design documents should establish system requirements, including control strategies, and specify the contractor's testing requirements.

During the implementation phase, testing should include a complete factory test of all hardware and software, field-testing of all process input and output (I/O) points to each end element (e.g., instrument or control element), and complete system testing for operation with the treatment process, including all control strategies.

Testing plans must be flexible, so testing will not unduly delay the project schedule. For example, one project called for a full-system factory test of both hardware and software, but delays in the control-strategy submittal process resulted in the hardware being ready well before the control-strategy configuration was complete. So, the project task force modified the factory-testing requirement to allow for two tests. The first only tested the hardware; once approved, the hardware was shipped and installed. The second only tested the software. This allowed more time to finalize and test the control configuration without delaying onsite hardware installation and wiring.

Also, all changes need to be fully tested before the control system is turned over to plant operators.

CONTINGENCY PLANS. Despite the best intentions, not every aspect of a project goes exactly as planned. Task force flexibility, contingency plans, and creative solutions are essential to getting derailed projects back on track. Flexible budgets and schedules, in particular, can best absorb the results of unexpected changes.

TRAINING. The best control-system projects will be pointless if utility staff do not know how to use the system as intended. Training should be scheduled before the system is turned over, and the training program should include multiple approaches—train-the-trainer, formal classroom training, hands-on, self-directed, on-the-job, etc.—to maximize results by allowing for different learning styles.

The training budget should include both the cost of training itself and the arrangements for staff to attend. Training for small systems or system changes can be done in the control room during normal shifts. Large systems or system changes—especially those involving a major process upgrade—may require dedicating a shift to staff training. If so, plans should be made for temporary personnel to handle the workload while staff is in the classroom.

In addition to basic training, the program should include refresher training after the system has been operating for several weeks. Follow-up training also should be periodically available thereafter to refresh staff skills and train new employees.

GOALS. Before starting any project, the task force should put together a written plan. This plan should state the tangible and intangible goals for the entire project. These goals need to define the specific benefits to be derived from implementing the control system. The plan should be authorized by managers and key personnel and referenced whenever questions of intent, goals, schedule, and priorities arise. It will be the project's justification and a measure of its success.

Consulting engineers use various methods to create these plans. The method used is not important, so long as the plan defines the objectives.

REFERENCES

Hill, R. D.; Martin, J. (1994) Measurement and Automatic Control of Chlorination. In *Critical Issues in Water and Wastewater Treatment, Proceedings of the 1994 National Conference on Environmental Engineering,* American Society of Civil Engineers: Reston, Virginia, 718–725.

Stire, T. G. (1983) *Process Control Computer Systems Guide for Managers*; Butterworth Publishers: Boston, Massachusetts, 170.

Texas Administrative Code (1990) Title 30 Environmental Quality, Part 1 Texas Commission on Environmental Quality, Chapter 317 Design Criteria for Sewerage Systems, 317.4 Wastewater Treatment Plants.

Water Environment Research Foundation (2002) *Sensing and Control Systems: A Review of Municipal and Industrial Experiences,* WERF Project 99-WWF-4; Water Environment Research Foundation: Alexandria, Virginia.

Water Environment Research Foundation (2003) *Efficient Redundancy Design Practices,* WERF Project 00-CTS-5; Water Environment Research Foundation: Alexandria, Virginia.

Chapter 3

What Is a Complete Automation Design?

The Design Process	20	*Vendor-Supplied Packaged-Equipment Panels and Controls*	31
Project Structure	20	Process-Control-System Configuration and Communication Schematics	32
The Predesign Phase	23		
The Detailed Design Outline Phase	24	Control-Panel Designs	32
The Final Design Phase	24	Instrumentation- and Panel-Location Drawings	32
Design Documents	25	Instrument Loop Diagrams	32
Process and Instrumentation Diagrams	25	Instrumentation Installation Details	33
Elementary Wiring Diagrams	25	Instrument and Input/Output Lists	33
Process-Control Narratives	25	Specifications and Bid Documents	33
Tag Numbers	28	Sources of Technical Design Information	33
Monitored Variables	28		
Valve-Position Switches and Indicators	29	Technical Societies	33
Alarms	29	Vendors	34
Reset Function	30	*Selecting Vendors and Technologies*	35
Remote Indicators	31		
Local Pushbutton Stations	31		
Uninterruptible Power Supplies	31		

Incorporating Selected Technologies into Design	35	*Witnessing Factory Acceptance Tests*	37
Training	35	System Performance	38
Web Sites	36	Hardware Testing	38
Books and Periodicals	36	Software Testing	38
Previous Design Experience	36	Simulation Software	39
Post-Design Phase	36	Operator Interface Terminals	40
Design Services during Construction	36	Construction Management	41
Reviewing Shop Drawings	36	*Field Tests*	41
Processing Change Orders	37	*Loop Testing*	41
Responding to Requests for Information	37	*Integrated System Testing*	41
Meeting with Contractors and Suppliers	37	*Final Acceptance*	42
		References	42
Making Regular Site Visits	37	Suggested Readings	44

THE DESIGN PROCESS

A typical control-system design project for a medium-sized wastewater treatment plant has three phases: predesign, detailed design outline, and final design (Figure 3.1).

PROJECT STRUCTURE. Wastewater treatment projects can be cost-driven, time-driven, or both. A cost-driven project aims to limit budget overruns by enabling the schedule to be adjusted for optimum design efficiency. A time-driven project places more emphasis on deadlines, limiting the engineer's ability to make any schedule changes.

Wastewater utilities typically pay for project design via fixed-cost (lump-sum), cost-plus, or design–build fee structures. In the fixed-cost approach, the utility pays

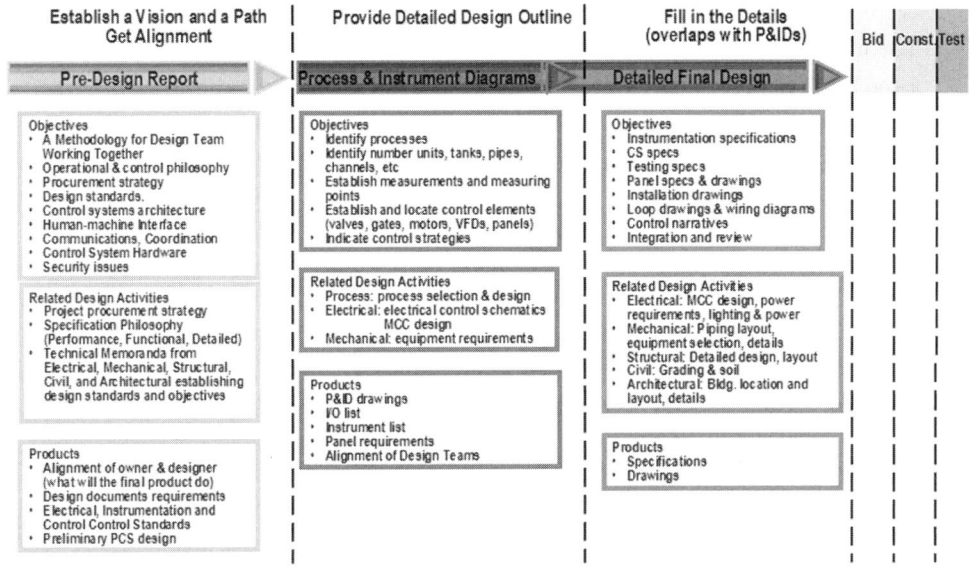

FIGURE 3.1 Typical automation project execution.

an engineering firm a predetermined fee to provide all of the project's design documents, from the request for proposal (RFP) to final design. In the cost-plus approach, the utility agrees to pay the engineering firm at an hourly rate for clearly defined services. In the design–build approach, the owner either prepares the contract documents (or hires a design engineer to do so) and then hires a project team—an engineering team, subconsultants, a general contractor, and subcontractors—to completely design and construct the treatment plant. The approach may also be mixed; for example, the utility may pay one engineering firm a flat fee for the design and pay another an hourly rate to oversee construction.

The desired level of detail can also vary. Fewer details are needed for RFP documents, for example, than those needed for system-integration specifications. In the first, the designer should define the criteria for programming the control system's programmable logic controllers (PLCs), while in the second, the designer should note the actual programming logic (e.g., ladder logic and, sequential function charts, etc.) required to implement the control strategies.

Some utilities prequalify control-system integrators for their automation projects. Koons and Conley (2004) recommend using a decision matrix to select a system integrator. When creating this matrix, utility staff should assign weights to all the decision criteria, including

- Eligibility,
- Unique features,
- Project understanding,
- Hardware and software recommendations,
- Project approach, and
- Price.

The Control Systems Integrators Association publishes free guides to help engineers and utilities document criteria for selecting system integrators (see reference section at end of chapter).

Additional factors to consider in making the selection include

- Field service capabilities to support project during startup, commissioning, and after system acceptance;
- Project management capabilities, including cost-control methods and scheduling;
- Problem-resolution approach and documented record of how problems were addressed, including any pending litigation;
- Affiliation with control-system equipment manufacturer;
- Insurance limits and bonding capacity;
- Project backlog, including other projects being handled concurrently; and
- Ability to staff project adequately.

The final selection process is based on consideration of financial, technical, and managerial capabilities. For the wastewater treatment industry, experience with wastewater-related projects should be considered more than industrial-related project experience.

THE PREDESIGN PHASE. The predesign phase has a major influence on the entire project and the design process itself—especially when the project is time-driven and involves major subconsultants. The goals of this phase are for all stakeholders to agree on project objectives and the necessary steps to achieve them. The participation of utility engineers, project managers, and plant operations staff (e.g., plant supervisors, chief operator, etc.) is critical because they will be using the final product.

This phase is also the time to choose design team members and determine a method for working together. Ideally, the design team should include the project manager(s), the owner's representatives, representatives of all related engineering disciplines, and any major subconsultants. This team will establish the project's operational and control philosophies; design standards; and the architectural, civil, electrical, instrumentation, mechanical, and structural memoranda on which the design documents will be based.

A project procurement strategy also should be established so owners can be confident that they are "getting what they paid for". This strategy also helps determine the specification philosophy (the required design documents), because the documents required for a low-bid project are different than those needed for a design-build project.

This phase is also the time to begin addressing security issues. The team should start by reviewing the standards and guidelines for supervisory control and data acquisition (SCADA) system security that have been developed since the Sept. 11, 2001, terrorist attacks on New York City and Washington, D.C. Such guides include *21 Steps to Improve Cyber Security of SCADA Networks* (U.S. Department of Energy and the President's Critical Infrastructure Protection Board, 2002), *Security Technologies for Manufacturing and Control Systems* [Instrumentation, Systems, and Automation Society (ISA), 2004b], and *Integrating Electronic Security into the Manufacturing and Control Systems Environment* (ISA, 2004a).

At the end of this phase, the team should write a predesign report that includes the project objectives, necessary project steps, the required design documents, the standards for these documents, and perhaps include some preliminary design documents [e.g., technical memorandum describing the proposed instrumentation and control (I&C) system] and prepare a budget estimate based on the available information and project scope. The owner should approve this report before the design team begins the next phase.

THE DETAILED DESIGN OUTLINE PHASE. During this phase, the design team should develop the process flow diagrams (PFDs) and resulting process and instrumentation diagrams (P&IDs). Process and instrumentation diagrams typically are the first design drawings completed. They serve as blueprints for the more detailed work in the final design phase. As *Automated Process Control Strategies* (Water Environment Federation, 1997) notes,

> "An excellent tool and starting point for any design is the process and instrumentation diagram… a schematic diagram showing major components of a process or processes, interconnections among them, instruments, measuring points, and control elements. If the designers can reach a consensus on these major design components, much of the rest of the design is straightforward and simplified."

Owners must be involved in P&ID development because the control strategies shown on these drawings will directly affect treatment plant operations. (For more information on P&IDs, see Chapter 4.)

Electrical designers typically also are involved in this phase because of the numerous interactions between the control and the electrical systems. The elementary wiring diagrams (EWDs) often are developed simultaneously with the P&IDs.

Other design documents, such as the input/output (I/O) and instrument lists, are generated directly from the P&IDs. They might be completed now or during the final design phase.

The owner should approve these documents before the start of the final design phase.

THE FINAL DESIGN PHASE. If the project standards have been established and the design activities have been outlined, then this phase should proceed smoothly. The goal is to produce the remaining I&C drawings and specifications needed to complete the final design.

The control-system specifications—the written details on the equipment, methods, and functions that contractors must supply or implement—are especially important in this phase. These specifications typically are organized according to the format established by the Construction Specifications Institute (CSI; Alexandria, Virginia). The automation and instrumentation industry standards, which were completely revised in 2004, organize specifications into sections that correspond to major construction categories. The CSI format is structured so there is minimal overlap with other sections. The completed specifications are then submitted to the end user or owner for their comments and approval. (For more information on the CSI format, see Chapters 8 and 15.)

DESIGN DOCUMENTS

The following are brief introductions to the major documents used to design control systems (Table 3.1).

PROCESS AND INSTRUMENTATION DIAGRAMS. Process and instrumentation diagrams typically are the most important design documents for a control system. They define the process, the physical facilities, the interconnections between units, the measurement types and points, the control elements, and the control loops. These documents reflect the consensus of designer and owner, and help design teams coordinate their efforts.

A typical wastewater project requires between 40 and 50 P&IDs—all meeting ISA S5.1, S5.3, S5.4, and S5.5 standards. Smaller plant upgrades or retrofit projects require fewer drawings, depending on the extent of process modifications mandated by the project. These documents typically are printed on 11 in. × 17 in., 22 in. × 34 in., or larger sheets so the lettering is easy to read. (For more information on P&IDs, see Chapter 4.)

ELEMENTARY WIRING DIAGRAMS. Elementary wiring diagrams illustrate how a specific piece of equipment will be controlled electrically. The design team should create a separate wiring diagram for each piece of control-system equipment. The diagrams should include switches, push buttons, lamps, relays, fuses, solenoid valves, and flow switches. They also may show inputs to and outputs from the control system.

PROCESS-CONTROL NARRATIVES. Before preparing detailed process-control narratives (PCNs; also called control strategies), the design team should put together a process-function description that notes the owner's goals for the control system, explains the team's intentions, and promotes discussion with the I&C designer.

Among other details, the process-function description should note how the control system and its components will be started manually, be started automatically, be

TABLE 3.1 Various design documents and industry standards.

Design document	Industry standard
Computer specifications	Provide the latest computer processors available and generously sized hard drives to ensure the longest life for the I&C system.
Conduit, wiring, and panel-board schedules	Provide a table of all cables and conduits, complete with sizes, numbers, and routing.[a]
Control-panel design	Provide full designs of all control panels, properly sized with external and interior views.
Control-room layout	Provide a drawing showing the location of all control-room equipment.
Control-system specifications	Provide functional specifications and performance standards on all hardware, software, configuration services, and testing (factory and site) in CSI format.
Elementary wiring diagrams (EWD)	Provide complete EWDs for all new equipment and at least partials for all existing equipment involved. Provide a table of all cables and conduits, complete with sizes, numbers, and routing.[a]
Instrumentation- and panel-location drawings	Provide drawings showing the location of all instruments and control panels.
Instrumentation-installation details	Provide the installation details for all major types of instruments.
Instrumentation specifications	Provide specifications in CSI format, with general conditions and 30 to 35 individual instrument specifications.
Instrument list	Provide a complete list of all instruments with all required details, including description, tag number range, specification number, P&ID number, and remarks.
I/O list	Provide a list of all I/O points with all required details, including I/O type (digital input, analog input, etc.) and engineering units for minimum and maximum values.
Loop diagram	Provide loop diagrams for each control loop with all required details.[b]
Panel list	Provide a complete list of all control panels, with NEMA rating, drawing locations, and related specifications sections.

(continued)

TABLE 3.1   Various design documents and industry standards. *(continued)*

Predesign report	Provide a predesign report, with options, recommendations, and design standards (e.g., the type of drawings required and the standards for them).
Process and instrumentation diagram (P&ID)	Provide P&IDs for each process or subsystem and meet ISA standards.
Process-control narrative	Provide specifications, with process-control narratives for each control loop.
Process-control-system configuration and communications schematics	Provide complete configuration drawings, including all required communications.
Process-flow diagram (PFD)	Provide simplified schematics of the process system's mechanical configuration.
Security issues	Meet ISA-TR99.00.01 standards.
UPS distribution-block diagram	Show location of all UPSs, with ratings and connected loads.

[a] These typically are not part of instrumentation-design documents. They are included with electrical-design documents and are listed here only for reference.

[b] These typically are not part of design documents. Loop diagrams are typically completed by the system integrator during the shop-drawing phase of the project and are listed here only for reference.

stopped under normal conditions, be stopped in emergencies, and react to malfunctions and alarms. All starts and stops should be presented in order of occurrence. If permissive interlocks are required for startup, they should be presented in the order in which they must be satisfied—along with associated monitoring, verifications, and timing constraints. The design team should be explicit, using tables as much as possible to simplify this description.

No matter how detailed, though, a process-function description is not a substitute for the PCNs, which the I&C designer should prepare as part of the instrumentation specifications. When dealing with a large or complicated system, the designer should divide it into subsystems and produce a control strategy for each one. However, the overall flow of the main presentation should not be interrupted. Subsystem details should be presented after the main sequence is completed, and interlocks between subsystems should be noted, or else the presentation will be confusing and disjointed.

The *process-control narratives* describe each control loop and its strategy. Each narrative must include

- Related equipment and instruments;
- Referenced drawings;
- A process description;
- Local controls;
- Interlocks;
- Motor-control center (MCC) functions;
- PLC/remote terminal unit (RTU) functions;
- Automatic and manual control functions, including control modes, proportional–integral–derivative (PID) algorithms, cascade controls, etc. (Metcalf & Eddy, 2003); and
- Discrete components.

The contractor or system integrator uses this information to program the correct control sequences and loops so the equipment will operate as designed.

Once the PCNs are drafted, the I&C designer should submit them to the project-coordination engineer for review and approval. The following paragraphs note some key details to consider when writing PCNs. (For more information on PCNs, see Chapter 13.)

Tag Numbers. Each instrument, piece of equipment, etc. should be described in the PCNs by its identification or tag number, if available.

Monitored Variables. A list of each system's monitored variables should be compiled in a table (Table 3.2). The table also should note each variable's anticipated range, whether a set point is required, and whether it is fixed or adjustable. (All set points are adjustable, unless otherwise specified by the equipment manufacturer, warranty, or other applicable authority.) In addition, the table should note the initial setting of each set point, if available; who (e.g., operator, supervisor, plant superintendent, etc.) can change it; and whether it should be password-protected.

Valve-Position Switches and Indicators. When developing a control system for a shutoff valve, the I&C designer should be aware of the significance of its position (fully open, fully closed, or somewhere in between). The narrative should note whether the valve position must be confirmed directly or can be inferred from a position switch. The importance of valve position depends on the treatment process. Suppose, for example, that the control system is supposed to turn off a centrifugal pump and its discharge valve simultaneously, and the sequence is timed so the pump motor will not shut off until the system has confirmed that the valve is closed. What happens if the valve fails to close? Does it matter how far open the valve is?

Designers should indicate in the PCN whether a valve modulates or simply opens and closes. All automatic valves include two-position limit switches (open and closed). Modulating valves come with a continuous valve-position transmitter that will signal the local control station (LCS), remote-control panel, or operator workstation how far open the valve is (0 to 100%).

All motor-operated valve actuators include local, manual controls. If a valve is inaccessible from the ground or a platform, however, then a separate, accessible LCS must be provided for it. This provision must be noted in the PCN because it affects both system wiring and equipment supplies.

Alarms. The process-control narratives should include a list of possible malfunctions and alarm conditions, along with descriptions of the responses to these events. This information should include where and how such problems will be indicated, which alarms will sound in which locations, and which control functions will be activated. Designers also should note how alarms will cascade from local to remote locations

TABLE 3.2 A sample list of monitored variables.

Variable	Engineering units	Range	Readout locations	Initial setpoint	Describe function; alarm or control

(e.g., when one alarm sounds at a local control panel, if and when another will sound at the central remote-control panel or operator workstation). The cascade description should specify whether two specific alarms will sound at each location or whether the local alarm will be specific, while the remote one will be a common alarm for the panel or system. If common alarms will be used at the central control site, the PCN should note whether, if a second alarm sounds at the local control panel before the first one is cleared, the common alarm at the central control site will be reactivated.

Designers also should note whether each alarm is a warning-condition or latching alarm (Table 3.3). A *latching alarm* requires an operator to reset the system after correcting the problem—even if the condition corrects itself. The latch can only be released when the system is reset.

If the control system includes human–machine interfaces (HMIs)—color graphic screens on personal computer (PC)-based operator workstations—the alarms will be recorded (with a time stamp) in the computer's hard drive. If the hard drive is full, then new alarms will overwrite old ones unless provisions are made to store alarm logs on removable media [e.g., compact discs (CDs), digital versatile discs (DVDs), backup tape drives, etc.]. In addition, alarms for PC-based systems must be acknowledged at the operator workstation (an industrial-grade computer mounted on a control panel or a desktop PC in a control room). Some systems also include an alarm printer in the control room that prints out all alarms as they occur. The printouts include an alarm description, the time it happened, the priority level (e.g., critical, warning, equipment status), and whether and when the alarm was acknowledged by an operator.

Reset Function. If the system must be inspected after a malfunction or alarm before it can be restarted, then the system will need a reset function to clear it after the problem is corrected. Designers have several reset options. For example, they could provide a "reset" button or combine a reset button with the control system's "stop" button. The subsystem or equipment could be momentarily disengaged from the

TABLE 3.3 A sample description of an alarm control strategy.

Alarm description	Function or control	Annunciation location	Latching or self-clearing	Reset function

automatic control system and then reconnected. (A motor overload, however, can only be reset at the MCC starter—typically via a reset pushbutton mounted on the MCC starter compartment's front door that pushes the spring-loaded overload relay contact back into position.)

When writing the PCN for system resets, the designer should describe when resets are necessary and how each one functions—especially for complicated systems with multiple subsystems. Typically, the reset function releases all latched alarms and associated permissives so the automatic control system can function normally. *Associated permissives* are all conditions—monitored by sensors whose signals are wired to the control system—that must be satisfied before the equipment can be started.

Remote Indicators. The process control narrative should include a list of all remote indicators (e.g., alarms, monitored set points, status lights, etc.). A *remote indicator* is any signal at a site other than the equipment's local control panel or station.

Local Pushbutton Stations. The process control narrative should include local pushbutton control stations for manual operations, maintenance, and testing. Designers should specify the type of controls and indicators needed [e.g., start/stop, local/remote, speed, and status lights (run, stop, fault, etc.)].

Uninterruptible Power Supplies. The process control narrative should specify which equipment must have an uninterruptible power supply (UPS) in case the utility's main power source fails. Designers not only should note which equipment must have a UPS and which can withstand a short power interruption (minutes) until the generators are fully on-line, but also calculate the power load and duration (in minutes) required.

This information should be shared with the project's electrical engineer.

Vendor-Supplied Packaged-Equipment Panels and Controls. The process-control narrative should provide simple descriptions of vendor-supplied packaged controls, their functions, and which parameters must be monitored or controlled remotely. Designers also should provide all related vendor information (e.g., functional descriptions, control-panel drawings, wiring schematics, and equipment catalog cuts) to the I&C and electrical engineers as soon as possible. In addition, designers should let the I&C engineer know what instruments the packaged equipment will need that are not provided by the vendor so these instruments can be included in the I&C specifications. Finally, to ensure that the design is well-coordinated, the I&C and electrical engineers should be asked to review the packaged-equipment specifications.

PROCESS-CONTROL-SYSTEM CONFIGURATION AND COMMUNICATION SCHEMATICS. The system configuration is a basic drawing of control-system components and their locations. The system configuration and communication schematics show all of the process control system's major components, including workstations, printers, modems, RTUs, PLCs, interfaces between units, networks, communication media, and communication protocols.

CONTROL-PANEL DESIGNS. This is a basic component of control-system engineering. Without detailed designs, the I&C contractor often has no guidelines on how to size and lay out each control panel.

A good control-panel design must

- Size the panels appropriately (to ensure that they will fit in the allocated spaces);
- Provide exterior views of the panels that illustrate doors and hinges; and
- Provide interior panel layouts of all components, along with the naming and numbering standards.

For more information on control-panel design, see Chapter 8.

INSTRUMENTATION- AND PANEL-LOCATION DRAWINGS. These drawings should show the physical location of all I&C equipment. They also can be included with the electrical-conduit layout drawings, or they can be included with the mechanical-equipment layout drawings (in which case, separate electrical drawings should be made based on the mechanical drawings).

INSTRUMENT LOOP DIAGRAMS. ISA Standard S5.4 details the content and format standards for loop diagrams. Each diagram traces the complete route of a control loop's wiring. They are useful during installation and critical for system maintenance.

Loop diagrams cannot be prepared during the design phase because they involve actual vendor shop drawings and documentation. Instead, they are created during the shop-drawing phase by the design engineer or the contractor's system integrator.

Typically, it is better to have the design engineer make the loop diagrams, but it depends on whether the owner is willing to pay for this service.

INSTRUMENTATION INSTALLATION DETAILS. Whether instruments will operate properly at a treatment plant often depends on how they are installed. Installation drawings show how instruments should be mounted and installed. When creating these drawings, designers should make sure each instrument will be accessible for maintenance. Designers also should review specific manufacturer installation details for any special requirements that should be included with these drawings.

INSTRUMENT AND INPUT/OUTPUT LISTS. Complete listings of instruments and I/O locations will help the contractor build and program an effective control system.

SPECIFICATIONS AND BID DOCUMENTS. To keep the project on track, the design team should establish milestone dates for all design documents (Table 3.4). How many documents will be due on a given date depends on the contract between the owner and the design engineer, but at least one milestone date should occur before the owner releases the bid documents to ensure that the designer is making timely progress and fulfilling the project's objective(s).

For more information on bid specifications, see Chapter 15.

SOURCES OF TECHNICAL DESIGN INFORMATION

Several sources of information, including technical societies, vendors, classes, Web sites, books, and periodicals, are available to automation-system designers.

TECHNICAL SOCIETIES. Several technical societies offer a wealth of information—CDs, conferences, online forums, periodicals, standards, textbooks, videos, workshops, etc.—on instrumentation and control-system design. Of all these organizations, the Water Environment Federation (WEF; Alexandria, Virginia) and the Instrumentation Testing Association (ITA; Henderson, Nevada) tailor their information to wastewater treatment plants, and ITA even provides test reports on instrumentation used at treatment plants.

For more information on technical societies, see the end of this chapter.

TABLE 3.4 An example of a milestone schedule for a typical control-system design for a wastewater treatment plant.

Milestone	Design documents due
30% complete	• PFDs • Drawing list • Specification list • Area classification • Cost estimate
50 to 60% complete	• PFDs • P&IDs • Drawing list • Specification list • Construction cost estimate • Process-control network and system-architecture diagrams • UPS power-distribution block diagrams • Control-room layouts
90 to 100% complete	• Contract bid documents • PFDs and P&IDs • Construction cost estimate • Instrument installation details • Schedules (I/O, instrument, and panel lists) • Complete specifications

VENDORS. Because control systems must be built using commercially available products and services, vendors are an important source of information. Design engineers should keep up to date on vendor information to ensure that their control systems make use of the best technologies available.

Getting this information is not a problem. Design engineers typically are bombarded with vendor information—via brochures, catalogs, CDs, Web sites, etc.—that they must sort through and evaluate. Fortunately, the Internet makes this task easier.

After a preliminary review using Internet sources, design engineers still need to contact vendors directly for more project-specific information. Then they can request a meeting with the vendor (or vendor's representative) for a more in-depth presentation.

An excellent opportunity to meet vendors and compare products is the annual Water Environment Federation Technical Exhibition and Conference (WEFTEC), the largest water and wastewater treatment conference in North America. Other conferences worth investigating include the ISA Expo and the American Water Works Association's (AWWA's; Denver, Colorado) IMTEch Conference for Water and Wastewater Information Management Technology and Applications.

Selecting Vendors and Technologies. Design engineers should thoroughly review product literature, specifications, references, and detailed manuals before selecting equipment and vendors for a wastewater treatment project. Of course, previous experience with particular products and their manufacturers should influence the final decisions.

Incorporating Selected Technologies into Design. When designing a control system for a wastewater treatment plant, engineers must accommodate multiple vendor and equipment variations. Public works projects typically are awarded to the lowest bidder, which often chooses the least-expensive instruments that meet project specifications. So, engineers must allow for such eventualities and adjust the design accordingly. For example, space must be allocated to accommodate the differences in dimensions among various vendors' equipment. Fortunately, the wiring for most treatment-plant instrumentation has been standardized {e.g., 4- to 20-mA outputs, standard protocols [i.e., highway addressable remote transducer (HART) and Fieldbus]}.

TRAINING. Appropriate training in instrumentation and control systems can be obtained from many sources, including universities, technical societies, vendors, and experienced colleagues.

For example, some technical societies now offer Internet-based training, which allows individuals to earn professional-development or continuing-education credits without having to travel. In addition to online presentations, this training approach allows students and the instructor(s) to interact via a phone link.

Vendors often will offer introductions to their product line via free seminars that can be held at either the engineer's or the vendor's facility. Most vendors also offer some form of training on their products and services that typically is tailored to users.

WEB SITES. The Internet itself is a vast pool of information that sometimes can be overwhelming, but it can be harnessed with the right tools and some patience and perseverance. A good place to start is a general search engine; some of the currently popular ones are listed at the end of this chapter.

Useful information is also available on professional societies' Web sites, including those of WEF (www.wef.org), ISA (www.isa.org), and ITA (www.instrument.org). For more information, see the end of this chapter.

BOOKS AND PERIODICALS. "Low-tech" media, such as books, journals, and magazines, are great sources of technical information. Many instrumentation, control, and automation trade magazines also publish annual vendor guides. For more information, see the end of this chapter.

PREVIOUS DESIGN EXPERIENCE. Of course, there is no substitute for actual experience. Many engineering firms offer a wealth of experience and can be a tremendous asset to users, who may lack the time or resources to undertake the work themselves.

POST-DESIGN PHASE

DESIGN SERVICES DURING CONSTRUCTION. The design team's job is not over when the design documents are completed. During project implementation, the team should be involved in shop-drawing reviews, change orders, contractor and vendor meetings, requests for information, and other construction activities to ensure that the final control system will function as intended.

Reviewing Shop Drawings. Contractors and system integrators typically prepare detailed drawings of instrumentation systems before installing them. These drawings include loop diagrams, panel drawings, manufacturers' data, PLC programs, distributed control system (DCS) configuration, interconnecting wiring diagrams, etc. The design team should thoroughly review these documents to ensure that they comply with the design's requirements.

Processing Change Orders. No design is perfect, and a certain amount of change orders is inevitable. Changes can be caused by user requests, design omissions, design errors, field conditions, or unforeseen situations. When such changes happen, design engineers should evaluate them, estimate their costs, and produce the documents needed to support them.

Responding to Requests for Information. Contractors and system integrators often will have questions for the design team during the construction phase. Such questions, typically called requests for information (RFIs), and their responses must be documented and filed with other design and project documents for future reference. Electronic documentation systems can make it easy to archive and retrieve this information.

Meeting with Contractors and Suppliers. The design team will often meet with the contractors and system suppliers to clarify design issues that are not easily addressed in other forums. During the early phase of the contract, design engineers should schedule monthly meetings with the system integrator. Later meetings should include electrical; heating, ventilation, and cooling (HVAC); and other subcontractors for coordination purposes.

Making Regular Site Visits. Although a construction management team (i.e., resident engineer, field inspectors, etc.) typically handles daily construction activities, the design team should visit the job site periodically to address and resolve implementation issues, such as alternate installation procedures, that could affect the design and overall project success.

Witnessing Factory Acceptance Tests. Contracts for I&C systems typically specify that these systems must be fully tested at the manufacturer's facility before shipment. Depending on the size and type of control system, the factory tests may involve simulation software (Dougall, 1997; Blevins et al., 2000).

The design engineer, system user, or other designated parties typically witness this control-system testing to ensure that the system is fully debugged, complies with the design requirements, and is free of internal-wiring errors when the equipment leaves the factory. This reduces the control-system startup time and confines most of the remaining errors to field wiring or other field-related issues.

At least two months before testing begins, the contractor must submit to the engineer and owner a schedule of test dates and a factory test procedure. The factory test procedure must include all testing steps, checklists, settings, and a list of all test equipment to be used to verify that the scheduled testing will fulfill the contract requirements.

When the factory tests have been successfully completed, the contractor's control-system supplier should submit a test report to the design engineer with the test results and a clear indication that all deficiencies found during the factory test were corrected. The equipment should not be shipped until the contractor's control-system supplier receives a notice from the engineer that the test results were acceptable.

System Performance. Before the witnessed factory test, the instrumentation system supplier should perform a full-system test, during which the entire system must operate continuously for 100 hours without failure, in accordance with the requirements of the specifications and drawings. The purpose of this test is to ensure that any component failures are discovered before the witnessed factory test begins.

This test should verify that all the components work properly as an integrated system. It involves assembling the entire control system—hardware, software, peripheral devices, and interconnecting cables—on the factory test floor and using a workstation loaded with simulation software (Simple control systems usually can be verified without simulation software) to operate the control system and simulate both treatment processes and signals from field devices that cannot be connected to the panel. If any component fails during this test, it should be replaced and the test restarted.

While the entire system is assembled, the supplier should inspect all panels, consoles, and cabinets to ensure that the following are correct and consistent:

- Nameplates and tags;
- Wire sizes and color coding;
- Terminal block contract ratings and number;
- Annunciator and terminal block spares;
- Proper wiring practices and grounding; and
- Enclosure flatness, finish, and color.

Once the 100-hour test has been completed successfully, the contractor should certify this success in writing, submit this document to the design engineer, and then schedule the witness testing.

Hardware Testing. Each hardware component should be tested to verify that it operates properly by itself. Such tests should include

- Alternating current (AC)/direct current (DC) power checks,
- Power-failure and -restart tests,
- Diagnostics checks, and
- Proof of all specified functions.

Also, all I/O devices and their components should be tested to verify operability and basic calibration.

Where applicable, proper communication among system components should be verified. Communication between each remote I/O and control-panel PLC and the master control-panel PLC also should be demonstrated.

Software Testing. The contractor should demonstrate that all system software—especially its security components—works as intended. For small or simple systems, this test may involve wiring some simulated signals (toggle switches, lights, analog signal from a signal generator, etc.) to the control panel. For larger or more complex systems, this test should involve the use of simulation software installed on a separate computer. During this test, the contractor should demonstrate the operation and display of all software based on a simulation involving 100% of I/Os, both analog and discrete.

Simulation Software. Many system integrators now use simulation software to test the program logic before scheduling a factory test. The contract should specify that simulation software be used during testing as well. This software should have the following features:

- Process-specific libraries of unit operations and physical properties;
- The ability to define characteristics (e.g., flow, level, temperature, and pressure) of discrete devices, loops, and objects in the control system model;
- The ability to simulate all process-control-system signals, alarms, and shutdown scenarios in real time;
- The ability to model all process feedback using graphics and symbols;
- Full, workstation-based control over the behavior of each object in the model;
- The ability to develop custom graphics;
- Point-and-click access to object parameters as they are being executed in the background;

- The ability to communicate directly with operator interface terminal (OIT) software;
- The ability to communicate with specified PLCs;
- Operator-training tools that meet Occupational Safety and Health Administration (OSHA) 1910.119 recommendations; and
- The tag numbers for panels, equipment, and instruments used in simulation software should be identical to those in the contract P&IDs.

Also, the contractor should be required to retain the services of the simulation software supplier's factory-trained technician for at least 2 days. These services should be provided during factory testing, startup and field testing, and training. The contractor should bear all costs (including transportation and lodging expenses) related to these services.

Operator Interface Terminals. Before the control system is staged and tested, the OIT displays must be configured as agreed (e.g., layout, passwords, and security), loaded onto the process control system server or standalone PC OIT), and tested.

The control-system's OIT software should include the following:

- Graphic symbols of the project's process schematics, as shown on the contract P&IDs;
- Graphic symbols of each field panel, as shown on the contract P&IDs;
- Graphic symbols of each field instrument, as shown on the contract P&IDs;

During OIT testing, testers should

- Review the main menu display contents and demonstrate how an operator will navigate within the overall display structure (the main menu display should provide a list of all available displays and a link to the various subsystem displays);
- Demonstrate which displays are assigned to which keys on the workstation keyboard;
- Confirm that all graphic display components (e.g., layout, symbols, and color scheme) are correct;
- Prove that standard-alarm management displays (e.g., current alarm display and alarm history) function as intended; and

- Show that each specified type of report can be generated and printed.

CONSTRUCTION MANAGEMENT. Once the new control system is delivered to the site, the construction management team is responsible for installation, field tests, and other related activities.

Manufacturers may need to certify the installation of specialized equipment (e.g., process analyzers). They also can test equipment, provide startup assistance, and train users.

Intelligent field instruments and asset-management software can simplify commissioning activities, such as system configuration and loop checkout. Intelligent field devices or instruments generate equipment status and diagnostic information that are gathered by the asset-management software at a central site (e.g., a control room or maintenance shop), where the installation team can review it all. This approach is more efficient than the traditional method of having team members travel throughout the plant to test each device individually (Johnson and Bailey, 2000; Raven, 2001).

Field Tests. The following field tests (site acceptance tests) should be done after the new control system has been installed and wired, and each instrument has been verified and calibrated. Also, an electrician should have double checked the integrity of all wiring, including terminations (panel and field), insulation, continuity, and grounding.

Loop Testing. Loop testing confirms that each "control loop" in the system functions as designed. A *control loop* is a combination of one or more interconnected instruments that are arranged to measure or control a process variable (ISA-67.02.01-1999; ANSI/ISA-5.1-1984).

All loop testing should be overseen and approved by the site engineer. The resulting loop sign-off sheets should be kept as test records.

Integrated System Testing. After each control loop has been tested, the construction management team should test the entire control system, including signals to and from controlled field equipment (e.g., motors, sensors, transmitters, valves, and variable-speed drives). Before the test begins, the system integrator should write up the entire test procedure and submit it to the design engineer for approval. It should include a list of items to be checked, such as a database listing grouped by I/O for each area of the treatment plant. The database listing should include all wired I/O points, virtual (nonphysical) points, and derived points. Sorting the database I/O listing by process area helps making system checking easier.

During the integrated systems test, the control system must operate for a specified period without failures. If the factory acceptance test was well-executed and all control-system equipment from the factory was tested properly, then the integrated systems test should only be limited by field wiring issues.

The integrated system test should be overseen and approved by both the engineer and the owner. A log of all testing activities—with the system integrator's, engineer's, and owner's initials next to each functioning item—should be kept as test records.

Final Acceptance. The new control system should not be formally turned over to the owner until all startup and commissioning activities are complete. The contractor also should be required to meet the following two conditions before the utility formally accepts and pays for the system:

- All as-built drawings and operations and maintenance manuals must be approved and delivered to the owner and
- All user training must be completed (see Chapter 18 for a detailed discussion of training requirements).

These prerequisites ensure that the owner ends up with a fully documented system that staff is comfortable using.

REFERENCES

___(2003) *Second Generation Searches on the Web*. University of Albany Libraries, www.library.albany.edu (March).

Blevins, T.; McMillan, G.; Wheatley, R. (2000) The Benefits of Combining High Fidelity Process and Control System Simulation. *Proceedings of the International Society for Measurement and Control EXPO 2000*; New Orleans, Louisiana, Aug 21–24; Instrumentation, Systems, and Automation Society: Research Triangle Park, North Carolina.

Control Systems Integrators Association (2000) *Guide for Selecting & Working with a Control Systems Integrator*; Volumes I and II; Control Systems Integrators Association, Exton, Pennsylvania. www.controlsys.org (accessed June 2006).

Dougall, D. J. (1997) Applications and Benefits of Real-Time I/O Simulation for PLC and PC Control Systems. *ISA Transactions*, Issue 4; Instrumentation, Systems, and Automation Society: Research Triangle Park, North Carolina.

International Society for Measurement and Control (2004a) *Integrating Electronic Security into the Manufacturing and Control Systems Environment*; ISA-TR99.00.02-2004; Instrumentation, Systems, and Automation Society: Research Triangle Park, North Carolina.

International Society for Measurement and Control (2004b) *Securing Technologies for Manufacturing and Control Systems*; ISA-TR99.00.01-2004; Instrumentation, Systems, and Automation Society: Research Triangle Park, North Carolina.

Johnson, B. L.; Bailey, K. (2000) Leveraging Innovative Technology Delivers Benefits in Start Up and Daily Operations. *Proceedings of the International Society for Measurement and Control EXPO 2000*; New Orleans, Louisiana, Aug 21–24; Instrumentation, Systems, and Automation Society: Research Triangle Park, North Carolina.

Koons, D.; Conley, G. (2004) How to Select a Control System Integrator without a Dart Board. *Control*, June.

Metcalf and Eddy, Inc. (2003) *Wastewater Engineering Treatment and Reuse*, 4th ed.; McGraw Hill: New York.

Pinto, J. (2003) The Anatomy of Search Engines. *ISA Intech*, Apr 19.

Raven, B. (2001) Online Asset Management Software Facilitates Instrument Commissioning. *Proceedings of the International Society for Measurement and Control EXPO 2001*; Houston, Texas, Sep 10–12; Instrumentation, Systems, and Automation Society: Research Triangle Park, North Carolina.

U.S. Department of Energy; President's Critical Infrastructure Protection Board (2002) *21 Steps to Improve Cyber Security of SCADA Networks*, U.S. Department of Energy and President's Critical Infrastructure Protection Board. www.ea.doe.gov/pdfs/21stepsbooklet.pdf; September.

Water Environment Federation (1997) *Automated Process Control Strategies*, Special Publication; Water Environment Federation: Alexandria, Virginia.

SUGGESTED READINGS

American National Standards Institute Home Page. www.ansi.org (accessed June 2006).

American Water Works Association (2001) *Instrumentation & Control*, 3rd ed.; Manual of Water Supply Practices; American Water Works Association: Denver, Colorado.

Cockrell, G. W. (2001) Practical Project Management—Learning to Manage the Professional, *Proceedings of the International Society for Measurement and Control EXPO 2001*; Houston, Texas, Sep 10–12; International Society for Measurement and Control: Research Triangle Park, North Carolina.

Consulting Specifying Engineer Home Page. www.csemag.com (accessed June 2006).

Control Engineering Home Page. www.controlengineering.com (accessed June 2006).

Control Buyer's Guide Home Page. www.controlmag.com (accessed June 2006).

Engineering Search Engine. www.globalspec.com (accessed June 2006).

General Vendor Directories. www.thomasnet.com (accessed June 2006).

Institute of Electrical and Electronics Engineers Home Page. www.ieee.org (accessed June 2006).

Instrumentation Testing Association Home Page. www.instrument.org (accessed June 2006).

International Electrical Testing Association Home Page. www.netaworld.org (accessed June 2006).

International Society for Measurement and Control Home Page. www.isa.org (accessed June 2006).

International Society for Measurement and Control (2003) *The Automation, Systems and Instrumentation Dictionary*. International Society for Measurement and Control: Research Triangle Park, North Carolina.

McMillan, G. K.; Considine, D. M. (1999) *Process/Industrial Instruments and Controls Handbook*, 5th Ed.; McGraw-Hill: New York.

Meier, F. A.; Meier, C. A. (2004) *Instrumentation and Control Systems Documentation.* Instrumentation, Systems, and Automation Society: Research Triangle Park, North Carolina.

National Fire Protection Association (2003) *Standard for Fire Protection in Wastewater Treatment and Collection Facilities,* NFPA 820; National Fire Protection Association: Quincy, Massachusetts.

WaterWorld Home Page. www.pennnet.com (accessed June 2006).

Whitt, M. D. (2004) *Successful Instrumentation and Control System Design*; Instrumentation, Systems, and Automation Society: Research Triangle Park, North Carolina.

Chapter 4

Process and Instrumentation Diagrams

Introduction	48	*Control-System Architecture Diagram*	56
How To Create Process and Instrumentation Diagrams	49	Process and Instrumentation Diagrams as Communication Tools	57
Process-Flow Diagrams	50	Contracts	57
Symbols	51	User and Owner Feedback	57
Computer-Aided Design Software	53	Operator Training	58
Interactions with Other Design Documents	54	Operator-Interface Graphics	58
		As-Built Documentation	59
Instrument List	54	New Developments	59
Input/Output List	54	Intelligent Process and Instrumentation Diagrams	59
Panel List	55		
Cable and Conduit List	55	*Streamlined Diagram Development*	60
Process-Control Narratives	55		
Panel and Field-Instrument Specifications	55	*Object-Based Schematics*	60
		Modular System	61
Electrical Schematics	56	*Integration with External Data Sources*	61
Instrument Loop Diagrams	56		
Equipment-Location Drawings	56	*Interoperability*	61

Symbol-Management Tools 61	*Construction, Startup, Operation, Maintenance, and Asset Management Uses* 62
Review and Navigation Tools 61	
Component-Management Tools 61	Process and Instrumentation Diagrams for HVAC Systems 62
Automatically Generated Lists 62	References 63
Design History for As-Built Drawings 62	Suggested Reading 64

INTRODUCTION

When designing a new or upgraded treatment plant, instrumentation and control (I&C) systems must be an integral part of the process—not an afterthought. That's why process and instrumentation diagrams (P&IDs) are important.

Process and instrumentation diagrams are schematics that illustrate all the components of a treatment plant. Typically, a P&ID provides enough information for the project team and stakeholders to understand how the treatment process will be measured and controlled but leaves out the details that require a specialist's expertise. These details are covered in related specifications, data sheets, instrument lists, logic diagrams, and installation details.

Design engineers have various philosophies on how to use P&IDs in treatment-plant design. Some engineers, for example, use them to develop piping and equipment designs. However, this approach can make other design documents too dependent on the P&IDs and actually hinder progress.

A more common approach is to use the P&IDs to develop the instrumentation and electrical designs. In this case, the P&IDs would include all instrumentation connections, control panels, workstations, signal lines, and electrical power needs. For example, it would show which instruments and control panels need 120 V AC, and which control valves and other elements need air supply. It also might include other equipment's power needs, motor-control centers (MCCs), variable-frequency drives (VFDs), and power distribution. Electrical engineers would then use the P&IDs to develop power-distribution drawings for all equipment.

Mechanical engineers use the P&IDs to develop piping designs. So, they need the diagrams to show all wall-mounted, panel-mounted, and free-standing instruments, as well as the in-line devices (e.g., magnetic flowmeters and pressure taps) mounted on the process equipment (e.g., tanks, centrifuges, and scrubbers). Sometimes the P&IDs also include instruments that are part of a vendor-supplied skid-mounted package. These details are necessary to ensure that the piping, including required straight runs before and after a meter, is configured properly.

HOW TO CREATE PROCESS AND INSTRUMENTATION DIAGRAMS

When creating a P&ID, design engineers should follow these steps (see Figures 4.1 and 4.2 for examples of typical drawings):

- Start with a process flow diagram (PFD).

- Create a legend sheet specific to the project, defining all symbols, abbreviations, numbering scheme, and other conventions to be used.

- Develop instrumentation numbering system based on the Instrumentation, Systems, and Automation Society (ISA; Research Triangle Park, North Carolina) Standard ISA-5.1 and customize to adapt to specific project or end user requirements.

- Add the instrument bubbles for all process measurements.

- Add bubbles for process displays, including those for all panel-mounted and process-mounted field instruments.

- Add all signal lines showing interconnections between all field sensors, transmitters, panel-mounted instruments, and control-room equipment, including any basic interlocks using the appropriate ISA symbols.

- Show all inputs and outputs to and from each PLC and DCS.

- Show power requirements for all instruments and panels. Be sure to indicate where uninterruptible power supply (UPS) power is needed.

- Add any special notes needed to supplement the information shown.

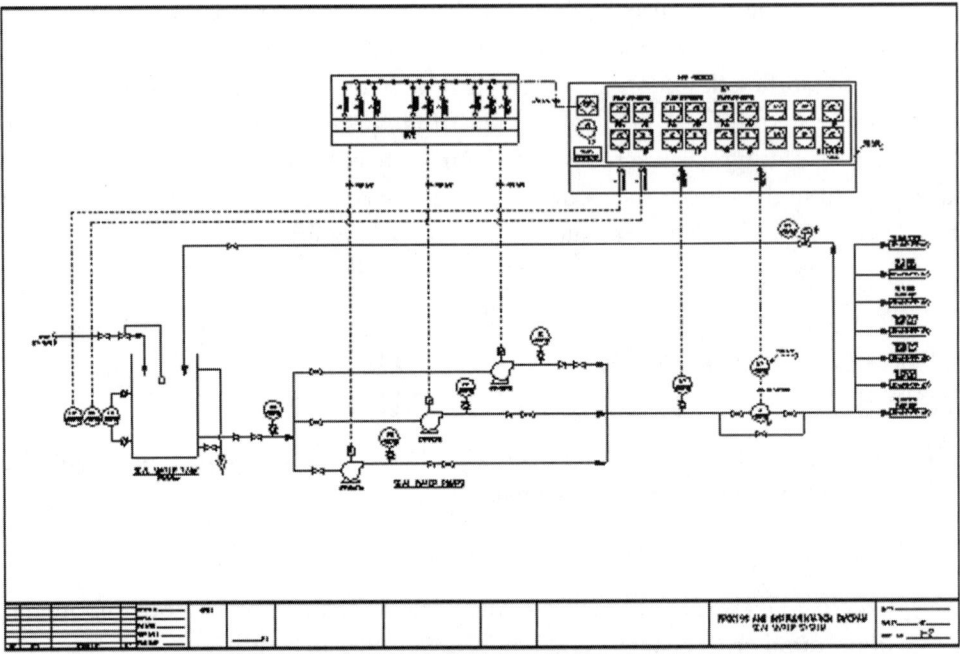

FIGURE 4.1 Example of a P&ID without operator interface.

- Complete drawing by making sure all devices are tagged according to the convention established in the legend sheet and any connections to other drawings are correctly shown.

PROCESS-FLOW DIAGRAMS. The process-flow diagrams that engineers create in the initial design phase establish the major equipment that will be used in each treatment process (Figure 4.3). These documents also would include basic materials-balance information for expected operating conditions (e.g., normal dry-weather flow and maximum flow).

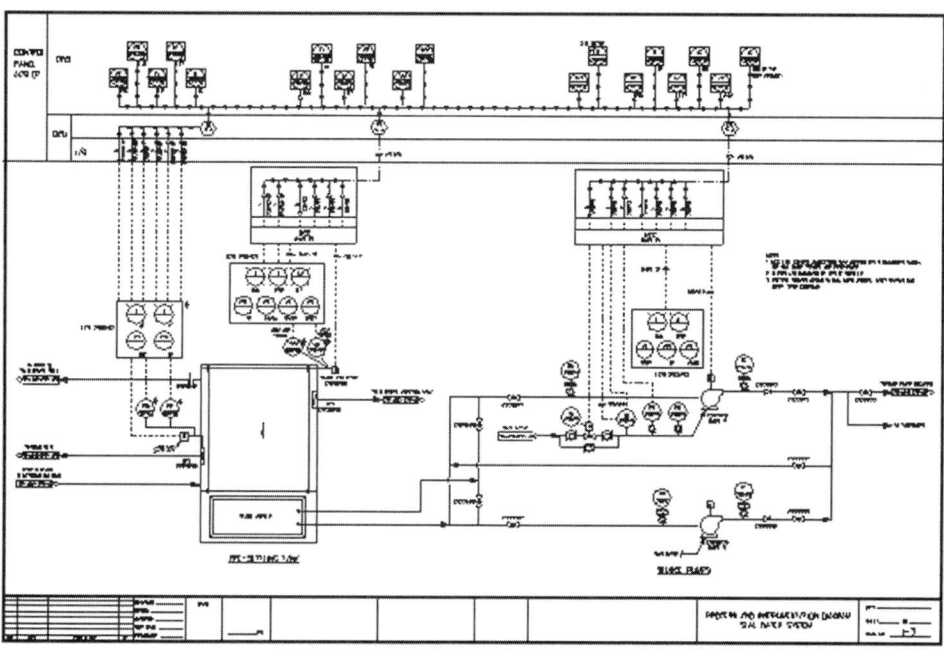

FIGURE 4.2 Example of a P&ID with graphic operator interface.

Once design engineers have completed the PFDs and incorporated all owner comments, then process, mechanical, or I&C engineers (depending on the organization) use them to develop P&IDs. These documents provide more details, such as the equipment, piping, instruments, signals, valves, and control panels needed for wastewater screening, pumping, primary treatment, aeration, and so on.

SYMBOLS. The symbols and identification codes used to identify instruments in water and wastewater plants typically are based on ISA standards, with modifications for the unique nature of water and wastewater treatment plants (Harrod, 2000; Figure 4.4). Standard S5.1 (ISA, 1992), which defines the symbols used in P&IDs, was created to help promote uniformity in the instrumentation industry. The symbols in

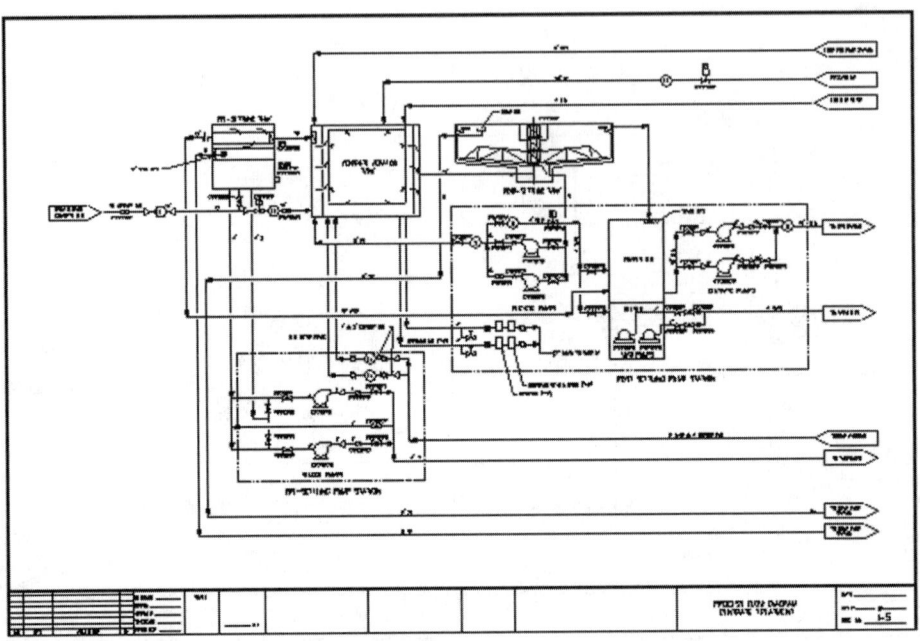

FIGURE 4.3 Example of a process-flow diagram.

this voluntary, consensus-based standard are adaptable and can be used in many applications. Through regular updates of the standard, ISA has incorporated other symbol standards endorsed by the American National Standards Institute (ANSI) and the Institute of Electrical and Electronics Engineers (IEEE). Another standard used in developing P&IDs is ISA-5.3-1983, *Graphic Symbols for Distributed Control/Shared Display Instrumentation, Logic and Computer Systems*. ISA also provides its library of standard symbols in an electronic format that can be easily imported into any computer-aided design (CAD) program.

The Society's standards only provide a framework for developing P&IDs, so documents prepared by different designers may vary significantly. Therefore, those responsible for developing P&IDs should include a legend defining all lines, symbols, abbreviations, instruments, and equipment-tagging number conventions used in the documents.

COMPUTER-AIDED DESIGN SOFTWARE. Today, most engineers use a CAD program to prepare P&IDs. The advantages of CAD software include

- A standard library of symbols for all project documents;
- The ability to reuse relevant P&IDs from previous jobs;
- The ability to reuse a template for a system-specific P&ID;
- A well-documented record of changes (in accordance with good information-technology management practices);

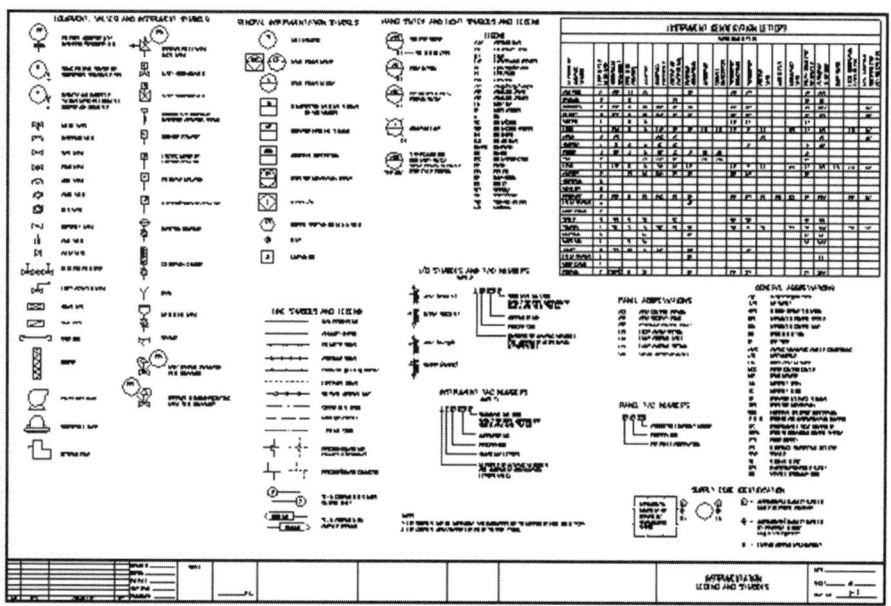

FIGURE 4.4 Example of a P&ID legend and symbols.

- Multiple options for archiving documents on removable media [e.g., recordable compact disc (CD), recordable digital versatile disc (DVD), or tape backup] or long-term storage [e.g., network server or personal computer (PC) hard drive]; and
- Several options for rapidly transmitting documents (e.g., e-mail, removable media, Internet sites, or intranet sites).

Commercial CAD software can produce two- or three-dimensional drawings for P&IDs and other engineering designs. The most popular programs use the Microsoft Windows operating system. Some manufacturers offer add-on software for existing CAD programs that can thoroughly check digital models for errors early in the design process. Such add-ons also can allow users to insert equipment "blocks" that automatically draw equipment, add connecting lines, and insert "text bubbles" identifying the parts. The blocks also contain equipment specifications, such as size and identification number. As each equipment block is placed on the drawing, the software creates a related database that can serve as an equipment list (AutoPLANT case studies, www.bentley.com).

For more information on CAD software, use any Internet search engine.

INTERACTIONS WITH OTHER DESIGN DOCUMENTS. P&IDs are typically supplemented by the following I&C design documents:

- *Instrument List.* This list provides more information on each instrument in the project P&IDs. It notes the instrument's tag number, process signal range (the minimum and maximum values that the instrument is to be calibrated for in engineering units; this is different from the instrument design range listed in the instrument specification), specification reference (detailed instrument specifications are located elsewhere in the project specifications and are referenced by their respective specification section), and drawing reference numbers (P&ID, location, etc.).
- *Input/Output List.* This list tabulates the actual input and output points for each signal shown on the P&IDs. It describes each point, noting its type (digital or analog input or output), associated programmable-logic-controller (PLC) or distributed-control-system (DCS) processor, and engineering units (units of measure that are useful for operators).

- **Panel List.** This complete list of all control panels is intended to aid contract bids. It includes each panel's location, National Electrical Manufacturers Association (NEMA) rating, and supplier. (For information on control-panel design, see Chapter 8.)
- **Cable and Conduit List.** This list tabulates the instrumentation-wiring requirements, including power, other electrical, and any special manufacturer's cables, as well as the quantity furnished with the instrument (e.g., magmeter's primary cable, float-switch cable, etc.). Depending on the organization, this list may be part of the I&C or electrical designs.
- **Process-Control Narratives.** Process-control narratives (PCNs) describe the strategies and exact sequence of events to operate all I&C equipment automatically or manually. This narrative can be an outline of the necessary steps (the most common option), a Boolean logic diagram, a sequential function chart, a flow chart, or a ladder logic diagram (Water Environment Federation, 1997).
- **Panel and Field-Instrument Specifications.** Detailed specifications of all panel and field instruments typically are prepared in Construction Specifications Institute (CSI) format by design engineers as part of the contract documents. The CSI specifications are not a substitute for the ISA specification forms, although there is some overlap in these documents. Owners and engineers should require that the completed ISA forms be provided by the system integrator as part of a complete set of submittals to ensure compliance with contract requirements and uniformity of the information provided. (See Chapter 8 for more detail). Instrument-specification forms are also available from the ISA in both paper and electronic formats. The ISA specification forms are typically provided by system integrators and those responsible for actual procurement of instruments and address such information as
    - Type of instrument;
    - Materials of construction for wetted and non-wetted parts;
    - Process and environmental conditions in which the instrument is used, (i.e., outdoors, corrosive area, or hazardous area);
    - NEMA rating;
    - Power requirements;
    - Signal output requirements;

- Alarm signal requirements, if any;
- Special mounting hardware required;
- All options to be furnished;
- Instrument span, calibrated range, and engineering units;
- Process connections [size and type of connections (i.e., flanged or screwed)];
- Transmitter requirements (i.e., whether remote or integral to sensor); and
- Manufacturer model numbers.

However, because most public works projects require competitive bidding for procurement, the design engineer's specifications should be general enough to enable several reputable vendors to bid on the project. More-detailed instrument specifications, using ISA forms, could be a bid-submittal requirement for I&C contractors.

- **Electrical Schematics.** Electrical schematics of some instrumentation systems may be needed to illustrate wiring details not apparent in the P&IDs. Consulting-engineer designs typically include these schematics with the electrical-design documents.

- **Instrument Loop Diagrams.** Loop diagrams are an important part of I&C design, but because they are specific to the actual devices used, they cannot be created until the system integrator has selected and the user or design engineer has accepted all instruments, control panels, control systems, etc. So, loop diagrams typically are prepared by the I&C contractor or system integrator, not the design engineer. (For specific guidance on preparing loop diagrams, see ISA Standard S5.4, 1991.)

- **Equipment-Location Drawings.** These mechanical or infrastructure drawings depict the actual locations of all I&C devices. Each device must be clearly identified by its tag number, as defined on the P&IDs. These drawings then enable the electrical engineer to create appropriate cable- and conduit-routing diagrams.

- **Control-System Architecture Diagram.** Control-system architecture diagrams are simple schematics of the control-network topology and interconnections among PLCs, DCSs, remote inputs/outputs (I/O), MCCs, VFDs, intelligent-valve networks, and other control components (Figure 4.5). They also illustrate any necessary network equipment (e.g., Ethernet hubs, switches, fiber transceivers, and routers).

For more information on all the documents included in a complete automation design, see Chapter 3.

PROCESS AND INSTRUMENTATION DIAGRAMS AS COMMUNICATION TOOLS

Process and instrumentation diagrams are the foundation of I&C engineering design. They are an essential element of the I&C design package. Process and instrumentation diagrams are multidiscipline drawings that present information from the mechanical, process, I&C, and electrical groups. They are also key documents used by system integrators in developing a bid for the I&C portion of a contract. Mechanical and process groups use P&IDs to show the various piping connections between process equipment, including main process equipment found in wastewater treatment facilities (i.e., process tanks, wet wells, channels, pumps, bar screens, grinders, and specialized equipment), manual and automated valves, pipe reducers, in-line instruments, special piping requirements (double wall pipes for chemicals lines, etc.), vendor supplied packages, and interconnections to the process, Electrical groups use P&IDs [supplemented by other documents (e.g., motor lists and equipment lists)] to determine the power requirements for instruments and motorized equipment, signal wiring requirements, networking requirements for the control system, interfaces to motor control centers and variable-speed drives, and so on. Process and instrumentation diagrams are also used by owners and end users for review of design documents to ensure that the design engineer has addressed their needs. They are also used by owners for training, operation, and maintenance.

CONTRACTS. Water and wastewater utilities use P&IDs when obtaining bids from contractors. Basically, "contract documents" consist of contract specifications, P&IDs, and their supporting documents.

USER AND OWNER FEEDBACK. During design reviews, control-system users and owners tend to base their written or oral feedback on the project's P&IDs. [Make sure they understand that other documents (e.g., written scope, design report, and standards) are used to implement the P&IDs.] Users and owners also use P&IDs to

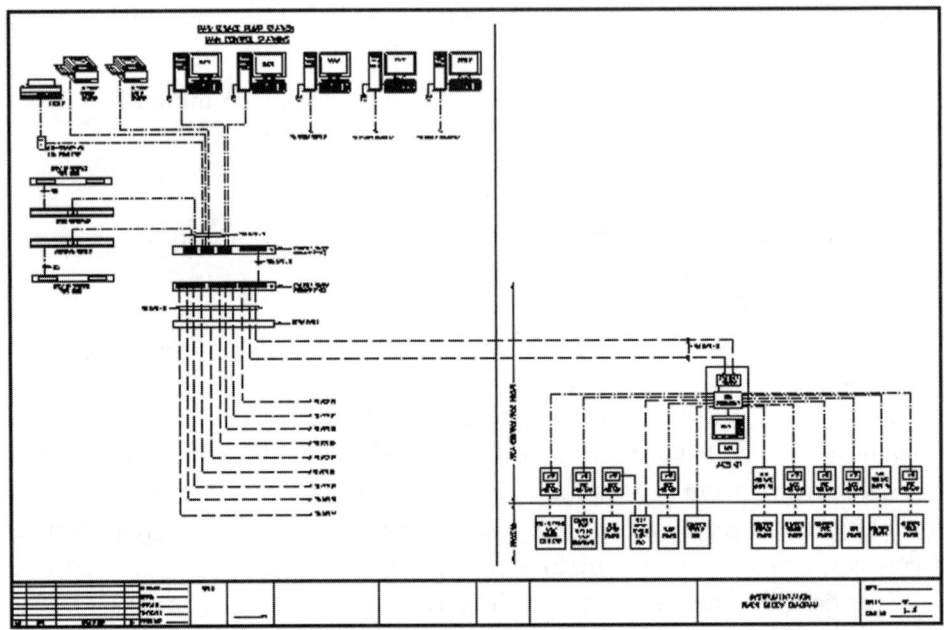

FIGURE 4.5   Example of a control-system block diagram.

verify that the concepts presented in the design report, flow diagrams, written scope, and standards are actually implemented in the new or upgraded system.

OPERATOR TRAINING. Because P&IDs show the process flow and all associated instruments, controls, panels, and major equipment, they typically are used when training plant operators. Process flow diagrams also are used because they can be easier to understand than P&IDs.

OPERATOR-INTERFACE GRAPHICS. Many treatment plants include graphics in their process-control or supervisory control and data acquisition (SCADA) system.

These graphics typically are displayed on operators' workstation screens. Depending on the organization's preferences and practices, these graphics can be developed by the

- Consulting engineer,
- Contractor's instrumentation system integrator, or
- User or owner.

Process and instrumentation diagrams and PFDs are used to develop process graphic displays. The graphics are designed to represent the process as closely as possible, as depicted on the P&IDs and PFDs.

AS-BUILT DOCUMENTATION. At the end of a project, users or owners typically require contractors to provide a set of "record" or as-built P&IDs—drawings that reflect the changes that occurred during construction and startup. They not only serve as an accurate record of the new or upgraded plant's actual equipment and system configurations, but also can be used as a training tool. As-built P&IDs also provide a basis for future renovations or upgrades.

NEW DEVELOPMENTS

The current P&ID trends include intelligent P&IDs and P&IDs for heating, ventilation, and cooling (HVAC) systems.

INTELLIGENT PROCESS AND INSTRUMENTATION DIAGRAMS. Today's CAD software packages allow design engineers to create intelligent P&IDs with minimal data. Users simply choose appropriate symbols and tags from pull-down menus and place them in the drawings with the help of toolboxes containing all symbol-placement and -modification capabilities. The software converts the drawing into an intelligent object model. Adding intelligence involves creating objects that include properties, methods, and relationships for each component (element) in the drawing. Thus, components that were symbols in a schematic drawing then become objects in an object model. Intelligent objects derive information from data entered with the object. Intelligent objects automatically inherit appropriate attributes from other objects in a process line (e.g., reducer size is based on the size of components on either side). Intelligent objects provide intelligent checking for data validity and consistent engineering practices. The software inherent with intelligent objects automatically

- Allows access to all pertinent component information by simply selecting the component on the P&ID. Software displays a list of piping, valves, instruments, instrument loops, and equipment in the P&ID.
- Reports conflicting flow, unconnected lines, and other inconsistencies.
- Checks for valid data input and inconsistencies between P&ID sheets.
- Provides integration with existing automation tools (design, engineering, construction, startup, and operation and maintenance).
- Provides links to project schedule, procurement databases and spreadsheets, and other information through a standard open database connectivity (ODBC) interface.
- Handles symbol orientation and line breaks, reducing the number of steps necessary to place all inline P&ID components. It also automatically manages all element symbology according to project standards, as defined by the user. In addition, it tracks all links on each drawing so, for instance, each arrow representing a link directing viewers to another drawing can actually send them to the correct drawing.

The CAD software also allows design engineers to integrate various databases and other data repositories (spreadsheets, text files, other drawings, etc.) into one I&C database linking the P&IDs with other supporting documents (Kelm, 2002; Knapp, 1999). This database contains all the data for every instrument, device, process line, and valve. Every change made to the database is instantly reflected in the P&ID. All documents are linked together via the equipment tag numbers.

The decision to use this software must be made early in a project because of the related learning curve. For more information on intelligent P&ID software, see the end of this chapter or use any Internet search engine.

Streamlined Diagram Development. Using intelligent P&ID software to create P&IDs and associated documents can reduce drafting time by 40 to 70%, compared to conventional methods. These savings are a result of the following:

Object-Based Schematics. The object-oriented software automatically reports conflicting flow, unconnected lines, or other inconsistencies. It can also check for valid data input and inconsistencies between P&IDs based on the engineer's or user's standards or project specifications.

Modular System. The software is modular and has an extensive symbol library based on ANSI standards. As a user develops the P&ID, each symbol automatically inherits data from related components. If, for example, the design engineer changes the size of a process line, the valves and other components on the line are automatically resized to match. It also generates detailed process-line, valve, instrument, and equipment lists automatically, improving accuracy and dramatically shortening the overall design time.

Integration with External Data Sources. The software can connect with external ODBC-compliant data sources, so front-end process design information and other valuable data are accessible as PFDs and P&IDs are being created.

Interoperability. The software can import and directly edit .dwg files (e.g., Microstation and AutoCAD files). It also can use block and cell libraries from other CAD programs and use them to create new drawings.

Symbol-Management Tools. The software's standard symbol libraries reduce the time needed to create new symbols. Its symbol-placement tools help users lay out diagrams faster.

Review and Navigation Tools. Drawing-review and auto-zoom features enable users to quickly locate specific components or lines anywhere in the entire set of project drawings.

Component-Management Tools. Software features for placing and manipulating components enable users to draw and edit P&IDs more rapidly. For example, the software automatically repairs line breaks when in-line components are moved or deleted. Component grouping enhances placement of commonly used components and provides more global control of symbol attributes. A group of components can be defined for various business or project purposes. For example, in the drawing, instrument engineers might want to give the instrument lines and symbols a group name for a specific use (i.e., odor control group). In this case, the group name exists only in the drawing for notational purposes and does not have anything to do with the tagging functionality and its stored values. The software's predefined grouping collections of components or assemblies provides the user the ability to assign logical collections of components for convenient group manipulations.

Automatically Generated Lists. An intelligent P&ID software program eliminates the tedious task of making the related lists—line lists, equipment lists, instrument lists, I/O lists, panel lists, and so on—because it automatically generates them as the P&ID is drawn. The lists can be formatted and printed in any OBDC-compliant spreadsheet or database program, such as Microsoft Access. Users can create custom lists by selecting and highlighting components of the software's master list of the piping lines, valves, instruments, instrument loops, and equipment in the P&ID.

Design History for As-Built Drawings. Documenting the design's complete history in a comprehensive log or journal enables users or project managers to understand the need for various design changes. The log or journal tracks each change, including what changed, when (date and time), who made it, and why.

Construction, Startup, Operation, Maintenance, and Asset Management Uses. During construction, the intelligent P&ID software can track which components have been purchased, ordered, delivered, installed, and tested. It also can highlight the construction status of the P&ID components as reflected by the project-scheduling software. After startup, it can highlight the equipment to be serviced, according to the maintenance-tracking software. The software also can enable treatment plant personnel to optimize the use of existing assets, while vastly improving access to and use of critical facility information.

Basically, the software can order or group information in any logical sequence for various uses. It can be linked to project-schedule and procurement databases and spreadsheets, plant models, and other data via a standard ODBC interface.

PROCESS AND INSTRUMENTATION DIAGRAMS FOR HVAC SYSTEMS.

Treatment-plant HVAC-control systems have gotten more complex. For example, boilers fueled by digester gas that use natural gas or other fuels as backup need an HVAC system that coordinates well with the digester gas-handling system. Because such HVAC-control systems use the same instruments, PLCs, DCSs, and control panels as the treatment-process I&C system, consulting engineers have begun using the same design tools for both systems.

Using P&IDs for all control systems, regardless of the application, simplifies the overall treatment-plant design. Design engineers typically can use the same specifications for both systems (except for system-specific requirements), so fewer types of hardware and software are needed. Also, the project documentation is easier to review and maintain.

REFERENCES

American National Standards Institute Home Page. www.ansi.org (accessed June 2006).

Bentley Systems, Inc. AutoPLANT Case Studies: *3D Model of Wastewater System Reduces Design Time*, Bentley: Exton, Pennsylvania. www.bentley.com (accessed June 2006).

Bentley Systems, Inc. (2004) PlantSpace P&ID Quick Startup Guide. www.bentley.com (accessed June 2006).

CATIA/CADAM P&ID by IBM. www.306ibm.com (accessed June 2006).

Harrold, D. (2000) How to read P&IDs. *Control Engineering*, Aug 1.

International Society for Measurement and Control (1992) *Instrumentation Symbols and Identification*; ISA-5.1-1984 (R1992); International Society for Measurement and Control: Research Triangle Park, North Carolina.

International Society for Measurement and Control (1991) *Instrument Loop Diagrams*; ANSI/ISA-5.4-1991; International Society for Measurement and Control: Research Triangle Park, North Carolina.

Institute of Electrical and Electronic Engineering Home Page. www.ieee.org (accessed June 2006).

Kelm, A. (2002) Using a Project Management Methodology for a Control System Project. *Proceedings of the International Society for Measurement and Control Expo 2002*; Chicago, Illinois, Oct 21–24; Instrumentation, Systems, and Automation Society: Research Triangle Park, North Carolina.

Knapp, R. (1999) Computer-Based Engineering with Integrated Systems. A New Approach in Electrical and Instrumentation Engineering. *Proceedings INTERKAMA 1999*; Dusseldorf, Germany, Oct 18–23, 1999.

Metcalf and Eddy, Inc. (2003) *Wastewater Engineering*. McGraw Hill: New York.

National Electrical Manufacturers Association Home Page. www.nema.org (accessed June 2006).

Smart Plant P&ID. www.intergraph.com (accessed June 2006).

Vantage Plant Engineering by Aveva. www.aveva.com (accessed June 2006).

Visio by Microsoft Corporation. www.microsoft.com (accessed June 2006).

SUGGESTED READING

Hordeski, M. (2000) *Control and Instrumentation Technology in HVAC: PCs and Environmental Controls*; Fairmont Press, Inc.: Lilburn, Georgia.

Chapter 5

General Characteristics

Background	66		*Dead band*	73
Instrument Classifications	67		*Linearity*	74
Signal Types	67		*Drift*	74
Sampling Method	68		*Repeatability*	75
Point Sensors	68		*Sensitivity*	75
Nonpoint Sensors	68		Dynamic Characteristics	77
Measurement Method	69		Value Conversions	77
Instrument Characteristics	69		*Linear Value Conversion*	78
Properties	69		*Nonlinear Value Conversion*	79
Range	69		*Nonlinear, Non-zero-based*	
Span	70		*Conversion*	80
Rangeability	70		Environmental Concerns and	
Overrange	70		Design Considerations	82
Zero Suppression, Elevation,			Temperature	82
and Offset	70		Moisture	83
Damping	70		*Short-circuits*	83
Static Characteristics	71		*Corrosion*	83
Zero error	72		*Swelling*	83
Span error	72		Corrosion	84
Hysteresis or hysteretic error	73			

Responses to Environmental Concerns	84	Internal Electronics	93
		Terminal Strip	94
Enclosures and Electronic Standards	84	Power Supply	94
		Printed Circuit Cards	94
Enclosed Spaces	86	Connectors	94
Temperature Controls	87	Calibration Points	94
Fire Ants	87	Display	94
Testing and Quality Assurance	87	Electrical Concerns	94
Need for Instrumentation Testing	87	*Voltage Spikes*	94
		Transient-Voltage Surge Suppressors	95
Instrumentation Test Reports	88	*Grounding*	96
Standards	91	*Shielding*	96
Sources and Types	92	Smart Process Instrumentation	97
Calibration Standards	93	Documentation	98
Primary standards	93	Suggested Readings	98
Traceable standards	93		
Portable standards	93		

BACKGROUND

Instrumentation is one of the most rapidly evolving segments of the wastewater treatment industry because

- Today's instrumentation typically is more reliable and accurate than that used in the 1980s and 1990s;
- The industry's emphasis has shifted from treatment plant construction to permit enforcement and reliable operations;
- Instrumentation and control (I&C) systems enable utilities to get more performance from existing treatment plant equipment;

- Construction costs make control-system improvements an attractive option; and

- More wastewater treatment professionals, including manufacturers, vendors, engineering consultants, contractors, and plant staff, know how to better design, apply, specify, install, and maintain instruments.

Wastewater treatment professionals used to think that instruments were fragile, unable to take the abuse that some process equipment can survive. Today's instruments, however, are more reliable and rugged because of new construction materials, coatings, and technologies that enable them to handle extreme conditions (e.g., overpressure, temperature, vibration, and corrosion).

Many modern instruments also have internal software burned onto their chips. Because of this software, complex microprocessor technology, and innovative new sensor technologies, these "high-tech" instruments are simpler to operate. They also need less maintenance due to built-in diagnostics, checkpoints, and auto-calibration features.

INSTRUMENT CLASSIFICATIONS

All instruments quantify or detect. They differ in several ways, however, and can be classified according to signal type, sampling method, and measurement method.

SIGNAL TYPES. Instruments emit either analog (continuously variable) or digital (discrete, two-state; on–off) signals, and the I&C system's computers must be able to handle both. *Input* is the signal entering a control device; *output* is the signal leaving it. These signals also can be described as *analog in* (AI), *analog out* (AO), *digital in* (DI), or *digital out* (DO), as appropriate. [Some engineers call these signals *analog receive* (AR), *analog transmit* (AT), *digital receive* (DR), and *digital transmit* (DT).]

Computers are digital, so even though an analog circuit's signal is continuously available, the computer can only extract data from this input periodically. A computer typically "takes samples" from the analog circuit, gives each sample a time tag, and stores each in its memory.

The relationship between sampling frequency and the analog signal's rate of change is important. The sampling must be frequent enough to follow signal changes and detect its maximum and minimum values. A simple linear interpolation of values between samples should be as accurate as the original signal. Modern digital multiplexing equipment can scan 5000 signal samples per second. Computers typically sample a given analog input once per second, and this rate can accommodate normal variations at wastewater treatment plants without degrading signal accuracy.

Not surprisingly, computers handle digital inputs more easily than analog inputs. Digital information about the state of equipment (e.g., valve open or closed, pump on or off, or conveyor belt direction) can be generated by digital instruments. Digital instruments also might indicate whether a specified tank level or bearing temperature has been exceeded.

Some digital instruments use pulse counting signals, in which each pulse indicates a specified quantity. For example, wattmeters use pulse counting to measure kilowatt-hours of energy. The computer accumulates and counts the pulses and periodically converts the count to kilowatt-hours.

SAMPLING METHOD. Instruments provide useful information based on samples of a large volume, area, or process.

Point Sensors. Many instruments provide information about a single point or small area of the overall process. For example, a pressure gauge tapped into a pipeline tracks data at that one point in the pipe. This may be desired when other sampling methods could be erroneous or confusing. Often, a single point is representative, either naturally or through mixing, of an entire section of a basin.

Nonpoint Sensors. Sometimes conditions are not uniform from one point to another because of non-ideal upstream and downstream conditions. For example, wastewater velocities in a pipe or duct may vary considerably from one point in a cross section to another. Analyzing one point for velocity might be correct under one flow condition but not another. In these circumstances, nonpoint analyzers attempt to gather information that is representative of an entire area. These instruments work by sensing

- Throughout a section, thus averaging that section (e.g., time-of-flight sonic flowmeter);
- Multiple points in a fixed array (e.g., velocity sensors in a large air line); or
- Multiple points via a moving sensor (e.g., reel-type, blanket-level sensor).

MEASUREMENT METHOD. Instruments measure variables in three ways: directly, indirectly, or inferentially. Direct measurements, such as temperature, pressure, pH, and dissolved oxygen, involve simply "reading" the substance being monitored.

Indirect measurements are more common and are effective if the instrument is accurate and its correlation to the measured variable is good. For example, pressure transmitters can be used to measure water level if the fluid density is constant. Differential pressure can be used to measure flow, and optical density can be used to measure solids concentration if a correlation exists and its variation is not too great. Engineers should keep in mind that the correlation between an instrument and the desired variable may be site-specific.

Inferential measurements involve using a given set of conditions to measure a variable. For example, a valve's position or pump's speed can be used as meters, given known pressures upstream and downstream. Inferential-measurement meters can be extremely accurate when in good condition (e.g., little wear or fouling).

INSTRUMENT CHARACTERISTICS

Below are some common characteristics of instruments. More complete descriptions of instrument terms and definitions may be obtained from related organizations, such as the American National Standards Institute Inc. (ANSI; www.ansi.org); the American Society of Mechanical Engineers (ASME; www.asme.org); the American Society for Testing and Materials (ASTM; www.astm.org); the Institute of Electrical and Electronics Engineers (IEEE; www.ieee.org); the Instrumentation, Systems, and Automation Society (ISA; www.isa.org); and the Measurement, Control & Automation Association (MCAA; www.measure.org).

PROPERTIES. The following terms describe an instrument's ability to measure the desired variable.

Range. The *range* notes the highest and lowest limits of a variable at which an instrument can function. For example, a range might be 0 to 20,000 m^3/d (0 to 5 mgd), 75 to 90 m (250 to 300 ft) elevation, or 700 to 1000 rotations per minute (rpm). Many instruments are zero-based (i.e., the lowest limit of their range is zero); others start at higher values. Some instruments can function in more than one range, depending on how they are calibrated or set up. For example, a range of 0 to 5/30 means the instrument's range can be either 0 to 5 or 0 to 30. A range of 0 to 25/150 means the instrument's range can be either 0 to 25 or 0 to 150.

Span. The *span* is the difference between the upper and lower values of an instrument's range. For example, a meter with a range of 3 to 6 m (11 to 21 ft) would have a span of 3 m (10 ft).

Rangeability. Is the ratio of maximum capacity to minimum capacity over which an instrument will operate within a specified accuracy tolerance? This can be achieved through electronic adjustment of the output signal through amplification or offset within that range, and the range can be calibrated to full scale at any point desired. In the example above, a meter could be calibrated between 0 to 5 and 0 to 30 or any range in between.

Overrange. *Overrange* or *overpressure* measures an instrument's ability to handle conditions outside its range without damage. This is important to know when extreme conditions occur. Severe overranging may damage meter accuracy or cause it to fail.

Zero Suppression, Elevation, and Offset. *Zero suppression* is the ability to set the lower limit to a positive value. This would be useful, for example, when a pressure transmitter is at the base of an elevated storage tank but is only measuring the water level in the tank.

Zero elevation is the ability to calibrate an instrument so its lower limit is a negative value (Table 5.1).

Zero offset is an instrument's ability to emit a "zero" signal when measuring a low level and emit actual measurements over the rest of its range. This feature is useful under zero-flow conditions, when a minor drift in calibration could cause totalizers to indicate a flow that does not exist.

Damping. Field measurements are typically noisy, and physically or electronically averaging (damping) these values can make an instrument's signals more useful. Damping may vary from fractions of an oscillation to several oscillations.

Damping can also refer to a fast or slow control-loop response. *Critical damping* is the smallest amount of damping to which a system can respond to a step function without overshoot. *Underdamped* is when overshoot has occurred, and *overdamped* is when the response is slower than critical.

TABLE 5.1 Illustrations of the use of range and span terminology.

Typical ranges	Name	Range	Lower range value	Upper range value	Span	Supplementary data
0 to +100	—	0 to 100	0	+100	100	—
20 to +100	Suppressed zero range	20 to 100	20	+100	80	Suppression ratio* = 0.25
-25 to +100	Elevated zero range	-25 to +100	-25	+100	125	—
-100 to 0	Elevated zero range	-100 to 0	-100	0	100	—
-100 to -20	Elevated zero range	-100 to -20	-100	-20	80	—

*The suppression ratio is calculated as the ratio of the lower-range value to the span. For example, if the range is 20 to 100, the span is 80 and the lower range value is 20, so the suppression ratio is 20/80 = 0.25, or 25%. Source: *Process Instrumentation*, 1979.

STATIC CHARACTERISTICS. Static characteristics apply to steady-state conditions. The most important of these characteristics are accuracy, dead band, drift, hysteresis, repeatability, and sensitivity.

Accuracy notes the difference between the instrument reading and the true value of the variable being measured. It typically is expressed as a percentage of either full-scale or the actual reading (Figure 5.1). For example, if a thermometer with a range of 0 to 100°C (32 to 212°F) has an accuracy of 1% full-scale, then each reading would be expected to be within 1°C (1.8°F) of the actual temperature. However, if its accuracy is 1% of the actual reading, then each reading's accuracy would depend on the temperature measured. So, if the actual reading was 50°C (122°F), the thermometer would register a temperature within 0.5°C (0.9°F) of that reading [i.e., 49.5 to 50.5°C (121.1 to 122.9°F)]. If the actual reading was 80°C (176°F), the thermometer would register a temperature within 0.8°C (1.4°F) of that reading [i.e., 79.2 to 80.8°C (174.6 to 177.4°F)].

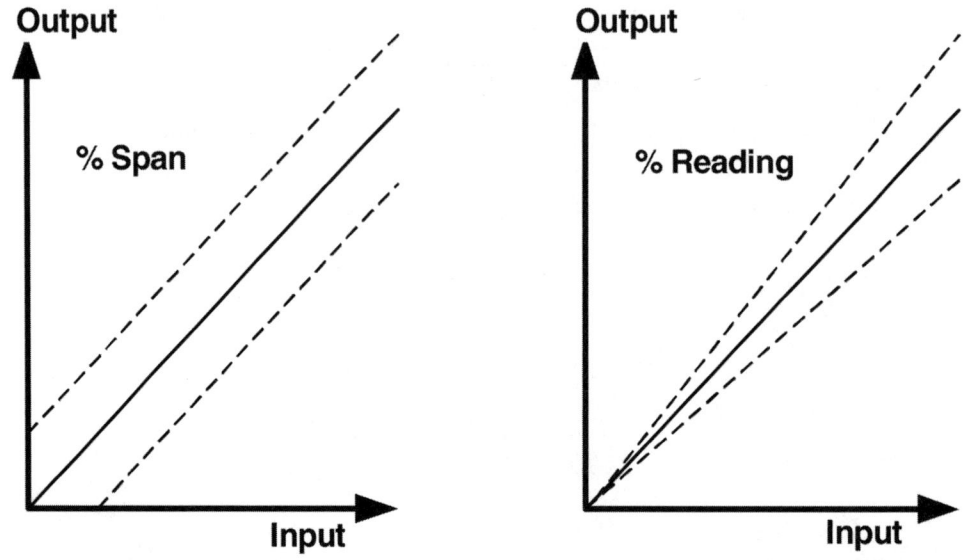

Figure 5.1 Accuracy. The true value is given by the solid line that lies within the band bordered by the dashed lines. Source: *Process Instrumentation*, 1984.

Accuracy is a measure of a device's total error. *Error* is the difference between the instrument's measurement and the actual state of the variable being measured. All instruments are subject to error for various reasons, including wear, internal friction, external disturbances, changes in ambient conditions, and the observer's limitations in reading or interpreting results. Some of these errors are consistent in direction and magnitude, while others occur randomly. Types of errors include

- **Zero error.** *Zero error* is the magnitude of an instrument's nonzero reading when there is no input.
- **Span error.** *Span error* is the difference between an instrument's calculated and actual spans.

- **Hysteresis or hysteretic error.** *Hysteresis*, which is typically expressed as a percentage of full-scale, is a measure of the variation in instrument readings for a particular input value, depending on whether it was approached from above or below (Figure 5.2).
- **Dead band.** *Dead band* is the range of a variable that does not initiate a response in the instrument (Figure 5.3).

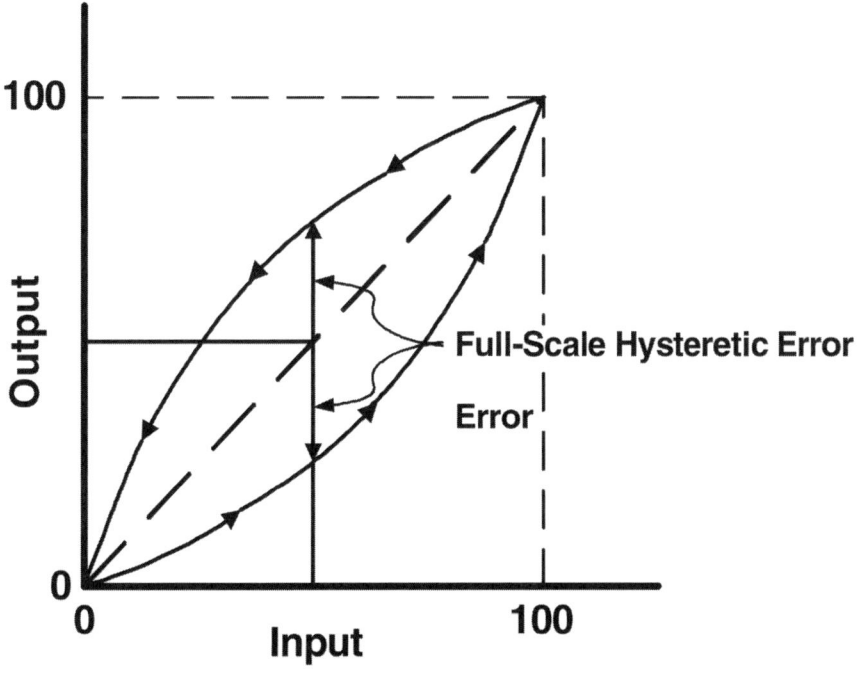

FIGURE 5.2 Hysteretic error only.

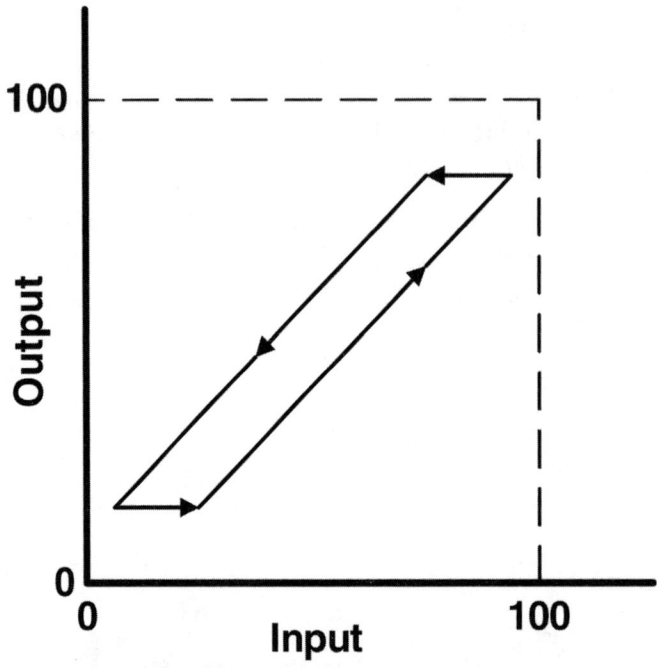

FIGURE 5.3 Dead-band error only.

- **Linearity.** *Linearity* is the degree to which a curve approximates a straight line. Manufacturers focus on producing instruments with linear outputs because they typically are easier to use and understand. Some transmitters are inherently linear, while others require complex schemes to produce linearity (Figure 5.4).

- **Drift.** *Drift* notes how much an output signal changes over time. Typically, it is expressed as a percentage of span over a specified time period. All output values are uniformly affected by drift, which typically is a linear function of time.

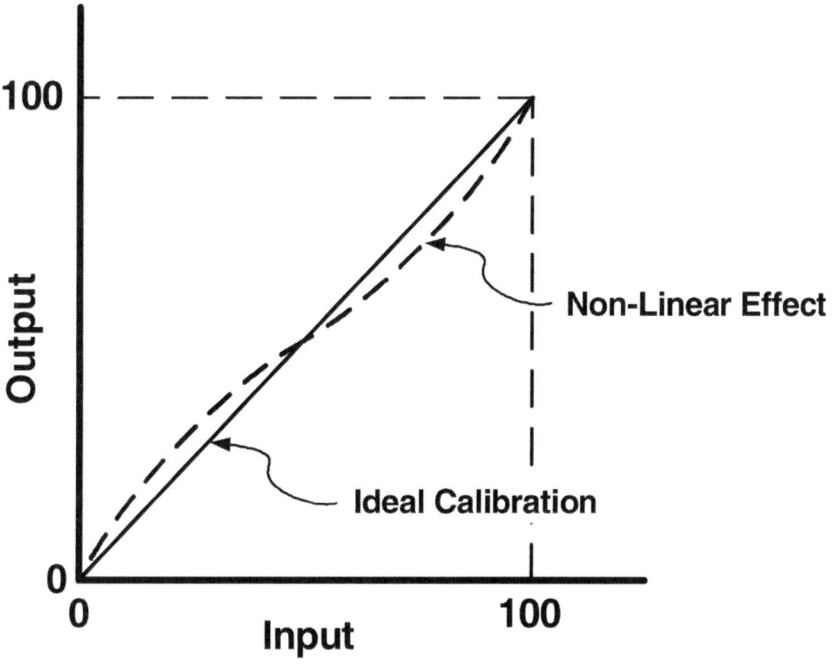

FIGURE 5.4 Nonlinear effects.

- *Repeatability.* *Repeatability* notes an instrument's precision when repeatedly measuring the same input value under identical operating conditions and approaching from the same direction, for full-range traverses (Figure 5.5).
- *Sensitivity.* *Sensitivity* is the ratio of the change in output to a given change in input. A high sensitivity is generally desirable because it provides high resolution with little amplification. This in turn results in a good signal-to-noise ratio. If the sensitivity is constant through the device's range, the device is linear; if the sensitivity changes, the device is nonlinear.

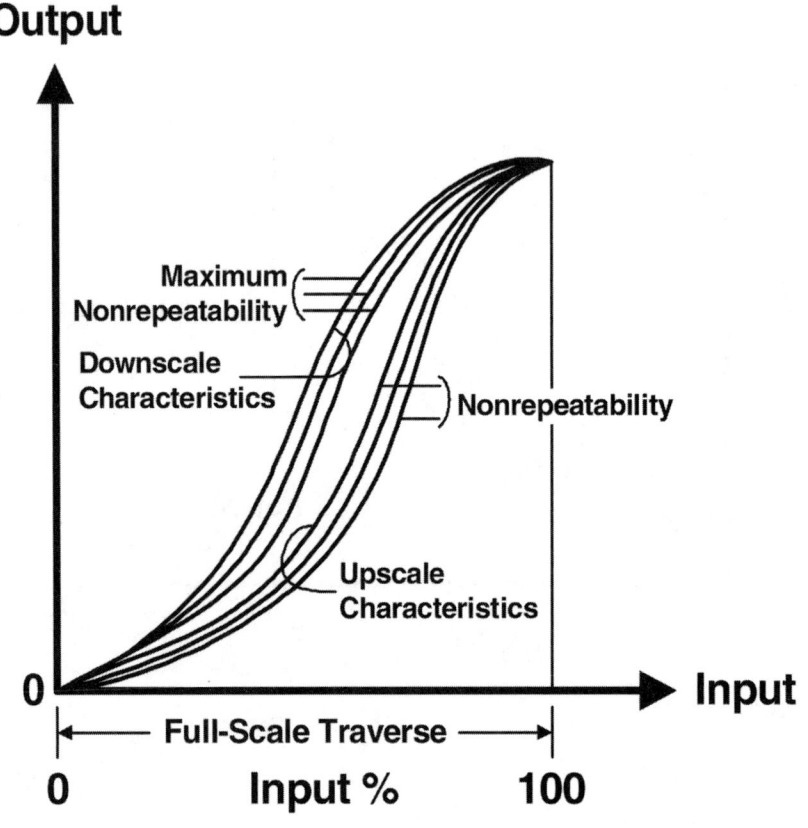

FIGURE 5.5 Repeatability. Source: *Process Instrumentation*, 1984.

Each instrument should be sized for both its initial and intended future uses. For example, a plant design may require a flowmeter to handle 375 000-m³/d (100-mgd) even though the present maximum flow is only 75 000 m³/d (20 mgd). So, the meter initially would use only a small portion of its range and could be significantly less accurate if this was not taken into account.

In multiple-component systems, each element introduces errors that collectively affect the entire system's accuracy. Errors from multiple units, however, are not additive. For example, if five flowmeters each had an accuracy of 5%, the error for the sum of the flows would not necessarily be 25%. Likewise, when two signals are subtracted, their combined error may be greater than that of each signal.

For example, the total probable error (TPE) for one measurement can be estimated by taking the square root of the sum of the squares of individual errors associated with it. For example, the TPE for a temperature measurement taken by a resistance temperature detector (RTD) connected to a temperature transmitter and data acquisition system can be calculated based on the errors associated with each device:

$$E_{TOTAL} = \sqrt{E_1^2 + E_2^2 + E_3^2} \tag{5.1}$$

Where

E_1 = RTD error,
E_2 = temperature-transmitter error, and
E_3 = data-acquisition error.

DYNAMIC CHARACTERISTICS. When instruments are used in dynamic processes, the most important characteristics are damping, dead time, decay ratio, decay time, overshoot, response time, rise time, step change, and step response time.

Because instruments cannot respond instantly to a change of input, response time is probably their most important dynamic characteristic. It typically is specified in percent of span per unit of time. Whenever a continuously monitoring instrument is placed on-line, staff need to know how quickly it will respond to changes in the variable it is measuring (Figure 5.6). Typically, an instrument's readings will at first overshoot and then oscillate around the new input value. Sometimes the instrument is damped to suppress the oscillations. The instrument's response time to a step change in the variable is usually measured as the time required for the readings to be within 10% of actual conditions.

VALUE CONVERSIONS. Although standardized transmission signals (e.g., 4 to 20 mA) exist, they are not transmitted directly as numbers and need to be converted into terms the user understands, such as gallons per minute (gal/min), million gallons per day (mgd), and milligrams per liter (mg/L). These units of measure are called *engineering units*.

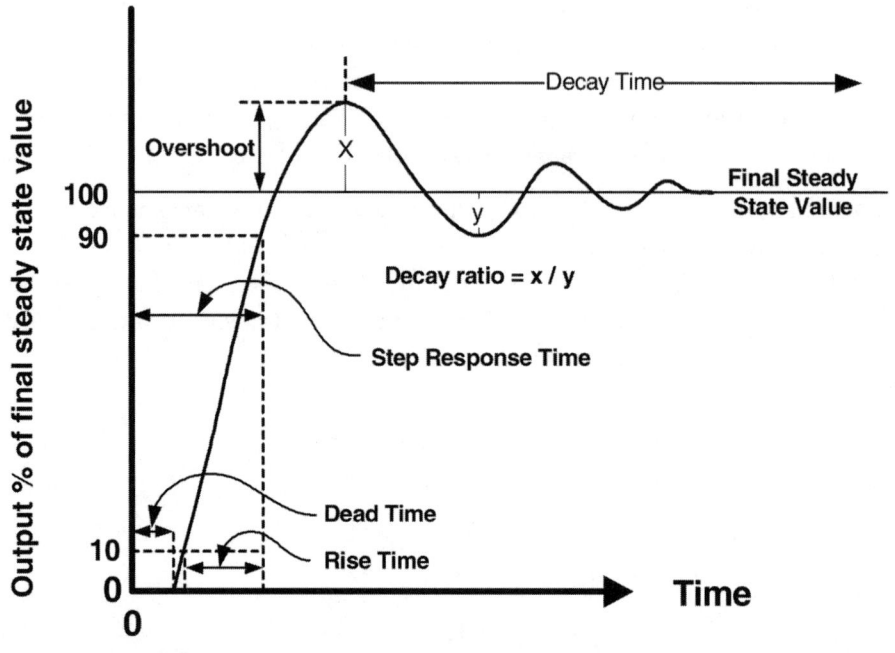

FIGURE 5.6 Typical time response of a system to a step increase of input.

All instrument technicians need to know how to convert transmission signals into engineering units and convert from one type of signal to another. This can be more challenging when the signals are not linear.

Linear Value Conversion. Technicians would use the following formula to convert a linear 4- to 20-mA signal into engineering units:

$$S_a = \frac{(L_a - M_a)}{(L_b - M_b)} \times (S_b - M_b) + M_a \tag{5.2}$$

Where

S_a and S_b are present values, M_a and M_b are minimum values, and L_a and L_b are maximum values for signals "a" and "b".

If a 4- to 20-mA signal is used for a 20-ft-deep (6-m-deep) wet well, the signal for 16 ft (5 m) would be calculated as follows:

$$S_a = \frac{(20 \text{ mA} - 4 \text{ mA})}{(20 \text{ ft} - 0 \text{ ft})} \times (S_b - 0) + 4 \text{ mA} \tag{5.3}$$

$$S_a = \frac{(16 \text{ mA})}{(20 \text{ ft})} \times (S_b) + 4 \text{ mA} = 0.8\, S_b + 4 \text{ mA}$$

Substituting the point in question (S_b = 16 ft) into this formula yields:

$$S_a = (0.8)(16 \text{ mA}) + 4 \text{ mA} \tag{5.4}$$

$$= 16.8 \text{ mA}$$

Nonlinear Value Conversion. Many signals have a square-root relationship, and in these cases, a square-root extraction would produce useful information. For example, to convert a Venturi or orifice-plate flowmeter's differential-pressure signal into a flow measurement, technicians would use the following formula:

$$Q = k1\sqrt{\Delta P1} \tag{5.5}$$

Where

$$k1 = \frac{Q_m}{\sqrt{\Delta P_m}} \tag{5.6}$$

reorganizing these equations yields

$$Q = Q_m \sqrt{\frac{\Delta P}{\Delta P_m}} \tag{5.7}$$

Where

Q = meter flow,
Q_m = maximum flow (i.e., 100% of scale),
ΔP = differential pressure,

ΔP_m = differential pressure at maximum flow, and
k1 = constant.

If a Venturi meter is rated for 5 mgd (19 000 m³/d) at 42.1 in. (1.1 m) of water-column differential, the flow at 12 in. (0.3 m) of differential is calculated as

$$k1 = \frac{Q_m}{\sqrt{\Delta P_m}} = \frac{5.0 \text{ mgd}}{\sqrt{42.1 \text{ in.} H2O}} = \frac{5.0}{6.48} = 0.771 \tag{5.8}$$

$$Q = 0.771\sqrt{12} = 0.771(3.46) = 2.67 \text{ mgd (10 105 m}^3/\text{d)} \tag{5.9}$$

Alternatively, the formula for nonlinear conversion may be written:

$$\Delta P = \Delta P_m \left(\frac{Q}{Q_m}\right)^2 \tag{5.10}$$

or

$$\Delta P = k2(Q)^2 \tag{5.11}$$

Where

$$k2 = \frac{\Delta P_m}{Q_m^2} \tag{5.12}$$

If an orifice-plate meter is rated for 500 gal/min (32 L/s) at 22.4 in. (0.6 m) of water-column differential, the differential for 200 gal/min (13 L/s) is:

$$k2 = \frac{\Delta P_m}{Q_m^2} = \frac{22.4}{500^2} = \frac{22.4}{250,000} = 0.0000896 \tag{5.13}$$

$$\Delta P = k2(Q)^2 = 0.0000896(200)^2 = 3.584 \text{ in. } H_2O \tag{5.14}$$

Nonlinear, Non-Zero-Based Conversion. Another calculation is common when transmitted signals are nonzero-based:

$$Q = Q_m \sqrt{\frac{\Delta P}{\Delta P_m}} = Q_m \sqrt{\frac{(S_a - M_a)}{(L_a - M_a)}} \tag{5.15}$$

The general equation for the square root of a 4- to 20-mA signal is:

$$Q = Q_m \sqrt{\frac{(S_a - 4)}{(20 - 4)}} = Q_m \sqrt{\frac{(S_a - 4)}{(16)}} \tag{5.16}$$

A Venturi meter with a maximum flow of 2 mgd (7570 m³/d) and a transmission signal of 4 to 20 mA has the following flow when the signal reads 13 mA:

$$Q = Q_m \sqrt{\frac{(S_a - 4)}{(16)}} = 2\sqrt{\frac{(13 - 4)}{(16)}} = 2\sqrt{(0.5625)} \tag{5.17}$$

$$= 2\,(0.75) = 1.50 \text{ mgd } (5677 \text{ m}^3/\text{d})$$

Another way to write the formula is:

$$S_a = \left[\left(\frac{Q}{Q_m}\right)^2 (L_a - M_a)\right] + M_a \tag{5.18}$$

If the transmitter is a 4- to 20-mA signal, the formula becomes:

$$S_a = \left[\left(\frac{Q}{Q_m}\right)^2 (20 - 4)\right] + 4 \tag{5.19}$$

$$S_a = 16\left(\frac{Q}{Q_m}\right)^2 + 4$$

If the flow for a Venturi tube is 80 mgd (303 000 m³/d), the 4- to 20-mA transmitted signal for 75 mgd (284 000 m³/d) is calculated as:

$$S_a = 16\left(\frac{75}{80}\right)^2 + 4 \tag{5.20}$$

$$= 16(0.9375)^2 + 4 = 14.1 + 4 = 18.1 \text{ mA}$$

ENVIRONMENTAL CONCERNS AND DESIGN CONSIDERATIONS

Wastewater treatment plant conditions can be harsh. Many instruments are installed outdoors, where they are subjected to sun, rain, snow, temperature variations, and airborne contaminants. The wastewater also contains various corrosives, debris, and gummy substances that find their way into instruments.

Nevertheless, delicate instruments are expected to work for extended periods with little or no attention. So, those specifying instrumentation for treatment plants must consider the environment and keep in mind that many common instruments are designed for less severe applications.

TEMPERATURE. Multiple problems arise when an instrument is put in a hostile or variable-temperature environment.

Excessive heat or cold can cause electrical or electronic components to fail. Temperatures higher than an instrument's rating may cause components to burn up, fail to perform properly for other reasons, or weaken and fail later. A 3-hour exposure to 50°C, for example, could reduce an instrument's life expectancy by 5 years. Solid-state components typically have problems at temperatures higher than 40°C (104°F), but technical developments now enable some to handle ambient conditions of 50 and 60°C (122 and 140°F).

Similar problems can occur when an instrument is exposed to temperatures below its rating. Freezing damages many instruments and sensing lines.

Variable temperatures cause two problems: cracking and faulty connections. More delicate electronic components swell and shrink, eventually cracking. Metal moving on edge-card connectors, pins, and contacts can cause fretting, which allows dust and corrosion flakes to build up on the metal surface, weakening electrical connections. Specially designed wax lubricants can minimize this problem.

Accelerated corrosion is a frequent byproduct of elevated temperatures, particularly when combined with moisture or high airflows. All instrument components—body, electrical components, connectors, cards, cables, and enclosures—are susceptible to corrosion.

Ideally, instruments should be put in areas with heating, ventilation, and air-conditioning (HVAC), but this is impractical at many wastewater treatment plants. So, design engineers should select instruments built to withstand extreme conditions; put instruments indoors or under a roof (with a heater, if required); or put them inside another enclosure with a heater, adequate heat dissipation, good ventilation (with filtered air), heat sinks to dissipate heat, heat pipes, and air conditioning.

MOISTURE. Moisture contributes to instrument problems in three ways: short-circuits, corrosion, and swelling.

Short-circuits. Water droplets from condensation or leakage can cause electrical short-circuits and associated errors or damage.

Corrosion. Moisture can combine with hydrogen sulfide or chlorine to yield acids that accelerate corrosion.

Swelling. Many electronic components—particularly solid-state devices—gain or lose water depending on the humidity. Continual changes in humidity can fatigue components and cause failures. Ideally, instruments should be put in areas where the humidity is between 30 and 80% with no more than a 2%-per-hour change, but this is difficult without using an excellent HVAC system. Fortunately, good manufacturers reduce such problems by using conformal coatings and airtight sealing.

TABLE 5.2 Maximum contaminant levels.

	Maximum level		
Contaminant	Mild	Moderate	Units
Dust	1000	5000	$\mu g/m^3$
H_2S	3	10	ppb
Cl_2	2	2	ppb
SO_2, SO_3	10	100	ppb
NO_x	50	125	ppb
Copper reactivity level	300	1000	Å

Source: Skrentner, 1988.

CORROSION. Corrosion damage is site-specific; it depends on the wastewater treatment processes and local climate involved. Hydrogen sulfide typically is the most common corrosive at a treatment plant because it forms an acid in the presence of water or water vapor—especially in pressure sewers, stagnant gravity sewers, solids-handling processes, or areas where the wastewater is more than 25°C (77°F).

Corrosive chemicals commonly used at wastewater treatment plants include alum, ammonia, chlorine, caustic (sodium hydroxide), ferric chloride, polymers, and sulfur dioxide. Table 5.2 lists the maximum contaminant levels to which instrumentation should be exposed. These levels are several orders of magnitude lower than the concentrations that may be present in the field.

RESPONSES TO ENVIRONMENTAL CONCERNS. Typically, most environmental problems with instrumentation at wastewater treatment plants have commonsense solutions. For example, instrument enclosures should have the National Electrical Manufacturers Association's (NEMA's) "X" rating for corrosion resistance (e.g., NEMA 4X rather than NEMA 4). Aluminum, stainless steel, and chemical-resistant paints have proven successful in corrosive areas. In sunny climates, plastic or fiberglass enclosures should be sunlight-resistant.

If an instrument will touch wastewater, solids, or treatment chemicals, it should be made of suitably resistant materials (e.g., polytetrafluoroethylene, 316 stainless steel, Iconel, Hastelloy, and Viton). Design engineers should ask manufacturers for recommendations or chemical-resistance charts. For example, engineers should use tinned copper wires rather than silver or copper ones to avoid corrosion problems with hydrogen sulfide, ammonia, and chlorine. Edge connectors, sockets, and rotary switches with gold contacts have fewer problems when low-voltage connections are used. These construction materials may be more expensive, but they ensure that the instruments will last longer.

Enclosures and Electronic Standards. Several groups publish instrumentation standards that are useful in the wastewater treatment industry. Such standards include NEMA Standards ICS-6 and 250; the National Fire Protection Association's (NFPA's) National Electrical Code (NEC) 70; and the Underwriters Laboratories' (UL©) UL508 and successors.

The National Electrical Manufacturers Association classifies enclosures for specific conditions; its designations range from 1 to 13 (defining suitable use environments, including indoor/outdoor, degree of protection from the elements, hazardous/nonhazardous environments, etc.) and may include modifying letters (e.g., R for rain, A for dust accumulation, and X for corrosion-resistant). The National Electric Code specifies where certain enclosures may be used, based on the potential for hazardous conditions. The enclosure codes are organized by class, division, and group. Areas determined to be nonhazardous are "not classified" (IEEE, 2002). Standard NEC 70 notes the minimum requirements for safe use of electrical devices.

Underwriters Laboratories tests equipment to see if it conforms to national codes and standards. Changing one component in a system invalidates its UL rating.

TABLE 5.3 National Electrical Code ratings for enclosures.

Code	
	Cause of hazardous condition
Class I	Flammable gas or vapor
Class II	Combustible dust
Class III	Ignitable fibers or flyings
	Conditions existing for hazardous environment
Division I	Normal operating conditions
Division II	Only during rupture or breakdown (or the area is close to a Division I area from which the hazardous environment could spread)
	Chemical group (similar explosive properties)
Group A	Acetylene
Group B	Hydrogen
Group C	Ethyl-ether vapors, ethylene
Group D	Gasoline, naphtha, natural gas
Group E	Metal dust
Group F	Electrically conductive dust, such as coal
Group G	Combustible dust, such as flour

TABLE 5.4 National Electrical Manufacturers Association's codes for enclosures.

Enclosure type	Definition
NEMA 1	Indoor, light indirect-splash protection. Its purpose is to prevent accidental personal contact.
NEMA 3R	Outdoor, rainproof, sleet-resistant. Without the "R" modifier, it is only rain-resistant.
NEMA 4X	Watertight, indoor/outdoor, can be hosed down. Without the "X" modifier, it may not have corrosion-resistant hardware.
NEMA 6	Submersible, generally rated for "X" feet for "Y" hours. The specification should state the desired level of protection.
NEMA 6P	Same as NEMA 6, including protection against water entry during prolonged submersion at a limited depth.
NEMA 7	Explosion-proof; designed to confine a spark or explosion generated inside the enclosure. These are required in areas where digester gas or explosive vapors could be present. [The NEC classification is more explicit: Class I, Division 1 or 2 (1 if normal, 2 if abnormal); Group D for methane gas.]
NEMA 12	Dustproof/industrial with flanged connections.
NEMA 13	Oil-tight, dust-tight, identical to NEMA 12 except cast construction with threaded conduit connections.

Enclosures are intended to minimize environmental problems, but they may create a harsh, high-temperature atmosphere, so the instrumentation's electronic components must be designed to handle such conditions. Codes related to enclosures commonly used in the wastewater treatment industry are listed in Tables 5.3 and 5.4. All treatment plant enclosures must comply with both the local code and the NEC.

Enclosed Spaces. Sensitive instruments have problems in confined spaces, where corrosives, dust, oil, moisture, and other foreign material can enter and cause damage. These materials creep into the enclosure during heating and cooling, when outside air is drawn in through cracks and conduits, and during periodic enclosure openings for readings or calibrations. To minimize such problems, engineers and plant staff should

- Drain and seal conduits entering the enclosure;
- Keep the door closed securely, except when servicing;
- Purchase instruments that require little or no service involving opening the door (some instruments provide for calibration by way of external connection, optical communication, or magnetic coupling);
- Provide cabinet heaters to prevent moisture buildup;
- Install passivating coupons to purge out contaminants or coat components with a protective layer;
- Purge the enclosure with clean and dry air or bottled gas; and
- Place in controlled atmospheric space-air conditioned and scrubbed.

Temperature Controls. Instruments used to be commonly rated for 40°C (104°F); now they are commonly rated for 50 or 60°C (122 or 140°F). Design engineers should select instruments that are rated for a higher temperature range than expected. They also should allow for heating via direct sunlight, if appropriate.

To shade instruments from the sun, engineers could provide for a small canopy. This may mean shifting the transmitter inside. If more temperature control is necessary, engineers could add a ventilation fan or convection, heat pipes, refrigerative air conditioning, chilled water, or vortex cooling.

Fire Ants. If fire ants are a local problem, design engineers should avoid rubber-insulated wire because the ants eat rubber and are attracted to electrical fields. Instead, they should seal conduits into instruments and enclosures to minimize the ants' access. They also should consider alternate gasket materials.

Other pests, such as rodents and insects, also can be problematic and are best dealt with by selecting appropriate enclosures.

TESTING AND QUALITY ASSURANCE

NEED FOR INSTRUMENTATION TESTING. Today, wastewater treatment plants use many types of instruments. Most instruments work well, but some are less than satisfactory. So, all instruments should be tested.

Design engineers can compare instruments by reading testing organizations' reports, evaluations in journals, and manufacturers' data or evaluations; they can talk to wastewater treatment professionals (users, consultants, and academia); and they can perform onsite testing. Testing reports contain essential accuracy, repeatability, and stability data, but these parameters have multiple definitions (e.g., as a percent of full-scale or actual reading) and the definition chosen will affect the results. Also, data obtained under controlled laboratory conditions may not be applicable to all field conditions, so testing conditions should be clearly defined. Engineers should scrutinize any data that does not cite a test standard or provide details of the conditions and methods used.

While instrument evaluations can quantify finite, measurable differences, they seldom measure site-specific factors as accurately as onsite testing. In-plant testing under typical maintenance procedures is the best way to ensure that the instruments will be effective.

The most important parameters for determining an instrument's success or failure cannot be found in a report. These intangible parameters include

- The utility's attitude, commitment, competence, and employee availability;
- The manufacturer's or representative's technical support and availability (i.e., help in understanding the instrument and how to get desired results with the least effort and confusion);
- The manufacturer's willingness to learn from field experience;
- Transferability of the level of operation from a test site to a given facility;
- The ease with which the instrument can be serviced and calibrated; and
- The quality of operation and maintenance manuals and identification of generic parts.

INSTRUMENTATION TEST REPORTS. A number of private organizations evaluate and calibrate instruments. Typically, they only evaluate the instruments submitted to them (usually by manufacturers) and only issue their reports to the submitters. Occasionally, the U.S. Environmental Protection Agency (U.S. EPA) publishes or sponsors reports on groups of instruments, but they conceal the brand names. Research may be necessary to relate the reported findings to a particular instrument.

Journals sometimes publish special surveys on one type of instrument that can be a useful starting point for comparisons. These surveys are based on market research and are limited to the manufacturers who respond to the journal staff's request for data.

TABLE 5.5 Typical protocol bench tests.

Wet tests
    Drift test
    Calibration
    Measured accuracy
    Linearity (independent, terminal-based, zero-based)
    Hysteresis
    Dead band
    Repeatability
    Reproducibility
    Ranging effects
    Response time
    Interferences
    Flow sensitivity
    Sample temperature effects

Power supply tests
    Steady-state supply voltage effects
    Transient-supply voltage effects
    Short-term supply interruption

Electronics tests
    Zero and span rangeability
    Warm-up effects
    Ambient temperature effects
    Common- and normal-mode interferences
    Output signal and noise
    Output-load effects

TABLE 5.6 Typical subjective evaluation areas.

Principles of operation
　Traceability to documented methods of measurement
　Susceptibility to interferences

Operating characteristics
　Clarity and resolution of indicators
　Sample conditioning requirements
　Level of skill required for normal operation

Maintenance characteristics
　Accessibility of components requiring service
　Frequency of cleaning and calibration
　Cost of expendables
　Operator exposure to injury during maintenance

Quality of construction
　Materials of construction
　Use of standard components
　Suitability of analyzer enclosure for outdoor exposure

System support
　Quality of documentation
　Distribution of service centers

Commissioning requirements
　Quality of shipping containers and packaging methods
　Availability of startup assistance

Manufacturer's policies and performance
　Warranty policy
　Policy for support or discontinued equipment

Instrument manufacturers publish data sheets that describe their products' capabilities. Design engineers should compare several manufacturers' specifications for I&C systems, making sure they use consistent terminology (e.g., accuracy and repeatability) and provide complete performance data.

The Instrument Testing Association (ITA; www.instrument.org) also publishes detailed reports of its testing results (Tables 5.5 and 5.6). Each report typically analyzes four to seven manufacturers' versions of one type of instrument (e.g., dissolved

TABLE 5.7 Standards-setting organizations.

ANSI	American National Standards Institute 1430 Broadway, New York, NY 10018
ASME	American Society of Mechanical Engineers 345 East 47th Street, New York, NY 10017
ASTM	American Society of Testing and Materials Race Street, Philadelphia, PA 19103
IEEE	Institute of Electrical and Electronic Engineers 345 East 47th Street, New York, NY 10017
ISA	Instrumentation, Systems, and Automation Society of America 67 Alexander Drive, Research Triangle Park, NC 27709
ISO	International Organization for Standardization 1, rue de Varembé, Case postale 56, CH-1211, Genève 20, Switzerland
MCAA	Measurement Control and Automation Association 2093 Harper's Mill Road, Williamsburg, VA 23185
NEC	National Electrical Code (by National Fire Protection Association) P.O. Box 9101, Quincy MA 02269-9959
NEMA	National Electrical Manufacturers Association 2101 L Street, N.W., Suite 300, Washington, DC 20037
NFPA	National Fire Protection Association 1 Batterymarch Park, Quincy, MA, USA 02169-7471
NIST	National Institute for Standards and Technology Gaithersburg, MD 20899
UL	Underwriters Laboratories 333 Pfingsten Road, Northbrook, IL 60062

oxygen or free chlorine residuals analyzers). ITA test reports can be purchased by the public and are offered to members at a discount.

STANDARDS. Worldwide, there are many types of standards and standards-setting organizations (Table 5.7). Instrumentation standards include

- Those used for comparison,

- Those that establish how things are done industrywide, and
- Those that clarify how an organization achieves more consistency.

Some of these standards are mandatory, and some are voluntary.

Sources and Types. For example, governmental and international standards organizations establish scientific or engineering standards for an instrument's basic physical, chemical, and electrical properties. They also distribute reference standards for calibrating instruments.

Industry standards for instrumentation establish terminology, function, customary methods, interchangeability, and connectivity. They typically are published by ANSI, ISA, and other technical organizations.

Most consulting firms develop design standards for themselves. These generally include:

- Instrument specifications,
- Typical instrument-installation drawings,
- Drawing formats for process and instrumentation diagrams, and
- Strategies for controlling certain tasks.

These standards ensure that designs are consistent and help service continually improve.

Wastewater treatment plants also develop standards for themselves to ensure that construction projects and ongoing operations are successful. The plant's terminology for equipment and numbering systems is especially important. Instrument tags, for example, should be uniform for the entire plant, not just one construction project. As-built information also should be kept current, even when multiple contracts or consultants are involved.

The National Fire Protection Association's Standard 820 lists requirements for avoiding fire and explosion hazards in wastewater collection and treatment facilities, as well as hazard classifications of specific areas and processes. The standard covers sewers and ancillary structures, pumping stations, wastewater treatment processes, solids-handling processes, and chemical-handling facilities.

Instrument manufacturers' operations-and-maintenance manuals provide invaluable information on application, installation, and maintenance. This information is usually general because the instrument may be used in many industries and environments, but the manufacturer typically can provide site-specific guidance.

Calibration Standards. Many calibration standards are available and are typically are defined as primary standards, traceable standards, or portable standards.

- *Primary standards.* Results related to a substance's physical or chemical property are referred to as *primary standards*. Science and engineering are built on the ability to communicate and compare such results.

- *Traceable standards.* A measurement or calibration that can be traced to a major standards laboratory, where environmental conditions (e.g., temperature, pressure, and humidity) are precisely controlled, makes an instrument's accuracy more credible. Otherwise, calibrations cannot be defended if a conflict occurs. Instrument distributors go to considerable effort to maintain this accuracy.

- *Portable standards.* Volt ohm meters, milliamp signal generators, portable dissolved oxygen probes, pH meters, color-wheel comparators, and other instruments used every day require calibrations based on working standards, not the conditions of a standards laboratory's controlled environment. However, periodic recalibration at a standards lab is essential.

INTERNAL ELECTRONICS

The entire electronics field has been changing rapidly for decades, and instruments have benefited from this evolution. Instruments began as mechanical devices. Then, resistive circuits were added to allow for signal adjustment, transmission, or display. Transistors then improved capabilities and reduced costs. Today, digital electronics are significantly changing the way instrumentation is designed, applied, and maintained.

Regardless of the technology involved, most electronics are designed to

- Convert a variable to an electronic signal,
- Amplify the signal so electronics can function,
- Linearize the signal,
- Scale the signal,
- Reject noise,
- Compensate for unrelated variables (e.g., pressure and temperature),
- Compensate for drift and environmental changes in the electronics or transducer,

- Smooth signal variations to give a steady output, and
- Indicate or transmit the signal.

To perform these tasks, an instrument typically has the following components:

- **TERMINAL STRIP.** This includes electrical connections for power, one or more signals from transducers that sense the necessary variables, and transmission signals (analog and/or discrete).
- **POWER SUPPLY.** Whether the instrument is two- or four-wire, its internal electronics must be controlled to a constant voltage and protected from voltage spikes by fusing and other methods.
- **PRINTED CIRCUIT CARDS.** Most instruments use electrical circuits built into printed circuit cards (or a roll of film, as digital electronics replace computer chips). If there are multiple cards, functionality is typically modular, with different cards for different tasks. This division of tasks is useful for calibration and troubleshooting.
- **CONNECTORS.** Components can be connected through terminal strips, but these connections take a lot of time and increase the probability of error. Edge-card and multi-pin connectors only require one component to be inserted.
- **CALIBRATION POINTS.** Instruments vary considerably in complexity, but most have adjustment "knobs"—typically, switches or multiple-turn potentiometers ("pots," which are variable resistors). Common calibrations include zero, span, and damping. Zero is used to set the instrument's minimum value. Span is used to set the instrument's maximum value. In older instruments, zero and span were interactive; that is, changing one affected the other, so calibrations could take a long time. Damping (smoothing) is used to slow signal responsiveness to provide an averaging effect. It minimizes meaningless values.
- **DISPLAY.** Most instruments include a local display, such as a galvanometer (pointer indication), light-emitting diode (LED), or liquid crystal, to indicate the value of the variable being measured and other data.

ELECTRICAL CONCERNS. *Voltage Spikes.* Although electronics are designed to handle a range of voltages and amperages (current), voltage spikes may weaken or destroy components, causing instruments to malfunction or fail. Voltage spikes are the result of lightning, power surges, and electrostatic discharge.

Lightning, which can contain 100 000 000 V, could be carried through power cables, signal-transmission pairs, or the instrument mounting (e.g., a handrail) and cause voltage spikes. Nothing can survive a direct lightning strike, but some instruments or special isolators can resist indirect strikes.

Voltage spikes caused by power surges during electrical-connection makeups or accidental short-circuits can be rendered harmless if an instrument is properly designed and fused. On the other hand, a poorly designed instrument can be damaged when simply connected or disconnected to a circuit.

Electrostatic discharge can cause spikes of 2000 V and damage instruments, even without physical contact or a noticeable spark. Certain components, such as digital electronics, are considerably more susceptible to damage than others. Some components may be damaged by multiple hits of small voltages (10 to 20 hits of 20 V). The damage may range from an instantaneous error in data manipulation to failure or reduced component life. Grounding mats, grounding straps, air deionizers, conductive storage containers, or a relative humidity between 65 and 90% can reduce such problems.

Transient-Voltage Surge Suppressors. An electrical transient is a temporary excess of voltage or current in an electrical circuit that has been disturbed. Transients are found in all types of electrical, data, and communications circuits, and typically last between a few nanoseconds and milliseconds. They can happen when

- A light, motor, or any other electrical device is turned on or off;
- High-amp loads, such as electric motors, are switched;
- Cloud-to-cloud lightning discharges or nearby lightning strikes occur;
- Tree branches or wet kite strings touch power lines and disrupt energy flow; or
- Noisy electrical uses, such as welding shops or manufacturing facilities, are underway.

Typically, the larger the current load, the greater the disturbance when it is switched off or on. If a high-transient voltage appears in an unprotected power, telephone, data, or coaxial line, it can destroy the system.

To avoid such problems, each circuit should include a surge-protection device. According to the UL 1449 *Standard for Safety Transient Voltage Surge Suppressors, Second Edition, transient-suppression voltage* is the maximum peak voltage occurring within 100 microseconds after a test wave is applied. In other words, it is the maximum amplitude of the voltage after the surge-protection device has done its job. All electrical products must comply with this safety standard.

Grounding. Electrical grounding protects the people using the instruments, gives the instruments a common voltage, and minimizes interferences and interactions between instruments. Antenna systems also need a properly functioning ground system to protect operators and efficiently radiate the maximum amount of radio frequency energy into the air.

Ground loops occur when there is more than one electrical connection path between two or more instruments (circuits between instruments) that are separated enough to be at different ground voltages, resulting in intersystem ground noise and potential damage. They can be created when ground connections on several pieces of equipment are connected in series, rather than to one centralized ground point. To avoid ground loops, each piece of equipment should be tied to one ground point.

Shielding. *Electrical induction* occurs when electrical voltages transfer from one pair of wires to another—even when they do not physically touch. Parallel wires will transfer some of their energy to each other, which is necessary for transformers but problematic in instruments. To avoid this problem, metal shields are used to break the magnetic fields and drain the inducted voltage to the ground.

Signal wires act like antennas. If their voltages are low, or their signals are run more than a few feet, induced voltages can damage signal accuracy. Twisted shielded pairs and, in extreme cases, coaxial cables have been used to minimize problems. Even with these precautions, signals should be kept away from power cables.

Radiation or radio-frequency transmission is an inadvertent byproduct of various electronic components. The Federal Communications Commission (FCC) controls radio-frequency transmissions, and instruments must comply with FCC regulations, such as certifying that the shielding will minimize communications interferences. Shielding integrity should be maintained to avoid violations.

Fiber optics eliminates electromagnetic induction, making it particularly useful in areas with several inducted voltages. This technology does not require separate conduits for power and communication cables.

SMART PROCESS INSTRUMENTATION

Smart process instrumentation is gradually entering the wastewater field as prices drop and industry standards develop. Smart instruments use two-way digital communications, so they can both send signals to and receive commands from a control system or host. These devices can record data and perform remote calibration and self-diagnostic routines. They also can be scaled, configured for various outputs and measurement ranges, calibrated, zeroed, and downloaded with process-loop variables.

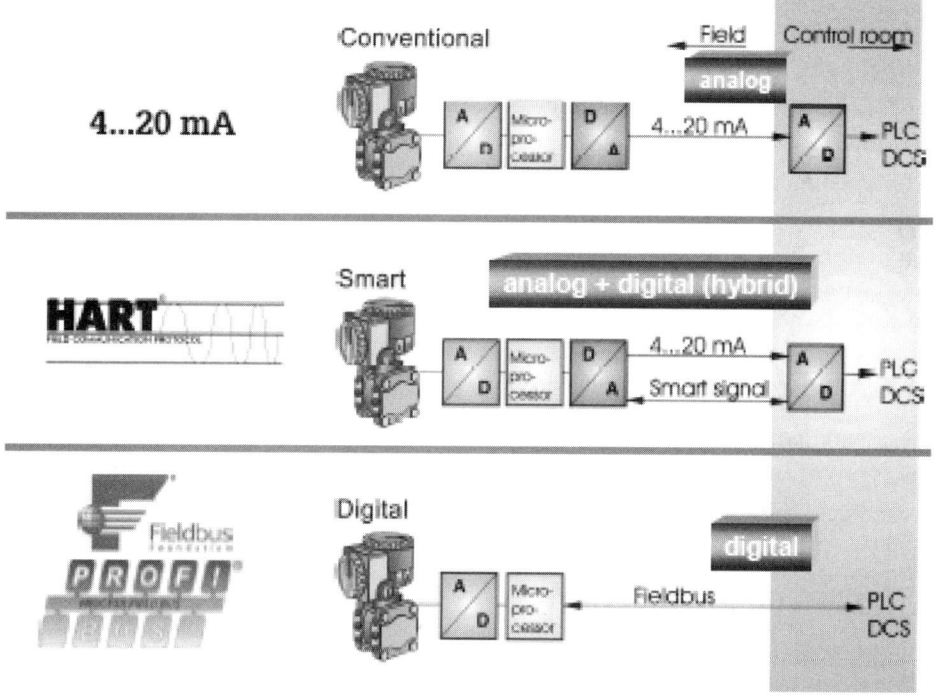

FIGURE 5.7 Comparison of analog, HART, and Fieldbus I/O.

The intelligent instruments communicate via one or more standard buses, such as HART, Profibus, and Foundation Fieldbus (Figure 5.7). The highway addressable remote transducer (HART) is a full-time, two-way digital communication method that is compatible with 4- to 20-mA analog instrumentation. It was developed to improve the startup and diagnostics of instruments and is the most common communication technology currently used in smart process instrumentation.

Currently, more water and wastewater treatment plants use Profibus than Foundation Fieldbus. Profibus consists of Profibus decentralized peripheral (DP) and Profibus process automation (PA). Profibus DP is designed for communications between automated control systems and distributed I/O at the device level, and Profibus PA is a lower-speed version for applications in process environments. The Fieldbus Foundation was formed when two rival Fieldbus groups merged in 1994 to develop a single interoperable fieldbus protocol. It is an open technology, and the Fieldbus Foundation establishes conformance and operability standards for manufacturers. Smart instruments are typically 5 to 10% more expensive than conventional analog instruments, but this higher capital cost is somewhat mitigated by the lower cabling (installation) cost associated with digital communications. They are also more complex and require knowledge of networking communications methods, which, combined with existing programmable logic controller and distributed control system plant standards, has made integration into the wastewater treatment industry somewhat slow. One difficulty with smart instruments is the built-in diagnostics, interpretation of equipment failure, and instrument response, which all add to the complexity.

DOCUMENTATION

At the very least, design documents should include basic installation details for various types of instruments. These details, which should include calibration, maintenance accessibility, etc., can be obtained from various instrument manufacturers.

SUGGESTED READINGS

American National Standards Institute; Instrumentation, Systems, and Automation Society (1993) *Process Instrumentation Terminology*, (R1993) ANSI/ISA-S51.1; Instrumentation, Systems, and Automation Society: Research Triangle Park, North Carolina.

American Water Works Association (2002) *Instrumentation and Control Manual*; American Water Works Association: Denver, Colorado.

Anderson, N. A. (1998) *Instrumentation for Process Measurement & Control*, 3rd Ed.; CRC Press: Boca Raton, Florida.

Boyes, W. (2002) *Instrumentation Reference Book*; Butterworth-Heinemann: Woburn, Massachusetts.

Institute of Electrical and Electronics Engineers (2002) *National Electrical Safety Code*.

Katebi, R.; Johnson, M. A.; Wilkie, J. (1999) *Control and Instrumentation for Wastewater Treatment Plants (Advances in Industrial Control)*; Springer: New York.

McMillan, G. K.; Considine, D. M. (2001) *Process/Industrial Instruments and Controls Handbook*, 5th Ed.; McGraw-Hill: New York.

National Board of Fire Underwriters (2002)) *National Electrical Code*; National Board of Fire Underwriters: Quincy, Massachusetts.

Seal, A. M. (1998) *Practical Process Control*; Butterworth-Heinemann: Woburn, Massachusetts.

Spitzer, D. W.; Boyes, W. (2002) *The Consumer Guide to Magnetic Flowmeters*; Copperhill and Pointer: Chestnut Ridge, New York.

Water Environment Federation (1996) *Operation of Municipal Wastewater Treatment Plants*, 5th Ed.; Manual of Practice No. 11; Water Environment Federation: Alexandria, Virginia.

Chapter 6

Sensors

Sensor Characteristics	105	*Accuracy and Precision*	116
Wastewater Characteristics	105	*Installation*	119
Plastic Products	106	*Maintenance Requirements*	121
Hair and Fibers	106	Weirs and Flumes	121
Solids	106	*Operating Principles*	121
Grease	106	*Construction Materials*	127
Color	107	*Accuracy and Precision*	127
Sulfides and Hydrogen Sulfide	107	*Installation*	127
Diurnal Changes	107	*Maintenance Requirements*	128
Flow Meters	107	Differential-Pressure	129
Magnetic	108	*Operating Principles*	129
Operating Principles	108	*Construction Materials*	130
Construction Materials	110	*Accuracy and Precision*	130
Accuracy and Precision	110	*Installation*	131
Installation	112	*Maintenance Requirements*	131
Maintenance Requirements	114	Mechanical	132
Ultrasonic	115	Rotary-Element	132
Operating Principles	116	*Operating Principles*	132
Construction Materials	116	*Construction Materials*	134

Accuracy and Precision	134	Installation	145
Installation	134	Maintenance Requirements	146
Maintenance Requirements	135	Capacitance and Impedance	146
Variable-Area	136	Operating Principles	146
Operating Principles	136	Construction Materials	149
Construction Materials	136	Accuracy and Precision	149
Accuracy and Precision	137	Installation	149
Installation	137	Maintenance Requirements	150
Maintenance Requirements	137	Differential-Pressure	150
Mass	137	Operating Principles	150
Coriolis	138	Construction Materials	153
Operating Principles	138	Accuracy and Precision	153
Construction Materials	139	Installation	153
Accuracy and Precision	139	Maintenance Requirements	155
Installation	140	Sonic and Ultrasonic	156
Maintenance Requirements	140	Operating Principles	156
Thermal-Dispersion	140	Construction Materials	158
Operating Principles	140	Accuracy and Precision	158
Construction Materials	140	Installation	158
Accuracy and Precision	141	Maintenance Requirements	159
Installation	141	Microwave (Radar)	160
Maintenance Requirements	141	Operating Principles	160
Level-Measurement Devices	141	Construction Materials	161
Continuous	141	Accuracy and Precision	161
Bubbler	142	Installation	161
Operating Principles	143	Maintenance Requirements	162
Construction Materials	144		
Accuracy and Precision	144	Point	164

Operating Principles	164
Construction Materials	165
Installation	165
Maintenance Requirements	165
Pressure	166
Operating Principles	166
Construction Materials	167
Accuracy and Precision	167
Installation	167
Maintenance Requirements	168
Temperature	168
Thermocouple	168
Resistance	171
Thermistor	172
Thermal Bulb	172
Accuracy and Precision	172
Installation	173
On-line Analyzers	174
Dissolved Oxygen	175
Typical Membrane Sensor Design	175
Conventional Galvanic Measuring Cell	176
Clark Polarographic Measuring Cell	177
Ross Polarographic Measuring Cell	179
Zullig Sensor	180
Luminescent Dissolved Oxygen Sensor	180
Accuracy and Precision	181
Installation	181
Maintenance Requirements	181
Chlorine Residual	182
DPD Colorimetric Method Analyzer	183
Amperometric Bare-Electrode Analyzer	186
Amperometric Membrane-Covered-Electrode Analyzer	186
Oxidation–Reduction Potential Probes	187
Accuracy and Precision	188
Installation	188
Maintenance Requirements	189
Polymer	189
Operating Principles	190
Construction Materials	191
Installation	192
Maintenance Requirements	192
Nitrogen	192
Ammonia and Ammonium	192
Operating Principles	193
Construction Materials	194
Accuracy and Precision	194
Installation	194
Maintenance Requirements	194
Nitrate and Nitrite	195
Operating Principles	195
Construction Materials	196
Accuracy and Precision	196
Installation	196

Maintenance Requirements	196
pH	196
Operating Principles	197
Typical pH Sensor	197
Ion-Sensitive Field Effect Transistor	199
Accuracy and Precision	202
Installation	202
Phosphorus and Orthophosphate	204
Orthophosphate	204
Operating Principles	204
Construction Materials	204
Accuracy and Precision	204
Installation	204
Maintenance Requirements	204
Total Phosphorus	205
Operating Principles	205
Construction Materials	205
Accuracy and Precision	205
Installation	205
Maintenance Requirements	205
On-Line Solids Measurement	205
Operating Principles	207
Near-Infrared Analyzers	207
Microwave Analyzers	207
Construction Materials	210
Accuracy and Precision	210
Installation	210
Maintenance Requirements	210
Interface/Sludge Blanket Level (ISBL) Analyzers	211
Operating Principles	212
Ultrasonic ISBL Analyzers	212
Optical ISBL Analyzers	213
Applications	213
Variable Blanket Height	213
Variable Blanket Height and Suspended Solids Profile	213
Blanket Height Alarm	214
Construction Materials	215
Accuracy and Precision	216
Installation	216
In-Line with Skimmers or Flights	216
Off-Line with Skimmers or Flights	217
Fixed Submerged Location	217
Maintenance Requirements	217
Vibration	218
Operating Principles	218
Velocity Transducers	218
Acceleration Transducers	219
Displacement Transducers	220
Construction Materials	220
Accuracy and Precision	221
Installation	222
Flush-Mount	222
Nonflush-Mount	223
Maintenance Requirements	224
References	224

Since competitive and regulatory pressures began increasing in the 1980s, instrumentation and control (I&C) systems have become more common at and intrinsic to wastewater treatment facilities. While not every instrument commercialized in the last 20 years has been successful, the overall evolution of I&C systems has made wastewater treatment more sophisticated. So, people responsible for designing, operating, and maintaining these increasingly automated facilities need to be familiar with the I&C features available.

This chapter focuses on the sensors used to monitor various wastewater treatment processes.

SENSOR CHARACTERISTICS

All sensors consist of two components: the sensing head and the transmitter. These components may or may not be in one casing. The sensing head is the "business end", which tracks the variable being measured and produces a signal that is sent to the transmitter. It sometimes includes amplification hardware to increase its signal strength.

The transmitter typically includes a signal converter, which changes the signal into a usable form (for more information, see Chapter 5). The transmitter displays the data onsite or sends the signal to another piece of equipment or the plant's supervisory control and data acquisition (SCADA) system. It also can generate discrete signals to indicate a high or low measurement and use calibration algorithms to observe field conditions and adjust sensor output according to a predetermined procedure.

Sensors can be divided into two categories: direct and indirect. Direct sensors typically have their sensing head in the process stream. They are often easier to install and start up, but must be made of a material that can resist the process environment, to minimize fouling. Also, the sensing head must be able to detect the required field conditions without interference from other process-flow characteristics.

Indirect sensors typically are put next to the process stream and do not actually touch the process flow. In fact, they often are shielded from it. The shield, which may be a diaphragm or air space above the process, must be able to resist fouling and avoid influencing measurements of the variable.

WASTEWATER CHARACTERISTICS

Wastewater is inherently variable and complex; it consists of thousands of waste products from houses and industries. Nevertheless, sensors must be able to continually and reliably report on the variable it was designed to measure. Sensors designed for other industries may have difficulty functioning in this harsh, ever-changing environment.

Wastewater characteristics known to be problematic for sensors include plastic products, hair, solids, grease, color, sulfides and hydrogen sulfide, and diurnal changes.

PLASTIC PRODUCTS. Wastewater may contain sandwich bags, condoms, tampon applicators, toys, and other plastic products. Wastewater utilities typically use screens to remove this debris before any biological treatment process, but some will always get though. Plastic is problematic for direct sensors in two ways. Heavy plastic debris, such as toys, may hit the sensor and damage its sensing head. Flexible plastic items, such as sandwich bags and condoms, may wrap around the sensor, effectively shielding it and reducing its ability to measure the intended variable. Either problem may be detected by monitoring sensor data and noting when it suddenly changes or fluctuates oddly.

HAIR AND FIBERS. Hair and other long stringy fibers can wrap around a direct sensor's sensing head and cause two problems. It can accumulate biological material, changing conditions around the sensing head. It also can shield the sensor from the process flow, reducing its ability to measure the intended variable.

Minimizing a direct sensor's projection into the process flow may minimize fiber accumulation. Some sensors are designed to resist fiber accumulation and others have a self-cleaning apparatus, such as a wiper (although a wiper's sharp edge may catch and collect fibers). However, even well-designed sensors will need weekly cleaning.

SOLIDS. Wastewater contains both colloidal debris and suspended solids, which can be problematic for sensors. In activated sludge processes, for example, suspended bacteria may attach themselves to the direct sensor's instrument head and begin to grow colonies. Solids accumulation around the instrument head may change the instrument reading, or it may shield the sensor from the process flow, reducing its ability to measure the intended variable. So, sensors must be able to accommodate or resist solids-related fouling. Even well-designed sensors, however, will need weekly cleaning.

GREASE. Colloidal grease is problematic because at normal wastewater temperatures, microscopic grease balls become sticky and tend to accumulate on all surfaces, including sensors. Grease accumulation on a semi-permeable membrane may affect the constituents' ability to pass into an instrument. Grease accumulation on an optical lens will change its optical properties and could affect the subsequent readings.

COLOR. Wastewater color depends on the local dischargers. Most treatment-plant influent ranges from dark to light brown, but if a local industry uses dyes, the treatment-plant influent may be another color. Color particles are often colloidal solids and may affect optical sensors. Optical sensors typically are designed to ignore minor color changes, but if a treatment plant receives discharges from a dye house, optical sensors may not be suitable. A facility which doses iron (ferric or ferrous) to the treatment process may also have dark brown or black solids not suitable for optical instruments.

SULFIDES AND HYDROGEN SULFIDE. Anaerobic, anoxic, and septic wastewater may contain sulfides because the lack of oxygen causes bacteria to release hydrogen sulfide. When hydrogen sulfide dissolves in water, it produces sulfuric acid, which can accumulate and corrode any surface. If a sensor's integrated circuit board corrodes at all, the sensor may fail. Sensor transmitters are often designed with National Electrical Manufacturers Association (NEMA) 4X enclosures, which are sealed to minimize the electronics' exposure to moisture and corrosive gases.

DIURNAL CHANGES. All wastewater characteristics fluctuate throughout the day based on the community's living and working patterns. So, sensors will be subject to different types of fouling as the wastewater characteristics change. Effective sensor troubleshooting often depends on 24-hour monitoring.

FLOW METERS

There are many types of flowmeters, including magnetic, ultrasonic, flumes and weirs, differential-pressure, mechanical, and mass-flow. They can be used in closed or open pipes.

All flow meters should be sized to handle the flow that will exist at treatment-plant startup. If they are sized based on 20-year flow estimates, they may be too large to register existing flow rates.

Flow meters should be on the discharge side of pumps and on the upstream side of throttling valves. Those in closed conduits need the pipe to always be full of liquid; mounting the meter in a vertical pipe section with upward flow may help. Horizontally mounted flow meters need air-bleed valves. Fittings (e.g., reducers and elbows) upstream and downstream of the flow meter must contribute to even flows. Putting multiple sharp bends before a flow meter typically reduces its accuracy.

All flow meters need enough space for calibration, in-line maintenance, or meter body removal for calibration or maintenance. Removing the meter body should not hurt the associated process or surrounding area. (This does not mean a flowmeter that typically has a long service life cannot be buried, just that engineers should consider maintenance requirements when installing flow meters.)

Flowmeters also need a means of testing the electrical ground and confirming hydraulic flow during acceptance testing and performance monitoring. In addition, flow meters used for billing need factory-calibration documentation and verification of meter accuracy.

MAGNETIC. Magnetic flowmeters ("mag" meters) currently are the most popular method of measuring flow through a pipe. They typically have a long life, are cost-effective, and are accurate and precise if installed properly.

Operating Principles. Magnetic flow meters are based on Faraday's principle of electromagnetic induction, in which the voltage induced by an electrical conductor moving through a magnetic field is proportional to the conductor's velocity (Figure 6.1). Electric power enables the flow meter's coil driver to energize the magnetic coils encasing its spool pipe, creating a magnetic field. If the liquid has enough conductivity, it will act as an electrical conductor and induce a voltage. This voltage is the sum of the voltage of each liquid particle in the magnetic field and is proportional to field strength, pipe diameter, and flow velocity.

The induced voltage is received by two electrodes mounted 180 degrees apart in the meter. The higher the flow rate, the greater the instantaneous electrode voltage. The electrodes send signals to the converter or transmitter, where they are added together, referenced to a zero value, and converted from a magnetically induced voltage to the appropriate flow measurement. The data then is sent to a control-panel transmitter, I&C system computer, or other appropriate receiver.

Magnetic flow meters used to be available in both alternating-current (AC) and direct-current (DC) versions. In AC meters, electricity typically is applied to the coils to create a continuous flux, inducing a continuous, low-level AC voltage. They typically are used on slurries or in areas with rapidly changing flow rates.

DC meters can be pulsed or non-pulsed. In most DC meters, the magnetic coils are energized periodically and produce two induced voltages: one when energized

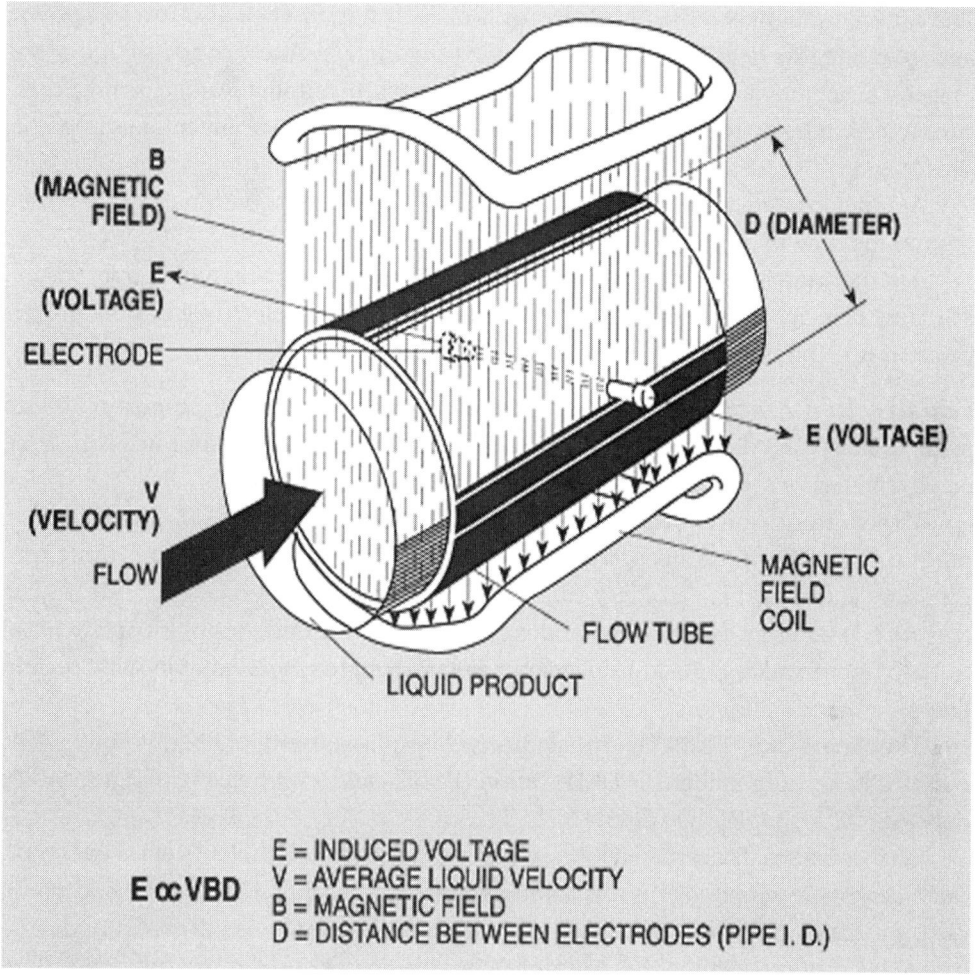

FIGURE 6.1 Faraday's principle of electromagnetic induction. (© Copyright 2006 Omega Engineering, Inc. All rights reserved. Reproduced with the permission of Omega Engineering, Inc., Stamford, CT 06907.)

and the other when not. The energized voltage is a combination of both true signal and noise. The de-energized voltage is only noise; it allows the meter to self calibrate to a zero value every cycle, which slows meter response time but improves accuracy.

The difference between the two voltages is a clean signal. They also use less power and generate less heat than other magnetic flow meters. Another advantage of the pulsed DC meters is that the current can be reversed to automatically clean the measuring electrodes so they will not be coated by ferric chloride or other charged materials.

AC meters and non-pulsed DC meters are no longer common because they require the flow to be stopped for periodic re-zeroing.

Another type of magnetic flow meter is the insertion probe, which operates on the same principle. Some probes have an alternate coil-and-electrode design and are inserted into the process pipe through a fully ported ball-valve tap.

Construction Materials. Magnetic flow meter bodies typically are available in primed-and-painted carbon steel, epoxy-coated carbon steel (best for buried meters), or 316 stainless steel.

The electrodes are available in platinum, Alloy 20, Hastelloy C-22, Monel, nickel, platinum-alumina ceramic, platinum-rhodium alloy, 316 stainless steel, Tantalum, titanium, and tungsten carbide. Some manufacturers offer optional electrode shapes (Figure 6.2) and self-cleaning electrodes (via heat, ultrasound, or polarity switching) to help avoid fouling. Field-replaceable electrodes are also available with certain liners.

The liners are available in ceramic, fiberglass, hard rubber, kynar, PFA, polyurethane, soft rubber (EPDM), Teflon (PTFE), and vitreous enamel; the choice depends on the application (Table 6.1).

Gasket materials are also chosen when a magnetic flow meter is specified; typically, gaskets compatible with the piping system are acceptable. Also, depending on the piping material and application, grounding wires may be required. Cathodically protected piping systems, for example, require special powering and grounding practices.

Accuracy and Precision. Under typical field conditions, pulsed DC flow meters may be accurate within ± 0.2% of full-scale and precise within ± 0.10% of full-scale. Meanwhile, AC flow meters should be accurate within ± 1% full scale and not exceed ± 3% of indicated flow when operating in the lower 30% of its range. Most flow meters can read flow equally accurately in either direction, although each manufacturer addresses flow reversal differently, and this must be accommodated in the design.

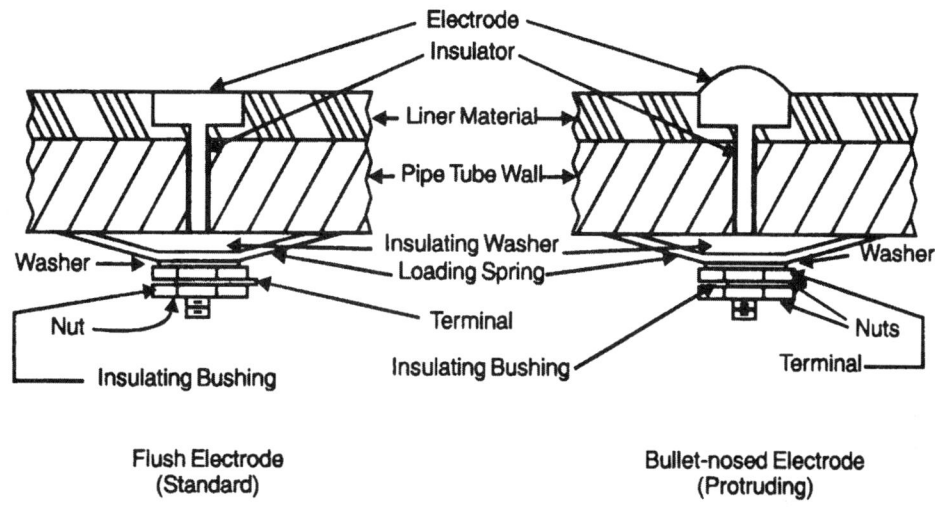

FIGURE 6.2 Optional electrode shapes.

TABLE 6.1 Construction material options for magnetic flow meters.

Service	Possible liner material	Abrasion resistance	Corrosion resistance	Maximum temperature
Raw wastewater	Polyurethane	Good	Poor	88 °C
Primary sludge	Ceramic	Excellent	Excellent	300 °C, but less than 90 change
Primary effluent	Polyurethane	Good	Poor	88 °C
Mixed liquor	Teflon	Poor	Good	
RAS/WAS	Teflon	Poor	Good	
Thickened WAS	Teflon	Poor	Good	
Digested sludge (warm)	Polyurethane	Good	Poor	88 °C
Digester supernatant	Polyurethane	Good	Poor	88 °C
Centrate/filtrate	Polyurethane	Good	Poor	88 °C
Process water	Polyurethane	Good	Poor	88 °C
Polymer solution	Teflon	Poor	Good	
Chemicals	Check material compatibility charts.			

Several circumstances can degrade a magnetic flow meter's performance level. For example, flow-disturbing obstructions too near the meter inlet and outlet may add another 1 to 10% of uncertainty to the measured flow.

Flow meter performance can be severely degraded if it is improperly oriented so the pipe only partially fills (trapping gases) or material builds up on the electrodes. If this happens, some flow meters will have jumpy readings and others will not report flow. Some flow meters can detect when a pipe is not full and notify operators so the relevant data can be ignored. Some are even designed to accommodate partially filled pipes; low-flow circuitry causes the meter to read "zero" when the flow rate is less than a specified amount.

Compared to AC and DC flow meters, an insertion probe is slightly less accurate and less suited for wastewater because its foil must be put in the center of the process stream. The probe typically is accurate within 1% of flow rate, depending on insertion depth, probe design, velocity profile, and debris or coating buildup on the probe.

Magnetic flow meters work best in streams where headlosses must be minimized, liquids with conductivity more than 5 μS/cm (5 μMhos/cm), corrosive or abrasive process streams, and liquid streams with a solids concentration of less than 10% (by weight). Most require 1 to 5 μS/cm of conductivity. Some probes require more, and some flow meters will function with as little as 0.05 or 0.1 μS/cm.

These flow meters are not recommended for non-conductive liquids, gas streams, streams with powdered or granular dry chemicals, or partially full pipes (unless the meter is specifically designed for such pipes).

Installation. Sizing is critical. Magnetic flow meters are sized for velocities of 0.3 to 10 m/s (1 to 33 ft/s) in non-solids-bearing liquids and 1.5 to 7.5 m/s (5 to 25 ft/s) in solids-bearing liquids. Appropriate reducers or expanders may be required to achieve recommended operating velocities. Slower velocities are possible for special applications, but higher velocities are preferred to keep the inside of the flow tube scoured clean.

The flow meter should be put in a straight run of pipe without valves or fittings (the minimum pipe length depends on the fittings and the meter manufacturer involved). Design engineers typically should avoid putting valves, gates, T connections, elbows, pumps, and severe (30-degree-angle) reducers and expanders less than five pipe diameters from the flow meter inlet or three pipe diameters from its outlet. However, fittings could be put as close as one pipe diameter from the flow meter if space is constrained; the added uncertainty of close fittings typically is less than 1% (Endress and Hauser, document TIF 008 3.98). So long as billing is not an issue, readings from such flow meters will still be precise enough for treatment-plant operations. Magnetic flow meters also can be installed on diaphragm metering pumps if the sensor is designed to compensate for the pulsating flow.

The decision to use bypass piping for flow meters depends on pipe size, available space, and the need to shut down the line while maintaining operations or shifting to a parallel treatment-process unit. Metered pipes should not automatically drain when shut down. Intermittently used pipes in sludge applications should be flushed and filled with clean water.

Magnetic flow meters typically are grounded with stainless steel grounding rings and straps supplied by the meter manufacturer. The inside diameter of the grounding rings should be 0.01 m (0.25 in.) shorter than that of the meter [for 0.1-m-diameter (4-in.-diameter) meters or larger]. The rings are installed on both flanges with grounding straps (Figure 6.3). Some magnetic flow meters use grounding electrodes in the meter body; if so, operators should periodically test the plant's electrical-system ground near the meter to see whether it is adequately grounded. If a plantwide grounding grid is available, the flow meter could be grounded to it instead.

Magnetic flow meters should be kept away from heavy induction equipment that could cause operating problems [no closer than 6 m (20 ft) for 75-kW (100-hp) motors and larger]. Also, before installing a flow meter next to large induction or radio equipment, consult the manufacturer to determine possible interferences. Some sensors will not have problems.

Magnetic flow meters typically are installed so the electrodes are parallel to the floor and the attached transmitter or signal converter is legible from the ground. If wall-mounted, the transmitter or signal converter typically is put within sight of the flow meter in a NEMA 4 enclosure (NEMA 6 if it may be submerged) or in a control panel if the cable is less than 60 m (200 ft) from the meter.

Driven-shield signal leads should be used between the transmitter and meter, typically in dedicated 0.02-m (0.75-in.) conduits. The same dedicated circuit typically is used to power both the coil driver and the transmitter or signal converter. If separate circuits are used, both must originate from the same phase of the primary power feed (follow the manufacturer's instructions).

Magnetic flow meters may include full-pipe detection, reverse-flow measurements, self-cleaning, and high-impedance electrodes that require less cleaning. Design engineers should discuss these options with the flow meter manufacturer. Manufacturers handle reverse-flow measurement differently with respect to totalizers and contacts, and self-cleaning probes are unusual and generally not recommended.

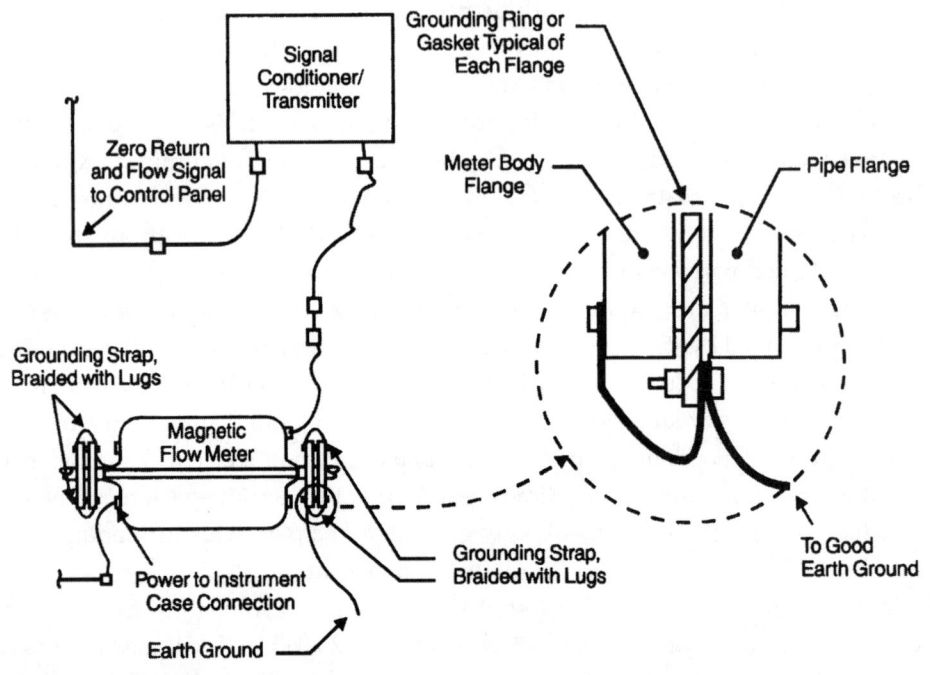

FIGURE 6.3 Magnetic flow meter grounding.

Flow meters are also available with all of the common communication protocols. Smart meters with digital bus technology can be coupled directly with flow-control valves to allow PID (proportional–integral–derivative) control algorithms to be used in the field instruments. They also incorporate several totalizers with enhanced capabilities, such as programmable intervals, so totalizers and counters can keep track of flow in each direction and the total combined rate.

Maintenance Requirements. Some magnetic flow meters need regular cleanings. Many have self-cleaning electrodes to prevent buildup. A cleanout T fitting can be installed when continuous electrode cleaning is not used or when the meter is used to measure flow in a pipe known to have a fouling problems. Possible problems with magnetic flow meters that would require regular maintenance include: rock-damaged electrodes in unscreened wastewater or accumulated grease in raw-wastewater, primary-sludge, and activated-sludge pipes that could eventually degrade accuracy. If buildup in the pipe is a concern, installing a smooth liner (e.g., Teflon or ceramic) and maintaining high-velocity flows can improve system performance. Also, installing an appropriately sized T with a blind flange before or after the flow meter allows the whole flow meter pipe to be wiped out (Figure 6.4) without having to remove the meter.

ULTRASONIC. There are two basic types of ultrasonic flow meters: transmissive (through-beam or transit-time) and reflective (frequency shift or Doppler). Both are suited for areas where headlosses must be minimized and the process pipe flows full.

Transmissive sonic flow meters are suited for liquids between 0 and 80 °C (32 and 180 °F) without entrained air bubbles. They work best in primary effluent, mixed liquor, secondary-clarifier effluent, final effluent, washwater, RAS, and WAS. They

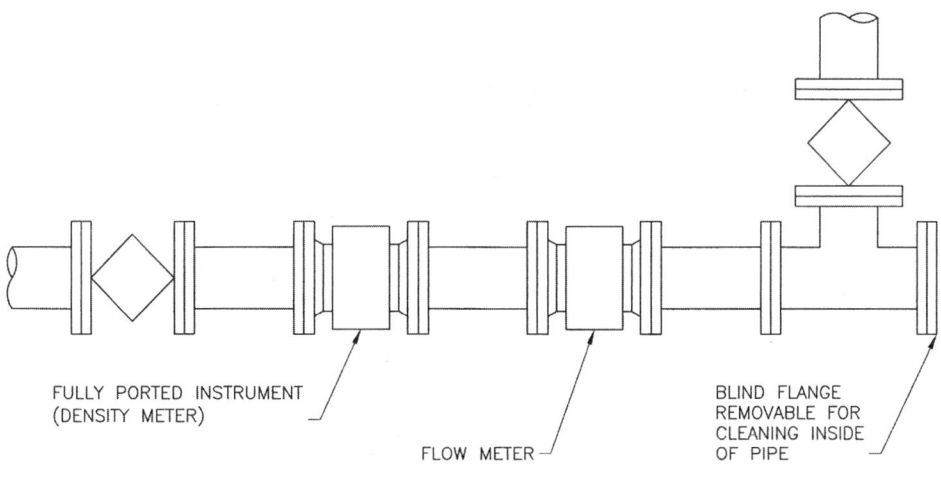

FIGURE 6.4 Cleaning T for pipeline known to have fouling problems. Courtesy of Carollo Engineers, Sacramento, California.

are not recommended for primary sludge, thickened sludge, nitrification RAS, or nitrification WAS. These meters may also be used in dirty or foul air (e.g., digester gas, or scrubber and biotower airflow).

Reflective sonic flow meters are suited for flows with air bubbles or suspended debris. They work best in raw wastewater, primary sludge, or thickened sludge. They are not recommended for primary effluent, secondary-clarifier effluent, final effluent, or washwater. Also, fluctuating conditions may cause intermittent metering problems.

Operating Principles. The transmissive sonic flow meter measures fluid velocity by calculating the difference in travel time for one sonic pulse to go a specific distance downstream and another to go the same distance upstream. Two transducers, alternately energized by electrical pulses, emit sonic pulses across the flow. The downstream pulse traverses the pipe in less time than the upstream pulse, and this difference is proportional to the flow velocity. The flow meter's transmitter computes an output signal linearly proportional to the flow rate.

The reflective sonic flow meter typically uses one transducer, and its transmitter can be separate from the sender (Figure 6.5). Mounted on the external wall of the pipe (or through the wall), the transducer sends a specific-frequency signal (between 600 kHz and 1 mHz) into the fluid, and suspended particulates or gas bubbles reflect it back to the transducer. Because the reflective matter is moving with the fluid, the frequency of the sonic energy waves shifts as it is reflected. The magnitude of the frequency shift is proportional to the flow rate and is converted electronically to a linear flow signal.

Construction Materials. The transmissive meter is available as an insertion meter (i.e., a pipe section with integral well-mounted transducers) or a spool piece (i.e., direct-mounted with the transducers mounted externally to an existing pipe). Both types operate on the same principle (Figure 6.6). Clamp-on meters are also available. The construction materials must be compatible with the fluid being measured.

The reflective meter is typically available as a clamp-on device that attaches to the process pipe with acoustical gel. It can only be calibrated if detailed information is available about the pipe diameter, pipe-wall thickness, process fluid, solids concentration, and temperature variation. Its construction materials only need to resist normal atmospheric conditions, which may include hydrogen sulfide and rainwater.

Accuracy and Precision. Transmissive sonic flow meters are more accurate than reflective ones. Transmissive meters should be accurate within ± 1% and precise within ± 0.5% of actual flow or 2% of full-scale. Reflective sonic meters should be

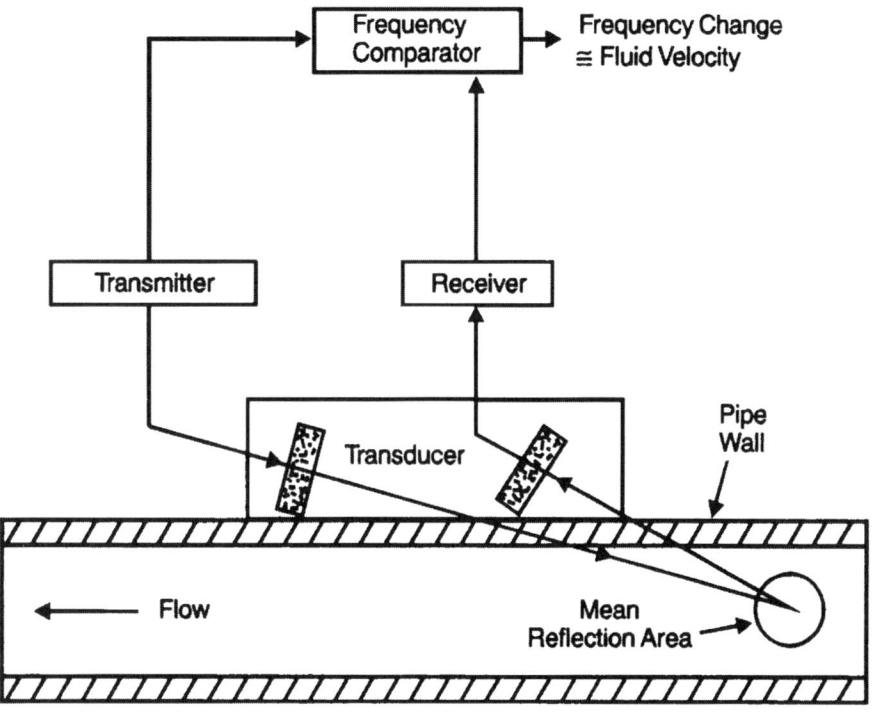

FIGURE 6.5 Doppler sonic flow meter operation.

accurate within ± 3% and precise within ± 1% of actual flow. Better performance is possible if the unit has two, four, or more pairs of transducers.

Purchasing sonic flow meters that meet these limits does not necessarily ensure that they will be this accurate in field conditions. Typically, installation issues reduce accuracy by 1.5%. To avoid degrading accuracy, design engineers should not put flow-disruptive obstructions (e.g., valves, gates, and pumps) too near the meter inlet. They can reduce accuracy by up to 10% (Table 6.2). Also, modulating and isolating valves, gates, elbows, T connections, pumps, and severe reducers and expanders

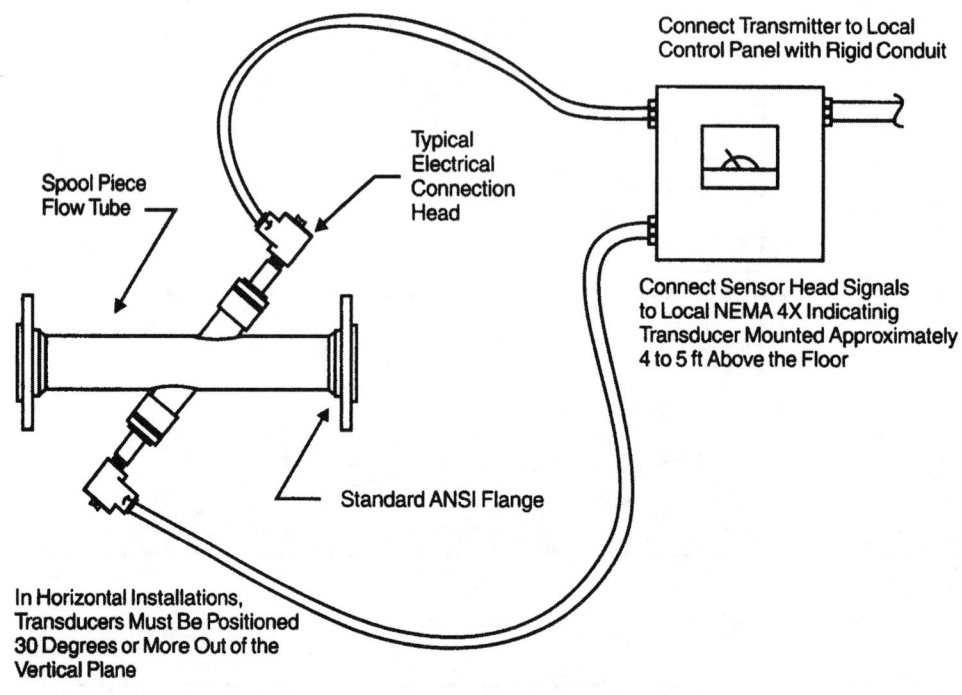

FIGURE 6.6 Transmissive sonic flow meter (ft 3 0.304 8 5 m).

(those with a 30-degree angle) should be at least seven to 10 pipe diameters from a transmissive meter's inlet or a reflective meter's external transducer and at least five pipe diameters from a transmissive meter's outlet.

Unless the manufacturer-recommended lengths of straight pipe are provided upstream and downstream of the flow meter, the velocity profile may be skewed, causing errors in flow measurement. Very disturbed flow profiles may require upstream runs to be straight for up to 20 pipe diameters. Also, the pipe must be completely full or else the flow velocity cannot be measured because air dampens sound waves.

TABLE 6.2 The effect of components on the accuracy of transit time flow meters.

Meter type:	Flow obstruction	Downstream position [% Error]			
		2D	10D	20D	40D
Single Channel	Globe valve 2/3 open	14.3	3.2	1.7	0.2
	Gate valve 2/3 open	12.4	4.0	2.6	0.5
	Butterfly valve 2/3 open	21.5	4.1	1.6	0.4
	90° Bend	22.3	7.0	3.4	1.0
	Two 90° bends in different planes	9.9	6.4	4.6	2.8
	Expansion	10.4	3.6	1.9	0.3
	Reduction	2.3	0.8	0.9	0.9
Dual channel	Globe 2/3 open	4.4	0.5	0.4	0.4
	Gate valve 2/3 open	14.9	1.0	1.1	0.1
	Butterfly valve 2/3 open	8.8	0.8	0.9	0.6
	90° Bend	6.7	0.3	0.6	0.6
	Two 90° bends in different planes	3.4	1.2	0.6	0.6
	Expansion	3.5	1.0	0.6	0.4
	Reduction	0.7	1.9	1.1	0.4

Installation. Transmissive sonic flow meters are typically wetted, but clamp-on models are available. The sensor can only measure clean liquids with less than 1% of bubbles and 5% solids (by weight). The pipe must be small enough so sonic-signal attenuation is not an issue. Design engineers should consult the manufacturer about applications in lines larger than 1 m (42 in.).

A spool piece is used for pipes between 0.08 and 0.9 m (3 and 36 in.) in diameter. Spool pieces are available for pipes up to 4.0 m (160 in.) in diameter. Figure 6.7 shows transmissive flow meters with either single or dual path measurement. For smaller pipes, transducers are mounted in an axial configuration.

Reflective sonic flow meters typically are clamp-on devices. The sensor can only measure flows with entrained air bubbles or solids concentrations between 0.2 and 60%, depending on particle size.

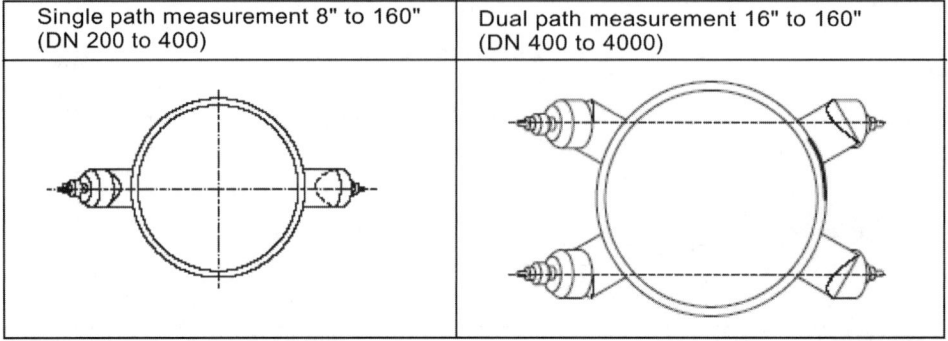

FIGURE 6.7 Single and dual path measurements for transmissive flow meters. (Courtesy of Endress + Houser, Greenwood, Indiana.)

Also, reflective sonic flow meters that operate at frequencies of 1 MHz or higher are available. Theoretically, the high frequency will operate in virtually clean liquids because the sound will reflect off swirls and eddy currents in the flow stream.

For both types of sonic flow meters, the fluid velocity at the transducer must be between 1 and 9 m/s (3 and 30 ft/s). If the water is clean and the sensor is relying on air bubbles, the velocity must be 1.8 m/s (6 ft/s).

Clamp-on meters are popular because they are portable. Although not as accurate as magnetic flow meters, these meters allow plant staff to verify the flow rate through a pipe. For these sensors to function effectively, the pipe wall must be less than 0.05 m (2 in.) thick, and the pipe must not be lined or constructed of aggregate material. Even a molecular layer of air can block the transmission of ultrasonic wavelengths. Clamp-on transducers should be installed according to the manufacturer's recommendations. Air bubbles should be removed from the epoxy sealant.

The transmitter is typically installed as close as possible [3.7 m (12 ft)] to the clamp-on transducer, or directly onto the meter tube.

Sonic flowmeters are available with all of the common communication protocols. Smart meters, which communicate directly with flow-control valves, allow PID control algorithms to be used in the sensors.

Maintenance Requirements. Ultrasonic flow meters typically do not require a lot of maintenance. The concerns about immersed sonic meters are similar to those for magnetic flow meters (i.e., material sticking to the sensor). However, they can be installed so the meters can be taken out of service for calibration or maintenance without disrupting the treatment process.

As long as the clamp-on meters are configured correctly, they are typically easy to use and do not require maintenance unless someone bumps the transceiver-mounting assembly or the pipe's wall thickness changes because of accumulated debris, grease, or chemical precipitation. Sometimes the sensor's acoustic-coupling gel must be replaced, depending on the environment.

WEIRS AND FLUMES. When instrumented, several types of flumes and weirs [e.g., Parshall flumes and flat, rectangular, Cipolletti (trapezoidal), and V-notch weirs] can be used to measure flow. Although typically used for regulatory purposes, weirs also can check the performance of other installed flow sensors. Flumes have historically been used to measure a plant's influent or effluent flows.

Weirs can accurately measure a wide range of flows (a maximum-to-minimum flow ratio of 20:1 is normal and ranges of 75:1 are reported for large weirs) in open channels. They should be put in areas where headloss is not a concern. Weirs require approach conditions that are tranquil and free of eddies or surface disturbances at all flow rates. At maximum flow, approach velocities in the upstream channel should not exceed 0.1 m/s (4 in./s).

Parshall flumes are open-channel, flow-measuring devices suitable for areas where headloss must be minimized. Typically, a rectangular weir's headloss is four times that of a comparable Parshall flume at the same flow. The flume is tolerant of solids and sediment, because fluid velocity tends to scour away deposits. Depending on flume size, a maximum-to-minimum flow ratio of 10:1 is practical, but tranquil and uniformly distributed approach conditions are necessary for accurate measurements.

Operating Principles. In this method, flow measurement depends on two elements: the Parshall flume or weir and a level-measuring device.

A weir is a dam or bulkhead placed across an open channel with an opening on the top through which the measured liquid flows. This opening is the weir notch; its bottom edge is the crest. The notch typically is cut from a metal plate (a sharp-crested weir) attached to the upstream side of the bulkhead. This plate prevents water from contacting the bulkhead (Figure 6.8).

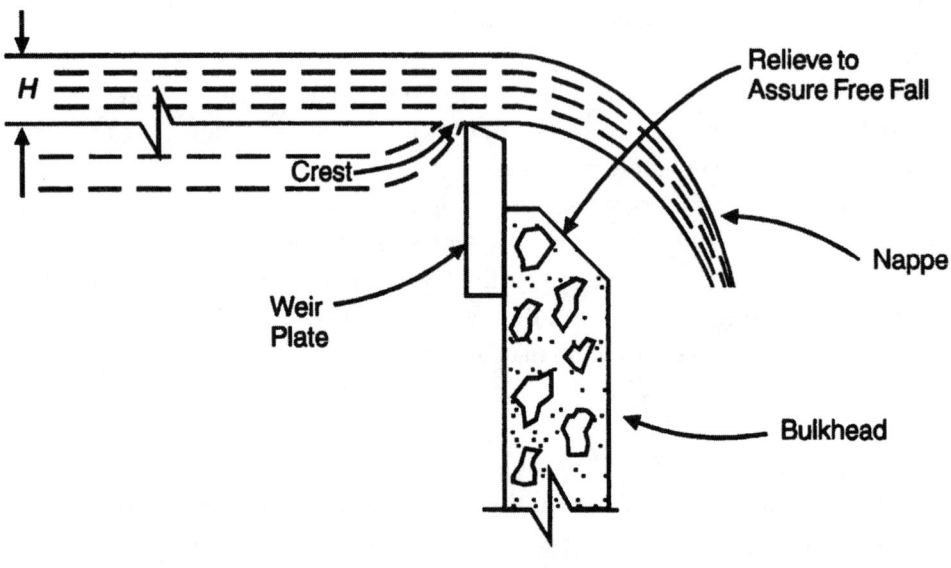

FIGURE 6.8 Sharp-crested weir.

The water depth is measured at a prescribed distance upstream (three or four times the head) to determine the discharge through the weir. *Head* is the measurement from water surface to crest of the weir elevation. Characteristic head-to-flow relationships are governed by weir geometry. All depth measurements are related to the crest elevation.

Weir openings are typically fabricated in rectangular, trapezoidal, or V-notch shapes (Figures 6.9 and 6.10). A trapezoidal weir with a side slope of 4:1 is a Cipolletti weir. Flow as a function of upstream head (H) is expressed by empirical equations, which depend on the weir shape (Table 6.3). A V-notch weir is the best choice for low flows.

Weirs work best on secondary and primary effluents (with provisions for sluicing). They are not recommended for raw wastewater, mixed liquor, or sludge.

A Parshall flume is a channel constriction that develops a hydraulic head proportional to flow (Figure 6.11). If the flume has been constructed to standard dimensions

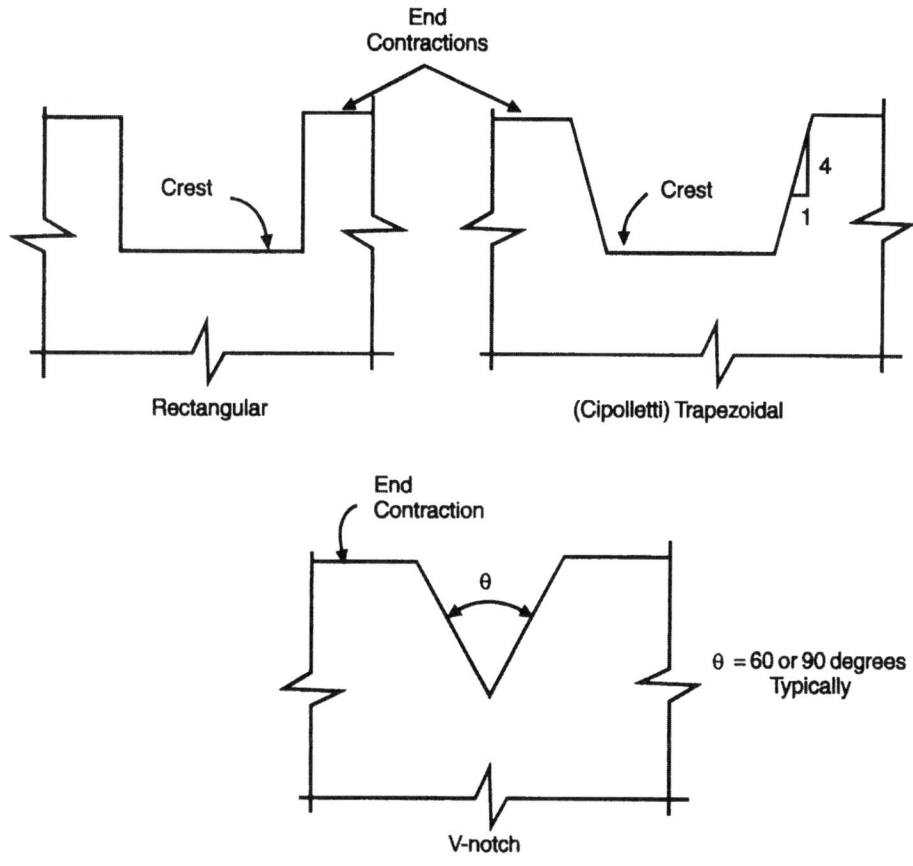

FIGURE 6.9 Weir shapes.

and properly set, flow through it can be calculated based on the depth measurement at one point (see point "A" in Figure 6.12).

Flumes are available in sizes from 0.025 to 15 m (1 in. to 50 ft); the size is defined by the width of the throat section. Large flumes are constructed onsite, but smaller flumes can be purchased as prefabricated structures or as lightweight shells set in concrete.

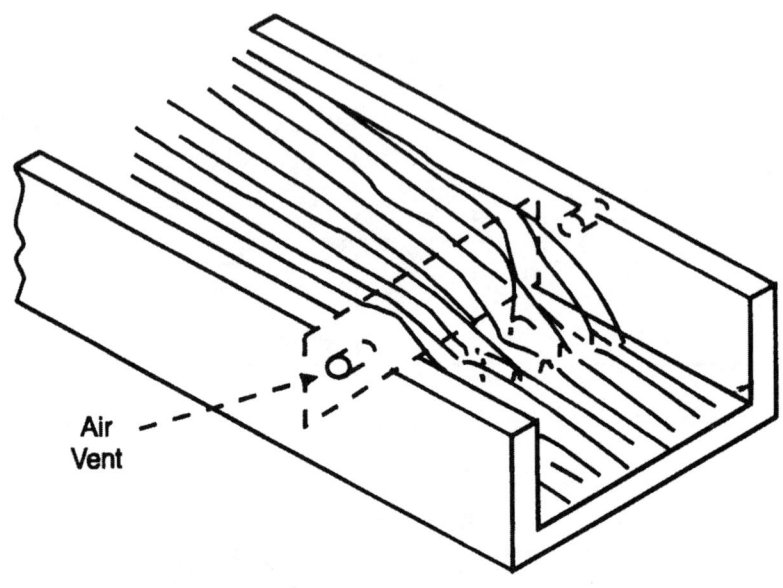

FIGURE 6.10 Suppressed rectangular weir.

TABLE 6.3 Weir equations.

	Customary (cfs)	**Metric (m³/h)**
Rectangular weir	$Q = 3.330(L-0.2H)H^{1.5}$	$Q = 6618(L-0.2H)H^{1.5}$
Cippoletti weir trapezoidal	$Q = 3.367LH^{1.5}$	$Q = 6692LH^{1.5}$
Rectangular (suppressed, without contractions)	$Q = 3.330LH^{1.5}$	$Q = 6618LH^{1.5}$
V-notch weir	$Q = 2.48 (\tan 0.5\theta)H^{2.5}$	$Q = 4929 (\tan 0.5\theta)H^{2.5}$

Where
 Q = rate of flow.
 L = crest length (ft)
 H = head of flowing liquid (ft)
 θ = V-notch angle in degrees for a V-notch weir.

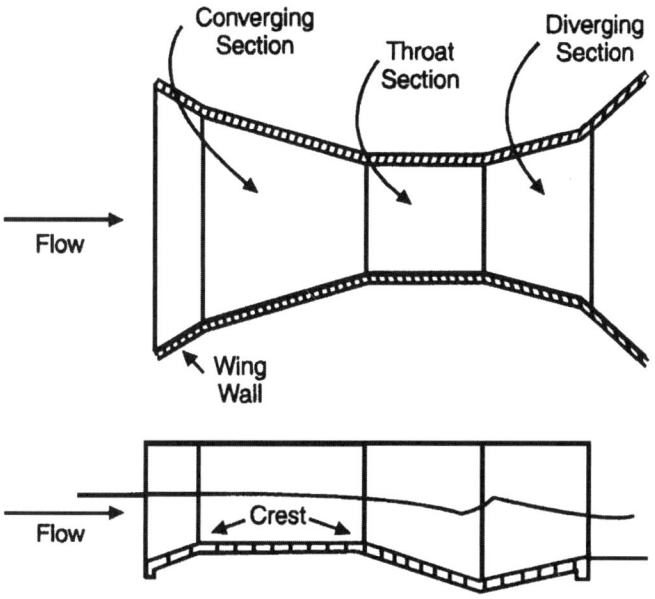

FIGURE 6.11 Shape and sections of a Parshall Flume.

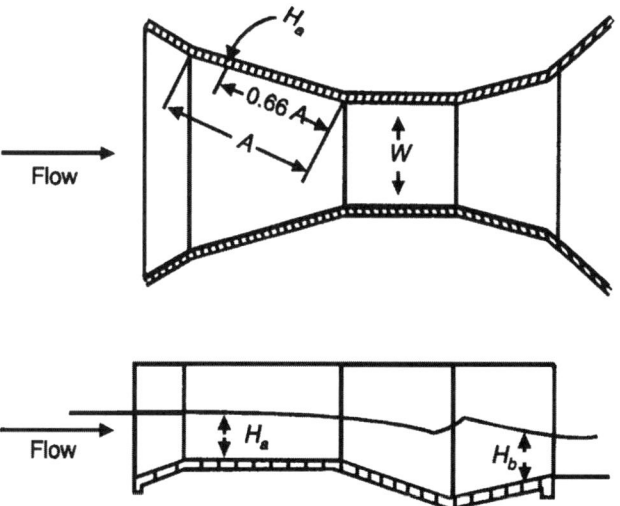

FIGURE 6.12 Head relationship for free flow (W 5 throat width, A 5 distance of converging section, Ha 5 hydraulic head upstream, and Hb 5 hydraulic head downstream).

Parshall flumes can function under either free or submerged flow conditions. *Free flow* is when fluid velocity is only restricted by throat width. Under this condition, upstream (H_a) and downstream (H_b) depth measurements can be used to calculate the flume's discharge flow (Figure 6.12). *Submerged flow* is when fluid velocity is restricted by both throat width and a rising downstream-channel level. Although Parshall flumes can operate under submerged flows, a second level measurement and a correction factor are required to calculate flow. The measurement is also less accurate. However, tests have shown that the flume's discharge rate does not change when the tailwater level (downstream of the flume) exceeds the crest elevation by a small amount. Free-flow conditions can still exist even with some degree of submergence (Table 6.4).

Parshall flumes work best on raw wastewater, primary effluent, secondary effluent, final effluent, and mixed liquor. They are not recommended for sludge or chemical applications.

TABLE 6.4 Parshall flume operating conditions.

Flume throat size "W"	Submergence [% (Hb/Ha)]	Minimum head	Customary (cfs)	Metric (m³/h)
0.075 m (3 in.)	50	0.03 m (0.1 ft)	$0.9920\,H^{1.547}$ 0.028-1.15 cfs	$0.9920\,H^{1.547}$ 2.80-125 m³/h
0.15 m (6 in.)	60	0.03 m (0.1 ft)	$2.060\,H^{1.580}$ 0.054-3.91 cfs	$2.060\,H^{1.580}$ 5.39-389 m³/h
0.23 m (9 in.)	60	0.03 m (0.1 ft)	$3.070\,H^{1.530}$ 0.091-8.87 cfs	$3.070\,H^{1.530}$ 9.01-882 m³/h
0.305 to 2.44 m (1 to 8 ft)	70	0.06 m (0.20 ft)	$4.0\,W\,H\,(1.522\,W)^{0.026}$ 0.120-16.1 cfs (1 ft)	$4.0\,W\,H\,(1.522\,W)^{0.026}$ 12.0-1610 m³/h (0.3)
More than 2.44 m (8 ft)	80	0.10 m (0.33 ft)	$(2.5+3.69\,W)H^{1.6}$ 5.74-199 cfs (10 ft)	$(2.5+3.69\,W)H^{1.6}$ 570-20,700 m³/h

The *Isco Open Channel Flow Measurement Handbook* (Grant and Dawson, 1997) provides a thorough discussion of the application of Parshall and other flumes, including different equations for flumes between 0.305 and 2.44 m (1 and 8 ft).

Construction Materials. Weirs and Parshall flumes may be fabricated out of any material that is stiff enough to keep its shape under the hydraulic load. Typically, fiberglass (FRP) or aluminum is used.

Design engineers should be aware that raw wastewater cascading over a weir will aerate, thereby converting dissolved sulfide into hydrogen sulfide gas. If a weir will be permanent, its construction material should be corrosion-resistant. Also, the surrounding concrete may need to be protected from water abrasion and sulfide attack, both of which will eventually expose aggregate and rebar.

Accuracy and Precision. The weir's head-flow relationship can be accurate within ± 2%. However, the accuracy of indicated or recorded flow also depends on the level sensor and flow converter used. So, if both the weir and level sensors are properly installed, the resulting flow measurement can be accurate within ± 5% of flow. (The total accuracy can be calculated as the square root of the sum of the squares of each measurement component's accuracy.)

Parshall flume accuracy also depends on the accuracies of both the flume and level sensors. If the flume is correctly fabricated and installed, its depth-discharge equation can be accurate within ± 3% of flow. If both the flume and level sensors are properly installed, the resulting flow measurement can be accurate within ± 5% of flow, and its precision can be within ± 0.5% of flow.

Flow-measurement accuracy can be degraded by the following errors (if uncorrected):

- Throat widths that deviate from standard dimensions;
- Any longitudinal slope of the floor in the converging section [tests on a 0.075-m (3-in.) flume showed that a downward sloping floor added errors of 3 to 10%, depending on flow rate];
- A transverse slope of the flume floor;
- Poor approach conditions;
- An incorrect zero reference during initial level sensor calibration; and
- If a stilling well is used, the connector hole is improperly sized.

Installation. In general, a weir's head height should be at least 0.06 m (2.5 in.) but less than 50% of the weir length. The approach channel should be straight and have a uniform cross-section for a length equal to at least 20 times the weir's maximum head. The distance from the weir's crest to the bottom of the approach channel should be either 0.3 m (12 in.) or twice the maximum head, whichever is greater. The free fall downstream of the weir should be at least 0.15 m (6 in.). The end contractions on each side should be at least 0.3 m (12 in.) or twice the maximum head—except for suppressed weirs. Suppressed weirs should be ventilated to prevent a vacuum from forming on the underside of the weir.

The sides of a rectangular weir must be vertical, while those of a Cipolletti weir should slope outward at a 1:4 horizontal-to-vertical ratio. Rectangular and Cipolletti weirs must have level crests that are at least three times as long as the maximum upstream head, while V-notch weirs must have a notch whose angle is accurately cut and that could be bisected by a vertical line. The bulkhead opening should be approximately 0.08 m (3 in.) larger on all sides than the weir notch. The upstream face of the bulkhead and weir plate should be smooth and vertical, perpendicular to the channel. The weir's edges should be straight and free of burrs, with the trailing edge chamfered to a crest thickness of 1 to 2 mm (0.04 to 0.08 in.).

The level sensor should be on the side wall upstream of the weir at a distance at least four times the length of the maximum head (to avoid drawdown effects). A depth gauge should be mounted next to it. Both the depth gauge and level sensor should be installed so their "zero" elevation is the weir crest's height. Level readings can be converted to flow by calculation or via a look-up table. Plant staff should visually inspect the level periodically to confirm that level-sensor and flow-converter readings are accurate.

A Parshall flume's approach channel should be straight for at least two channel widths or 10 throat widths to create a symmetrical, uniform velocity distribution and a tranquil water surface at the flume entrance. The flume floor must be level both longitudinally and transversely and high enough to prevent submergence conditions at maximum flow. If a fiberglass flume insert is used, the contractor should be required to pour the flume in stages to avoid bowing the insert.

A depth gauge should be mounted on the converging section so plant staff can confirm that flow indicators or recorders are accurate (Figure 6.12).

All system components should be easily accessible for maintenance. If necessary, work platforms can be installed for this purpose.

Maintenance Requirements. Weirs are simple hydraulic devices and have few maintenance requirements. They should be checked often to ensure that debris (e.g., rags and sediment) has not accumulated and interfered with the flow pattern through the structure.

A Parshall flume is typically an insert cast into a concrete channel, and its flow velocities are high enough to prevent sediment from accumulating. During installation, a ruler can be painted on the flume's wall so operators can monitor its performance.

If an ultrasonic level transmitter is used to monitor the water level, splashing water may eventually cause a deposit to form on its face that will interfere with distance measurements and flow calculations. Cleaning the transmitter face quarterly or annually with an appropriate cleanser should prevent such operating problems.

DIFFERENTIAL-PRESSURE. Differential-pressure flow meters are all based on the Bernoulli principle: in this case, when the material being measured passes through a pipe, it causes a predictable pressure drop that is proportional to the square of the flow rate. Differential-pressure flow meters with Venturi flow tubes and orifice plates used to be common for measuring flow through pressurized pipes. With proper flushing systems and component maintenance, they operate well in the hostile conditions typical at wastewater treatment plants. Now, however, many are being replaced by magnetic flow meters, which typically are less expensive and need less maintenance. Differential flow meters are discussed in the *Instrument Engineers Handbook* (Liptak, 1999) and the 1993 edition of MOP 21.

One type of differential-pressure flow meter still used in wastewater treatment plants is the Pitot gauge. These gauges are often installed in aeration air-drop legs to inexpensively measure airflow to various parts of aeration basins. When measuring gas flow, pitot tubes need clean (solids-free) gas or steam, minimal headloss, and a meter range of 3:1.

Operating Principles. A pitot gauge is geometrically constructed to measure average upstream pressure and compare it with the downstream pressure static port. It consists of an insertion probe with multiple upstream sensing ports and one downstream static port. It also includes a differential-pressure transmitter and a square-root extractor, which calculate the differential pressure and convert it into a voltage or current signal (Figure 6.13).

Pitot gauges work best on boiler steam; clean, compressed digester gas; natural gas; oxygen and blower air in activated sludge systems; draft or blower air in incinerators; and airflow in aerated grit chambers. They are not recommended for gas or steam with particulate solids, low-pressure (uncompressed) digester gas, or corrosive gases.

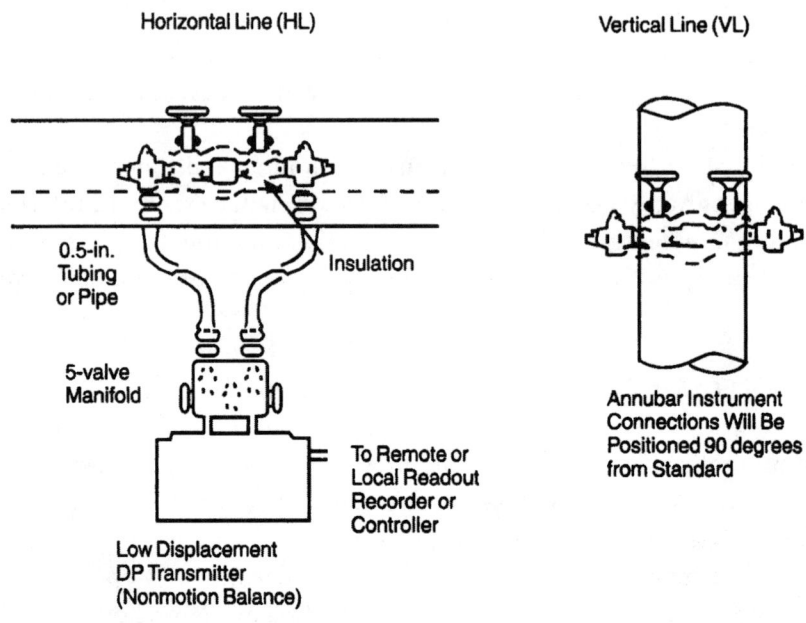

FIGURE 6.13 Pitot tube piping requirements (in. 3 × 0.025 4 = m).

Construction Materials. Pitot gauges are often made of stainless steel to avoid corrosion on the tip of the orifice, which would degrade the sensor's accuracy. Those used in aeration-drop legs, however, typically are measuring filtered air, so stainless steel or soft metals may be used.

Accuracy and Precision. Pitot tubes typically are accurate within ± 1 to 3% of full-scale and precise within ± 1% of full-scale, when properly installed and used. However, actual installed instrument accuracy will also be reduced by the accuracy of the differential-pressure transmitter and the square-root extractor. Other factors that can degrade accuracy include flows outside the expected range and piping elements that disrupt the velocity profile.

Installation. A pitot gauge should be installed with its multiple ports facing upstream (Figure 6.14). The piping upstream and downstream of the pitot tube should be straight and smooth. When this is impractical, straightening vanes may be used to condition the flow for more accurate measurement (Table 6.5).

Because the V-cone flow meter adjusts the flow regime through its head, it can be installed with only two pipe diameters of straight-pipe length upstream and five pipe diameters downstream.

Maintenance Requirements. Pitot gauges measuring filtered aeration air should need little maintenance, but they should be removed and inspected annually. Maintenance staff should test the pressure-differential element by attaching a portable monometer to the ports. They also should check whether high temperatures have warped the metal or debris has accumulated in the tip of the pressure-sensing tube.

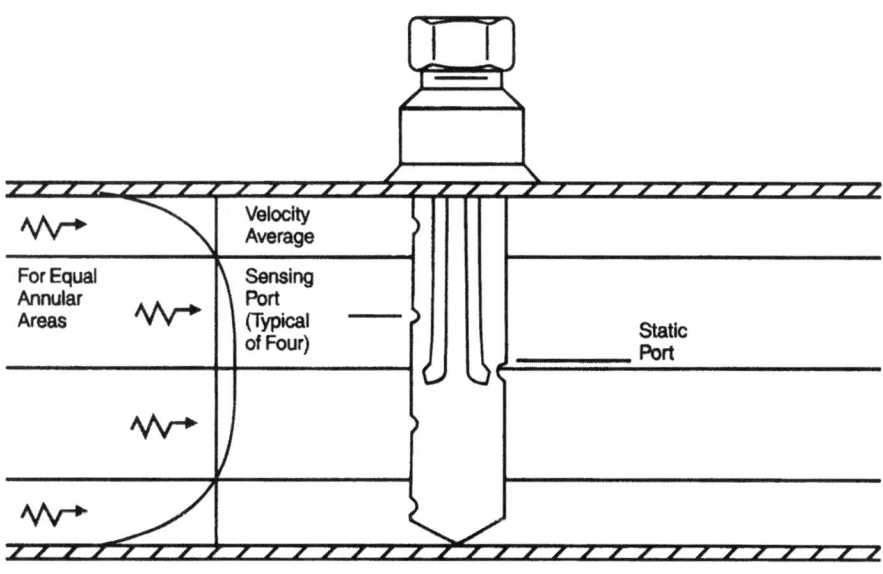

FIGURE 6.14 Pitot tube probe, multiple ports facing upstream.

TABLE 6.5 Pitot tube piping requirements.

Upstream fitting	In same plane	In different plane	With vanes
Expansion/contraction	8	8	8
Bend	7	9	6
Two bends in same plane	9	14	8
Two bends in different plane	19	24	9
Throttled valve	24	24	9
All downstream fittings	4	4	4

When reinstalling the gauge, staff should ensure that the hole in the pressure-sensing tip directly faces the flowstream.

MECHANICAL. Mechanical flow meters, such as rotary-element (e.g., turbine or propeller) and positive-displacement meters, operate on the principle that an object moved by water or air can be correlated to the flowrate of the fluid or gas.

Today, many mechanical flow meters are being replaced by magnetic flow meters, which are less expensive and need less maintenance. However, positive-displacement flow meters are still used for billing-related applications because they are very accurate, even if their headloss is high. This meter basically works by letting its internal cavity fill with fluid, discharging this fluid, and repeating the process.

Rotary-Element. Rotary-element flow meters work well in applications involving relatively clear liquids [i.e., solids concentration of less than 0.1% by weight (1000 mg/L) with no fibrous materials or debris] that can tolerate moderate headloss. They also can handle intermittent flows, so long as the pipe is full when flowing.

Operating Principles. Rotary-element flow meters consist of a pipe section in which a multi-bladed impeller is suspended in the fluid stream on a free-running bearing

(Figure 6.15). The impeller is perpendicular to the direction of flow, and its blades sweep nearly the full bore of the pipe. The impeller is driven by the fluid as it impinges the blades. Within the meter's linear flow range, the impeller's angular velocity is directly proportional to the fluid velocity, which is proportional to the volumetric flowrate.

An electromagnetic pickup coil, operating on either a reluctance or inductance principle, produces a pulse in proportion to the impeller's rotation speed. The flow meter's output signal is a train of voltage pulses, in which each pulse represents a discrete volume of fluid. The turbine output frequency is proportional to the volumetric flowrate at the actual operating temperature and pressure. When measuring gases, an appropriate temperature and pressure correction is required to convert the

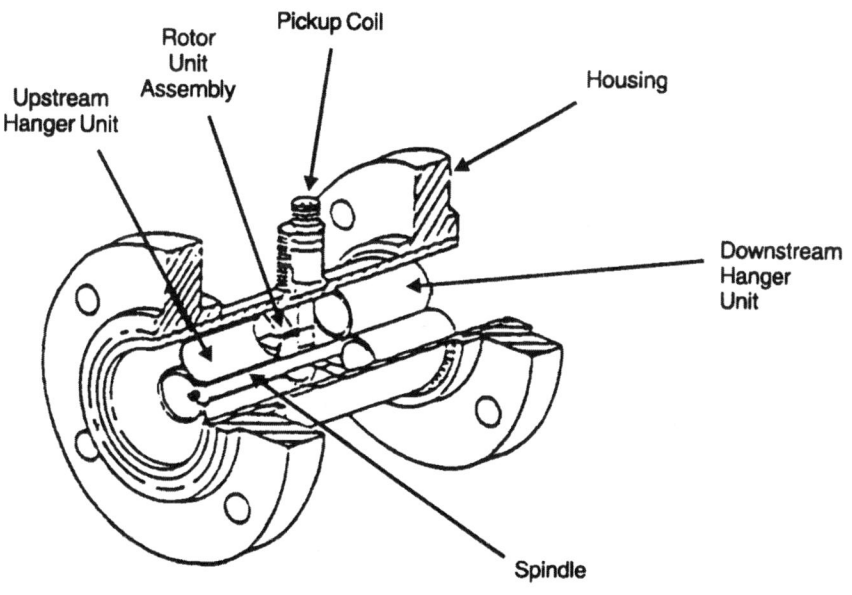

FIGURE 6.15 Turbine flow meter cutaway view.

meter output to a volumetric rate at standard reference conditions. Associated electronics units then convert and display the volumetric flow (flowrate) or total accumulated flow.

The meter's maximum range is 10:1 for liquids and 15:1 for gases in separate installations. Rotary-element flow meters work best on final effluent, secondary-clarifier effluent, washwater, steam, steam condensate, compressed digester gas, and natural gas. They are not recommended for raw wastewater, primary sludge, secondary sludge (RAS and WAS), mixed liquor, primary effluent, chemical slurries, low-pressure (uncompressed) digester gas, or corrosive gases because fibers or corrosion will slow the impeller, causing to meter to register less flow than actually passed through it.

Construction Materials. The flow meter housings typically are made of either stainless steel or coated carbon steel, so corrosion is minimized. The impeller's construction material depends on the type of meter: turbine meters typically have stainless steel impellers, and propeller units have plastic ones. The bearings need to provide a long service life with as little friction as possible.

Accuracy and Precision. A properly applied and installed rotary-element flow meter should be accurate within ± 0.25% and precise within ± 0.05% of indicated flow.

Each rotary-element flow meter has a unique "K" factor (number of pulses per unit volume), which is determined during factory calibration. This factor is degraded if the fluid viscosity is significantly more than that of air (for gas flow) or water (for liquid flow), or if the moving components become impaired by the buildup of solids or fibrous materials. Mechanical wear can also affect the K factor.

Installation. Rotary-element meters are available with 0.005- to 0.6-m (0.2- to 24-in.) diameters. They typically are smaller than the diameter of the process piping. The meters are sized by volumetric flow rate—not by reference or standard units—and each size has a specified flow range. The actual maximum flow rate should be between 70 and 90% of the meter's specified maximum flow rate.

Piping obstructions that disturb the flow profile can degrade the rotary-element meter's accuracy. For optimum accuracy, the minimum straight run upstream of the meter should be 25 to 30 pipe diameters (Figure 6.16). If necessary, this straight run may be only 10 pipe diameters if straightened vanes are installed. Pipe fittings—valves, gates, Ts, elbows, and severe reducers and expanders (30-degree included angle)—will produce flow disturbances that degrade meter accuracy if placed closer.

The meter may cavitate if the upstream line pressure is insufficient. To ensure sufficient upstream pressure, the downstream line pressure must be at least twice the

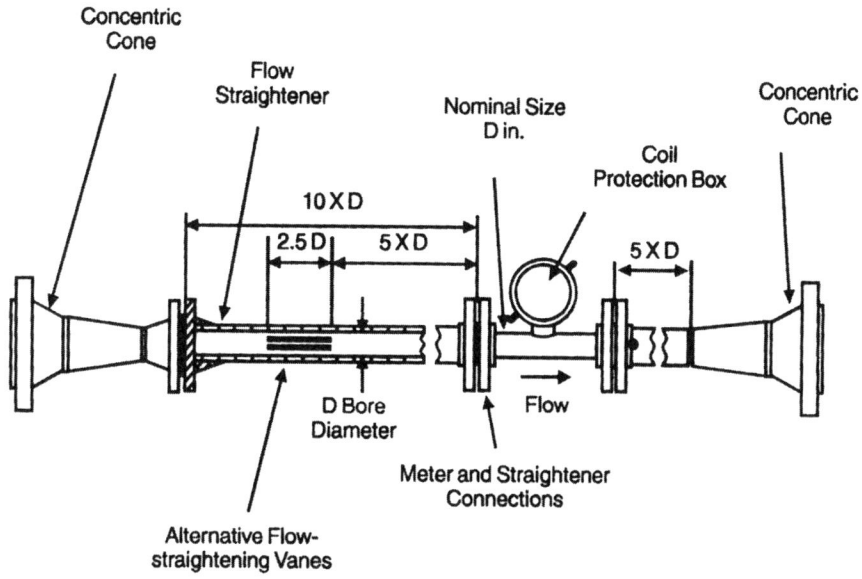

FIGURE 6.16 Turbine flow meter recommended piping installation (in. × 0.025 4 = m).

meter headloss plus 1.25 times the liquid vapor pressure. If this condition cannot be met, a larger meter with a smaller range is required.

Maintenance Requirements. Rotary-element flow meters typically measure the flow of clean water or a gas, so there is no debris to interfere with rotating equipment.

Meter maintenance may involve periodically removing the unit for cleaning and inspection to ensure that corrosion or debris is not interfering with rotation. Depending on the importance of accuracy, removal for recalibration may be required so new K values can determined. Also, the mechanical geared coupling between the turbine or propeller and the flow meter's display window may eventually need to be replaced to ensure that flow is displayed accurately.

Variable-Area. Variable-area flow meters (rotameters) are small floats often used in manifolds to verify that throttling valves are properly dividing a chemical among various injection points and that the proper amount of chemical is being delivered to each location. They are popular for measuring low flows because they are inexpensive and enable operators to physically see the status of a process.

Operating Principles. A variable-area flow meter typically consists of a glass flow tube and a stainless steel weight (float). To function properly, the tube must be vertical with the flow rising through it. The more flow through the instrument, the higher the float rises in the tube. A calibration plate mounted next to the tube allows operators to read the flow rate.

Because the fluid must rise past the small opening between the weight and flow tube, the meters can only be used in clean fluids, such as air, drinking water, or some chemicals. Also, because glass and stainless steel have different thermal-expansion properties, the meters are limited to 93 °C (200 °F).

Most variable-area flow meters are delivered to the project site precalibrated, assuming water or air with a specific gravity of 1.0 and a temperature of 20 °C (68 °F). To calculate actual flows of a substance with a different density, the conversion equations are:

$$\text{Actual flow (gpm)} = \text{Rotameter (gpm)} \times p_{\text{in rotameter}} \times 2.65 / [(p_{\text{float metal}} - p_{\text{in rotameter}}) \times p_{\text{in rotameter}}]^{0.5} \quad (6.1)$$

$$\text{Actual flow (lbm/min)} = \text{Rotameter (lbm/min)} \times 0.318 / [(p_{\text{float metal}} - p_{\text{in rotameter}}) \times p_{\text{in rotameter}}]^{0.5} \quad (6.2)$$

$$\text{Actual flow (scfm)} = \text{Rotameter (scfm)} \times p_{\text{gas@std}} \times 10.34 / (p_{\text{float plastic}} \times p_{\text{gas in rotameter}})^{0.5} \quad (6.3)$$

$$\text{Actual flow (gas lbm/min)} = \text{Rotameter (lbm/min)} \times 10.34 / (p_{\text{float plastic}} \times p_{\text{gas in rotameter}})^{0.5} \quad (6.4)$$

Construction Materials. Typically, the flow tube is made of glass and the float is made of stainless steel, but plastic flow tubes and exotic-metal floats are available. Some meters use metal flow tubes and a magnetic float. When specifying variable-area flow meters, design engineers must verify that the O-ring gaskets, flow tube, and metering float are made of materials that are compatible with the fluid to be measured. For example, glass and stainless steel are fine if the meter will measure 414 kPa (60 psi) carrier water, but if it will be measuring sodium hypochlorite, stainless steel cannot be used for the float or other internal components.

Accuracy and Precision. A properly installed industrial variable-area flow meter will perform within 1 or 2% of full-scale over a 10:1 range, so long as it is not fouled with debris or growth in the tube.

Installation. Variable-area flow meters typically are installed vertically with the flow rising through them. Many have small needle valves to control the flow through the tube and help prevent surges that can cause the metering float to hit the stop at the top of the tube. If a needle valve is not provided, one should be installed before the meter inlet. Installing it there will prevent the tube from overpressurizing and cracking if the valve is accidentally closed. Isolation and bypass valves also should be installed so the needle valve or flow meter can be removed from service without interrupting operations. Union fittings will make this project easier.

Maintenance Requirements. Because variable-area flow meters are such simple devices, they often are not included in preventive maintenance activities. However, the meter's small inlets and outlets can collect debris that alters the flow readings. Because the flow only travels one way, the units must be disassembled to remove such debris.

The tubes also can become habitats for algae and slime. If the meter is measuring unchlorinated water, for example, light can encourage algae to grow inside the tube. On the other hand, meters in dark rooms can develop a black slime on the inside of the tube. Either growth can obscure the calibration strip, making readings impossible. In severe cases, the growth can shrink the diameter of the tube, causing inaccurate measurements. The easiest way to clean the meter is disassembling it and washing it in bleach.

Sudden flow changes can cause the float to hit the stop at the top of the tube and become stuck or break the tube. Operators should be careful to avoid this when reinstalling the meter and opening the isolation valves.

If a variable-area flow meter has been taken apart several times, the metal calibration strip sometimes will become dislodged, causing operators to misread the flow.

MASS. Mass flow meters directly measure the mass flow rate (i.e., pounds or kilograms delivered in a unit of time) of a material. Some use density information to convert the output signal to a volumetric flow rate.

Mass flow meters are uncommon in the wastewater industry because the instruments are sensitive to process conditions and do hot handle fouling well. However, sometimes they can monitor difficult-to-measure process streams, such as polymer.

Coriolis. A coriolis flow meter is a small-diameter meter [up to 254 mm (10 in.) in diameter] that can measure liquids or gases by inducing a vibration and comparing it to the measured vibration of the material flowing through the meter to determine the material's mass. The meters often can measure multiple variables [e.g., mass flow, temperature, viscosity, and density (and, therefore, solids concentration or flow volume)] and read temperatures as high as 350 °C, but the small tubes inside them typically cause a large pressure drop and may be prone to plugging. However, the meters may work well when measuring process feed lines (e.g., chemicals). Coriolis flow meters can measure the flow of non-conductive liquids or gasses that cannot be measured with an electromagnetic flow meter. They can also measure gases, such as chlorine or ozone.

Operating Principles. These flow meters are based on the controlled generation of Coriolis forces, which are always present when both translational (straight) and rotational movement occur simultaneously. The Coriolis force is calculated as follows:

$$FC = 2\Delta m(\omega v) \qquad (6.5)$$

Where

FC = Coriolis force,
Δm = mass of moving body,
ω = angular velocity, and
v = radial velocity in a rotating or oscillating system.

Its amplitude depends on the moving mass (Δm), its velocity (v) in the system, and therefore, its mass flow.

The meter has two parallel measuring tubes. When fluid flows through them, they oscillate in antiphase, much like a tuning fork. When the meter vibrates the tubes, the resulting Coriolis forces cause a phase shift in the tube oscillation (Figure 6.17). This shift is proportional to the fluid density and mass flow rate. As the mass flow rate increases, the phase difference also increases. Electrodynamic sensors measure the tubes' oscillation at their inlets and outlets. The tubes' temperature is also measured to calculate the compensation factor due to temperature effects.

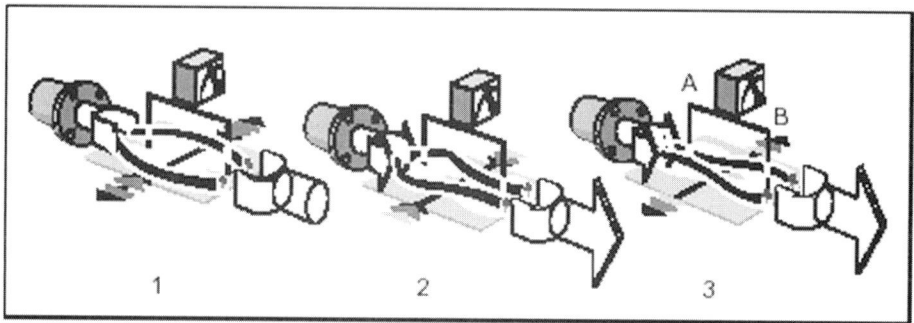

FIGURE 6.17 The Coriolis forces produced at the measuring tubes cause a phase shift in the tube oscillation. (1) When there is zero flow (i.e., with the fluid standing still), both phases are equal and there is no phase difference. When there is mass flow, the tube oscillation is decelerated at the inlet (2) and accelerated at the outlet (3).
(Courtesy of Endress + Hauser, Greenwood, Indiana.)

This flow measurement is independent of pressure, viscosity, conductivity, or flow profile.

Construction Materials. Coriolis flow meters basically consist of three parts: the flow tube, the driving mechanism, and the electronics. Because the flow tube will be in contact with the process flow, it should be made of stainless steel, Hastelloy, or titanium. Before specifying a Coriolis meter, design engineers should check a material's compatibility with the flow to be measured.

The driving mechanism uses a coil and magnet to vibrate the tubes. The transmitter, which is mounted remotely, is similar to those of other flow meters. Flange connections for this flow-through meter are commonly available.

Accuracy and Precision. Reported accuracies are ± 0.10 to 0.5% of flowrate, depending on whether the meter is measuring a liquid or gas. All Coriolis flow meters are factory-calibrated after zeroing to account for slight differences in manufacturing.

Coriolis meters are available in various ranges (one manufacturer offers a range up to 1000:1), so design engineers can upsize them to prevent plugging while retaining great control and accurate information. The meter also can be oversized to accommodate future expansions. However, design engineers should not size the meter too large, because plant startup flows are often 25% or less than initial design capacities.

Installation. Because of the small-tube design, the meter may not support the weight of a process pipe. Design engineers should include isolation valves, so the meter can be removed for maintenance. These valves may be put next to the meter because Coriolis meters are immune to most potential flow obstructions (e.g., upstream and downstream fittings).

Maintenance Requirements. Coriolis flow meters have small-diameter tubes that may be straight or bent. These tubes will need regular cleaning, and straight-tube designs are easier to clean. Flushing connections are a useful way to check meter zero calibration with water.

Thermal-Dispersion. Thermal-dispersion mass meters use heat transfer and known characteristics of the fluid being measured to determine the mass of material flowing through the meter per unit of time.

Operating Principles. A thermal-dispersion mass meter consists of a heater element and thermometer that fit together in the meter's head. It basically works by adding a known amount of heat to a fluid and then measuring how much its temperature increases (the actual temperature of the fluid is relatively unimportant). If the fluid's thermodynamic constants are known, then the mass of the flowrate can be calculated. If the fluid's physical characteristics are reasonably constant, the meter can also report volumetric flow.

These meters work well in an aeration air pipelines and digester gas pipelines if condensate is removed from the process flow.

Construction Materials. The meter's head may be made of stainless steel or another corrosion-resistant metal, as appropriate. Stainless steel works well in hot aeration air and in digester gas, which may contain hydrogen sulfide. A stainless steel thread-o-let and a full-ported-ball stainless steel ball valve with packing will allow the meter to be inserted and removed from the flow stream.

Accuracy and Precision. Like a pitot gauge, a thermal-dispersion mass meter measures flow characteristics in one part of the flow and then uses pipe geometry to report the total process flow. So, if the thermal head is not installed appropriately, there could be a large error in reported flows. The meters require a large number of upstream and downstream straight pipe diameters, but flow conditioners could reduce this requirement.

The meters are reported to be accurate within ± 1% of full scale and precise within 0.25%. This accuracy depends on whether the meter is fouled and how well the flow's physical constants are understood. Accuracy and precision also can be degraded if the pipe's flow regime is different than expected or if a drop of moisture or debris accumulates on the sensor tip.

When measuring digester gas, the meter's accuracy typically is within 5 to 7%.

Installation. Like pitot gauges, thermal-dispersion flow meters should have enough upstream and downstream diameters of straight pipe so a predictable flow profile can develop. When measuring aeration air, however, the meter may be installed wherever in the pipe is most convenient, so long as the flow profile is predictable.

If fouling is possible, design engineers should specify that these meters be insertion style, so they can be easily removed and cleaned. Insertion meters typically are calibrated for site-specific conditions before being inserted into the process pipe.

If measuring digester gas, the meters should be installed in the bottom half of the pipe (preferably at the 5 o'clock position), and their tips should point upward. So, when condensation forms, it will flow to the base of the meter rather than the tip and will not affect the readings.

Maintenance Requirements. All inserted instruments should be removed and cleaned regularly. When monitoring aeration air, the meters should be cleaned every 6 months; when measuring dirty digester gas, they may need cleaning monthly or more often.

LEVEL-MEASUREMENT DEVICES

Basically, there are two classes of level-measurement instruments: continuous and point. Figure 6.18 shows the many types of level instruments in each class.

CONTINUOUS. Continuous level-measurement instruments include bubbler, capacitance and impedance, differential-pressure, sonic and ultrasonic, and

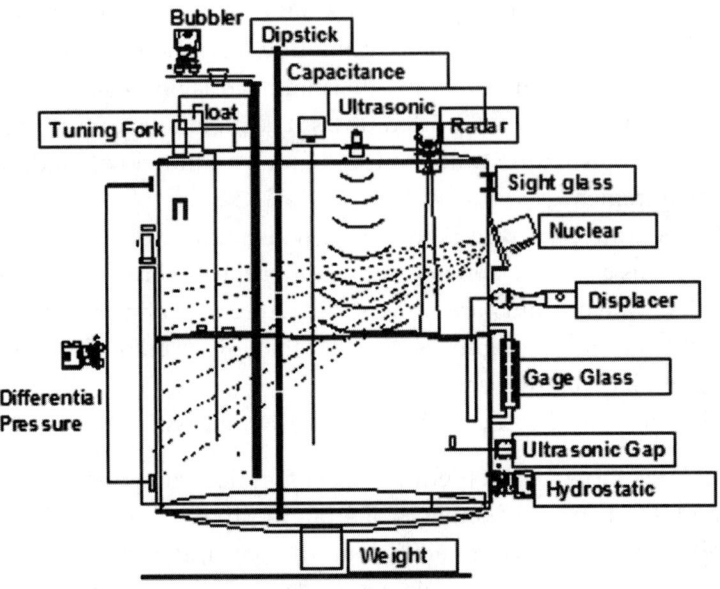

FIGURE 6.18 An example of the multitude of technologies available for level measurement.
(Courtesy of the City of Roseville, California.)

microwave (radar). Table 6.6 provides general guidelines for choosing a level-measurement instrument. Because the sonic, ultrasonic, and microwave devices do not involve direct contact with the product being measured, they typically are used to measure corrosive liquids or heavy sludge.

In addition to analog signals, most continuous level-measurement devices have options for discrete outputs meant for on–off control or alarms. (For more information on communication protocols and data transmission, see Chapter 9.)

Bubbler. Bubbler level-measurement instruments are used throughout the wastewater treatment process to measure liquid levels and level differentials.

TABLE 6.6 Level instrument application guidelines (+ = excellent, 0 = good, and - = poor).

Level measurement method	Freezing condensate on sensor	Light foam	Sludge >5%	Corrosive liquids	Conductivity <300 mho
Pressure-head-direct	+[a]	+	-	0	+
Pressure-head-bubbler	+[a]	+	0	0	+
Capacitance	-	-	-	0	0[b]
Impedance	-	+	0	0	0[b]
Sonic	c d	0	+	+	+
Ultrasonic	c d	-	+	+	+
Microwave	c	+	+	+	+

[a]Condition not applicable because process connection is typically submerged.
[b]Needs grounding strip or rod.
[c]Requires thermal protection.
[d]Instruments are not suitable in applications other than normal air or where conditions may cause vapor (clouds) to form in the air space. Vapor in the air changes the speed of sound.

Bubblers are frequently used to measure the hydraulic head created by flumes and weirs when measuring open-channel flows. Signal converters indicate flow based on the level sensed by a bubbler.

Operating Principles. An open-ended pipe, called a *bubbler tube* or *dip tube*, is connected to an air supply and positioned in the process so its open end is at a reference level (Figure 6.19). An airflow regulator maintains enough air in the tube so bubbles continuously trickle out the open end and the air pressure in the pipe equals the head of the measured liquid above the reference level. Meanwhile, a pressure transmitter measures the bubbler tube's pressure. (In closed, vented tanks, a differential-pressure transmitter is used, with its high-pressure port connected to the bubbler tube and its

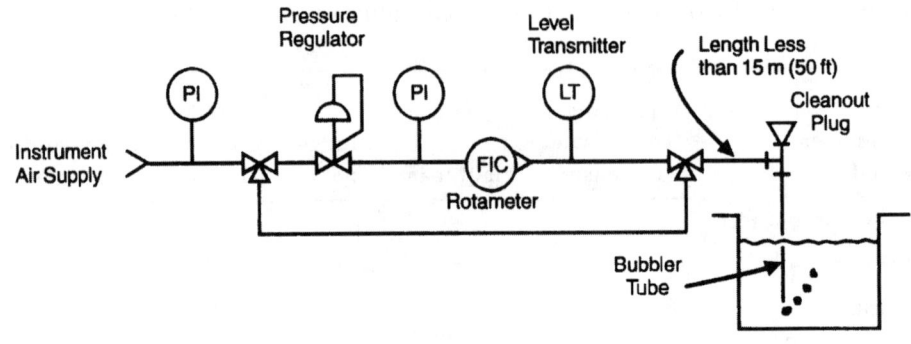

FIGURE 6.19 Schematic of open tank bubbler application. (Courtesy of the City of Roseville, California.)

low-pressure port connected to the tank's headspace.) If measuring water, the transmitter's measurement equals the water level. If measuring other liquids, the transmitter must be calibrated for the liquid's specific gravity.

Construction Materials. Bubblers can be purchased as complete systems or be fabricated using readily available materials. Ideally, they should be made of stainless steel tubing and fittings, and the joints should be welded to prevent leakage. Polypropylene tubing can be used, but it has a shorter lifespan.

The dedicated air supply can be bottled air, a compressor, or if extra reliability is required, two compressors with automatic switching. If using compressors, design engineers should specify intermittent-duty units capable of producing the required high-pressure purge range [about 500 kPa (60 psi)].

Accuracy and Precision. A bubbler's accuracy depends on headloss, process-specific gravity, barometric pressure, the uncertainty of the pressure-measuring device, and the temperature of both the fluid and air involved. While errors less than ± 0.1% can be achieved, the accuracy is typically ± 0.5 to 1% of full-scale. Precision depends on the variances from standard conditions.

Installation. The bubbler tube should be rigidly supported at a convenient location in the tank. A properly supported tube made of 12.7-mm-diameter (0.5-in.-diameter) stainless steel tubing or galvanized pipe can withstand process-related turbulence.

Because the tube's opening is the lowest level that can be detected, it should be set at or below the lowest measurement level needed. However, it should be at least 0.08 m (3 in.) from the tank floor to avoid solids buildup, and this offset must be compensated for in the flow calculation. On the other hand, if the bubbler is installed in a flume, its tip should be at the flume floor, or else the degree of elevation should be compensated for in the flow calculation. If installed in a weir, the bubbler tip should be at the same elevation as the bottom of the weir notch, or else the degree of offset should be compensated for in the flow calculation.

The bubbler's airflow rate is controlled by a pressure regulator, a flow-control valve, and perhaps a rotameter. Typical airflow rates are 0.008 to 0.03 L/s (1 to 4 cfh). The air can be supplied from instrument air, plant air, compressed gas tanks, or a dedicated oil-free compressor. Instrument air is preferable to minimize fouling the pressure-measuring instrument and other bubbler-system components.

Airflow head is affected by bubble formation, so the bottom of the tube should be notched to produce a continuous stream of small bubbles and minimize errors. Also, the open end of the tube should face downstream to avoid solids buildup and excessive pressure, which would affect bubble production.

Because of the headloss caused by airflow in the bubbler tube and connecting tip, the transmitter's pressure will not be identical to that at the tube's open end. To minimize headloss-related errors, the airflow regulator should be mounted as close to the bubbler tube as possible and connected with minimal fittings and tubing. For 10-mm (0.25-in.) tubing, the distance between the regulator and bubbler tube should not exceed 15 m (50 ft). The bubbler tube and pipe diameter should be at least 6.3-mm (0.25-in.) for distances less than 15 m (50 ft), and up to 12.7-mm (0.5-in.) for longer distances.

To prevent condensation from accumulating in the air purge tubing, the tubing should slope downward from the pressure transmitter and airflow regulator to the bubbler tube. If installation restrictions prohibit a continuous downward slope, condensate traps with blowdown valves should be installed.

In open tanks and flumes, for periodic reference checks, and to help recalibrate the tube if it is removed for cleaning or replacement, a depth gauge should be installed in the tank at a location visible from the bubbler tube. "Zero" should be the same elevation for both gauge and bubbler tube.

Maintenance Requirements. Maintenance access should be provided for bubbler-system components, tubing, and the cleanout T. A timer-and-valve package is available for periodic bubbler-tube cleaning.

Depending on the bubbler accuracy required, the pressure-measuring device should be calibrated annually. Once the pressure-measuring device is removed, full-pressure air [about 500 kPa (60 psig)] should be blown through the bubbler tube for 8- to 24-hour intervals to remove buildup from the end of the tube. Also, if dedicated intermittent-duty compressors are installed, periodic blowdown of condensate from the air receivers will be required via manual or automatic bleed valves.

Capacitance and Impedance. Capacitance and impedance probes can be used to measure the level of liquids, as well as some slurries and dry materials. They also can be configured to distinguish between interface levels in wastewater treatment plants. This is useful when a water surface level that requires monitoring is covered by a foam layer.

Some probes are modified to produce a signal proportional to flow in flumes and weirs. These probes have been characterized via electronic calculations or a probe-shape variation to produce a direct, linear relationship between capacitance and flow.

Operating Principles. A capacitor basically consists of two electrically conductive plates separated by a nonconductive material. A capacitance probe is typically constructed to act as one capacitor plate, while the tank wall or measured solution acts as the other plate (Figure 6.20). The tank's headspace acts as the nonconductive material. As the water level rises, so does the system's effective capacitance, which is linearly proportional to the level. In the capacitance instrument's transmitter electronics, the capacitance is measured by a wheatstone bridge circuit powered by a high-frequency oscillator. As the measured capacitance changes, it creates an imbalance in the bridge circuit, creating a voltage representative of the capacitance change that directly corresponds to the change in level. Capacitance probes are sometimes called *radio frequency (RF) probes* because of this measurement technique.

Because wastewater is a good conductor, the probe must be insulated. However, the insulation's exterior surface effectively becomes a third plate, complicating the probe's design.

The situation becomes more complex if wastewater solids coat the probe. The coating's capacitance and conductance effects can cause the probe to fail to respond to changes in level. So, some compensation method is necessary. The high-frequency

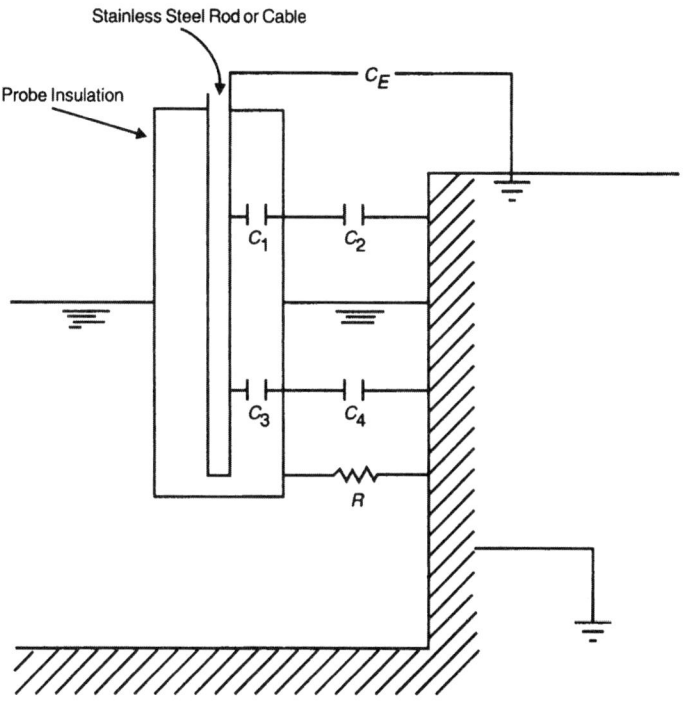

FIGURE 6.20 Capacitance probe electrical equivalent. (Courtesy of the City of Roseville, California.)

oscillator, for example, can reduce the errors that occur when conductive coatings short-circuit the capacitor.

Another method is to measure and subtract the coating's resistance from the effective capacitance. This method is based on the assumption that the coating's capacitive reactance is equal to its resistance, and that this resistance is the largest in the aqueous system. The result is proportional to the liquid level, but its degree of accuracy depends on how well the system matches the assumptions made.

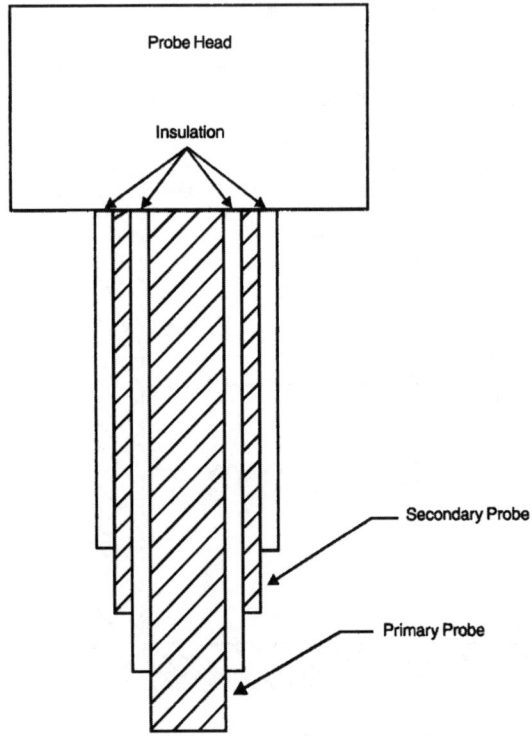

FIGURE 6.21 Impedance probe construction.
(Courtesy of the City of Roseville, California.)

Impedance probes involve two probes and more electronics to offset resistive paths for conductive coatings and capacitive paths for nonconductive coatings (Figure 6.21). Both probes are driven in phase at the same voltage. They typically are cylindrical rods or cables inserted perpendicular to the water surface. The probes used for open-channel, headloss flow meters are flat shapes designed to provide a signal proportional to flow (Figure 6.22).

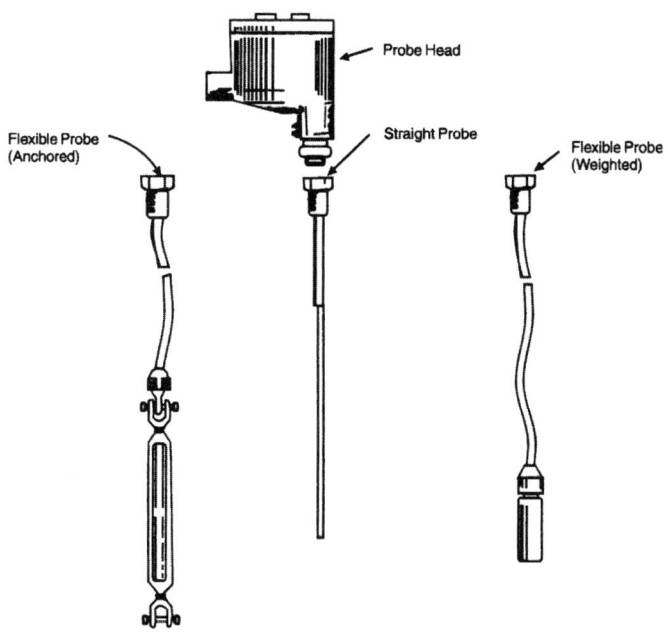

FIGURE 6.22 Capacitance probe variations.
(Courtesy of the City of Roseville, California.)

Construction Materials. Typically made of 316 stainless steel, the probes can be coated or uncoated rods or cables. Insulators can be made of Teflon or Delrin, depending on the maximum temperature of the process being measured.

Accuracy and Precision. A clean probe should be accurate within ± 0.25% and precise within ± 0.5% of full-scale. Improper installation can degrade the probe's accuracy.

Installation. The probe should be supported vertically at a convenient location in the tank that is accessible for maintenance, but not in the direct flow stream. There should be enough clear space around the probe so solids will not accumulate between the probe and the vessel.

If the process' conductivity is less than 7mS/m, grounding rods should be used. Also, a dual probe (ground reference) should be used when measuring anything in nonmetallic tanks or nonconductive media in spherical, horizontal-cylinder, or cone-shaped bins and tanks. The ground reference—a rod or a concentric shield—acts as a second capacitor plate in such circumstances.

All probes should be isolated from sources of vibration or insulation damage.

The probe is typically mounted within 75 m (250 ft) of the capacitance probe's transmitter. The probe manufacturer furnishes the cable, and its length should not be changed without recalibrating the instrument, because longer cables typically reduce the probe's resolution. The capacitance probe transmitter should be close enough to the probe so one person can see the level indication on the capacitance probe's transmitter change while monitoring the concurrent level change.

Also, enough extra cable should be provided in a junction box so the probe can be easily removed for maintenance (Figure 6.23).

Maintenance Requirements. The probe may need periodic cleaning, depending on how quickly a coating accumulates on its surface. It should be inspected and cleaned when personnel suspect that the calibration is off by about 5%. These instruments should not be used in sludge applications because excessive maintenance would be required.

Differential-Pressure. *Operating Principles.* Differential-pressure instruments are typically used in the wastewater treatment industry to measure levels of liquids and slurries; they are not effective for solids measurement.

The most common elements used to measure pressure are bellows and bellows' diaphragms. Bellows are deeply corrugated metal cylinders closed at one end. When process pressure or differential pressure is applied to the bellows, it expands. Restoring springs increase the bellows' maximum operating range and reduce wear on the bellows assembly by reducing the assembly's range of operation. As the bellows compresses, an electrical component (e.g., capacitor, strain gauge, resistor, or inductor) connected to it produces electrical signals proportional to the differential pressure, as reflected by the degree of diaphragm deflection. This changes the associated electrical property (e.g., the distance between the plates in a capacitor, piezoelectric response, or loop reluctance).

Diaphragms are flat or concentrically corrugated metal or ceramic disks that compress in response to pressure increases. As the diaphragm compresses, an electrical component (e.g., capacitor, strain gauge, resistor, or inductor) connected to it

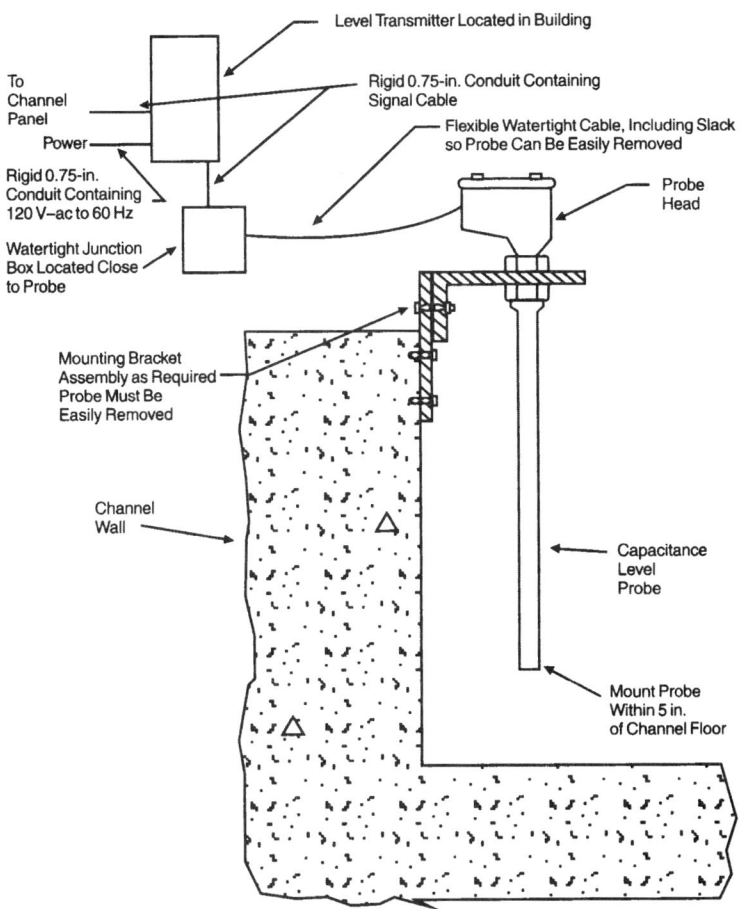

FIGURE 6.23 Typical capacitance probe installation.
(Courtesy of the City of Roseville, California.)

produces electrical signals proportional to the differential pressure, as reflected by the degree of diaphragm deflection. This changes the associated electrical property (e.g., the distance between the plates in a capacitor, piezoelectric response, or loop reluctance).

Diaphragm seal systems consist of a differential-pressure transmitter, a diaphragm seal, fill fluid, and direct-mount or capillary tubing (Figure 6.24). During operation, the thin, flexible diaphragm and fill fluid separate the measured substance from the pressure sensor. When process pressure is applied, the diaphragm is displaced, transferring the measured pressure through the fluid-filled system and the capillary tubing to the sensor. This displacement is proportional to the process pressure.

Submersible transducers come with a sealed cable assembly to prevent the measured substance from coming in contact with the sensor electronics. This assembly includes a breather tube that acts as the transducer's low-pressure reference leg. It is routed above the maximum height of the measured substance and either vented into

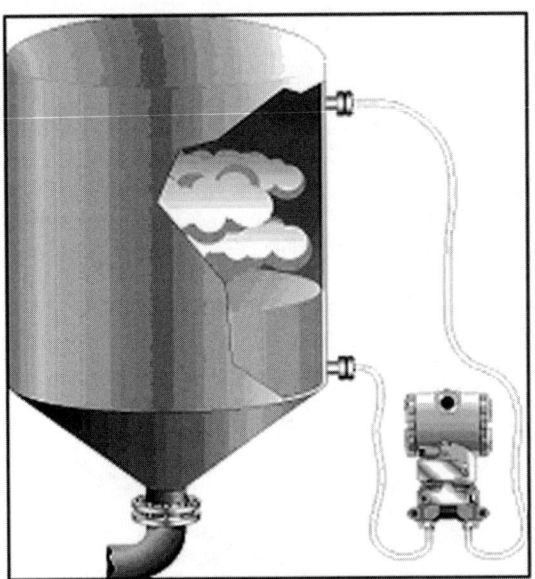

FIGURE 6.24 An example of a differential-pressure sensor with seal and capillary system.
(Courtesy of the City of Roseville, California.)

the enclosure or routed to a breather bag. When the tube is vented to the atmosphere, personnel should put a desiccant filter in the enclosure to ensure that humidity cannot condense into the tube and make level readings inaccurate.

Construction Materials. Differential-pressure sensors typically are stainless steel or ceramic. If diaphragm seals with either direct-connection or capillary tubing are used to extend the sensor, these connection materials should be made of 316 stainless steel, Hastelloy, Monel, or the appropriate material with the appropriate chemical compatibility.

Accuracy and Precision. Differential-pressure sensors typically are accurate to 0.75% of full scale. Those using diaphragm seal systems will be less accurate; engineers will have to add the inaccuracies associated with diaphragm responsiveness, the fill fluid's coefficient of thermal expansion, the capillary tubing's headloss, and other factors.

Installation. Installing the transmitter as close as practical to the process measurement site reduces response time for better level control. As with all instruments, easy access for maintenance should be provided, unless doing so would hinder the monitor's ability to meet process-control requirements. Using diaphragm seal systems can help reduce both operations and maintenance (O&M) requirements.

Differential-pressure transmitters are typically installed with valves that aid calibration efforts. Three-valve manifolds are used when staff will only calibrate the monitor by checking the sensor's zero point, and span checks or adjustments will be infrequent. Five-valve manifolds are used when the instrument will be used for billing or custody transfer of the measured substance.

Hydrostatic level monitors are used in enclosed tanks. The differential-pressure transducer is at the same level as the bottom tap, and the top tap is installed in the vapor space above the maximum level in the tank. The differential-pressure element uses the pressure in the vapor space as a reference, so the difference in pressure is proportional to the liquid level. If the tank is vented or open to the atmosphere, then the reference leg connection is unnecessary.

Differential-pressure transducers can be plumbed with a wet or dry reference leg configuration. The wet leg does not have to be filled with the same fluid as the product being measured, but compensation for the differences in the specific gravities would be required in the measuring instrument. In Figure 6.25 (wet leg drawing) the liquid in the wet leg prevents unwanted accumulation of condensate at the refer-

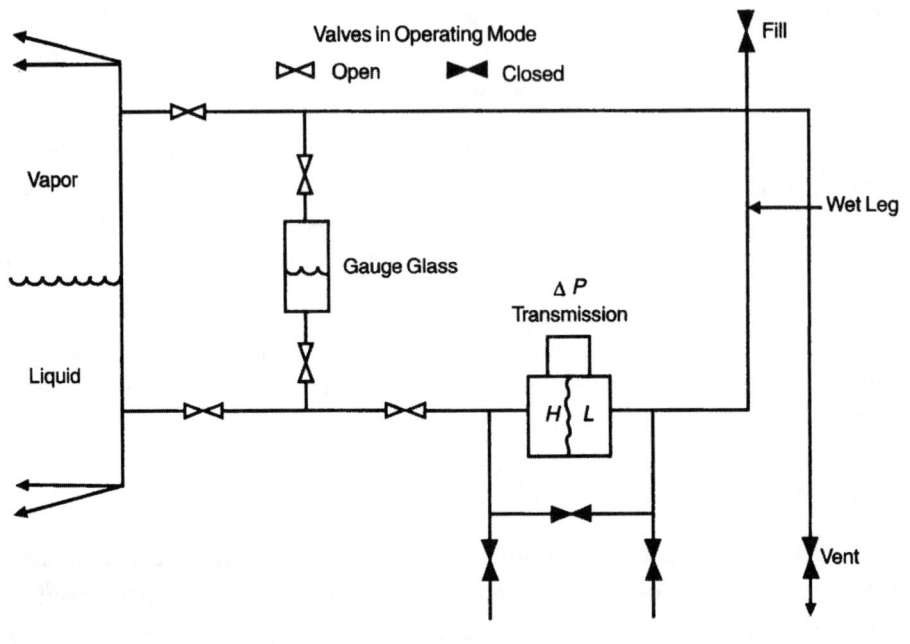

FIGURE 6.25 Differential-pressure wet-leg level system. (Courtesy of the City of Roseville, California.)

ence side. In this configuration, the reference (wet leg) is connected to the low-pressure side of the differential-pressure cell, just as in a dry leg setup [Figure 6.26 (dry leg drawing)]. As shown, the meter would read 100% at 0% liquid level. This can be corrected during calibration by suppressing the zero, thus providing the correct tank level. This configuration helps prevent reverse-locking bellows-type differential-pressure level transducers.

For applications involving solids-bearing liquids, flushing provisions or diaphragm isolation may be needed. Diaphragm connections to the process should be at least 25 mm (1 in.) for sludge lines. Diaphragms for solids applications are available in nearly every conceivable process connection. Diaphragm flanges can be acquired with integral flushing connections, or even with diaphragm extensions that will extend the sensing diaphragm into the process.

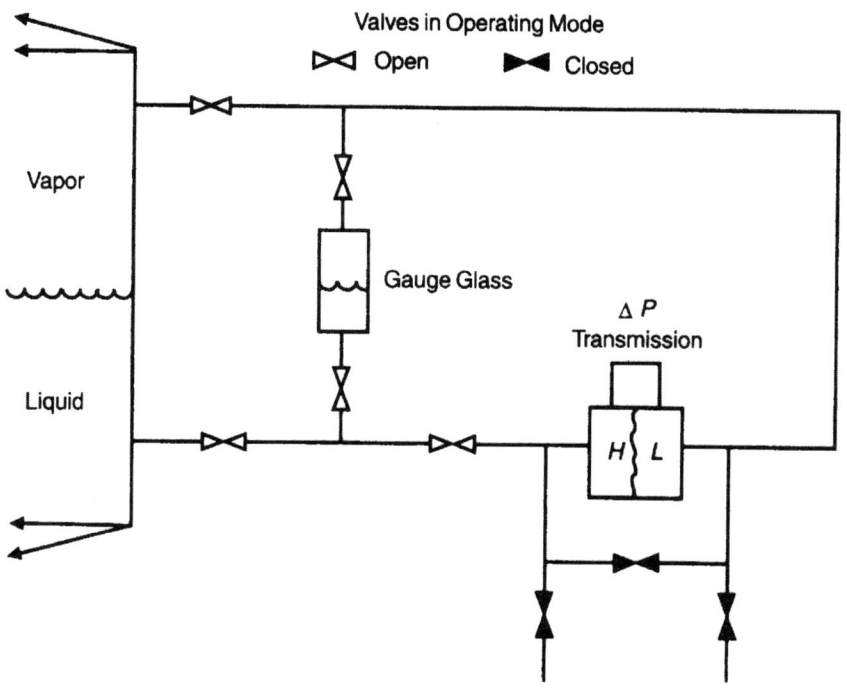

FIGURE 6.26 Differential-pressure dry-leg level system. (Courtesy of the City of Roseville, California.)

Special precautions are taken in steam applications to prevent overheating the manifold and transmitter. In steam applications, a condensate pot is installed to provide a water seal between the steam environment and the sensing element. Side-mounted pressure taps allow steam into the tap while still allowing drainage of excess condensate back into the process. Blowdown should not be done on steam level instrument piping to prevent overheating the sensor.

Maintenance Requirements. When used in substances that can coat its surfaces, the probe's sensing taps or flange faces should be manually or automatically flushed periodically. If the taps or faces cannot be flushed, they must be cleaned periodically.

The level sensor should be calibrated annually, unless it is used for compliance or custody transfer, in which case calibrations should be semiannual or quarterly.

Another form of a differential-pressure transducer is the submersible pressure transducer. These sensors use diaphragms similarly to differential-pressure transducers, but they do not require special process connections because they are installed directly into vented channels or tanks.

Submersible transducers are available with sealed cable assemblies to prevent the measured substance from touching the sensing-element electronics. The cable assembly has a breather tube that the transducer uses as a pressure reference. The breather tube is routed beyond the maximum level of the process and is either vented into the enclosure or routed to a breather bag. The breather bag is typically made of polypropylene and provides a humidity-free environment for the breather tube to vent into. When the tube is vented to the atmosphere, a desiccant filter should be installed in the enclosure to ensure that humidity cannot condense into the breather tube and make the pressure readings inaccurate or unstable. Many submersible transducers also have a Kevlar string in the submersible transducer cable assembly to help prevent elongation.

Submersible transducers typically are accurate within ± 0.1% of full scale, although a directly connected transducer can be accurate within ± 0.04% of span.

Sonic and Ultrasonic. Sonic and ultrasonic level transducers do not touch the process fluid, so they can be used in nearly every wastewater treatment process. They are suitable for and frequently used as secondary elements in flow measurement systems.

Operating Principles. Sonic and ultrasonic level sensors typically are based on the time-of-flight principle. A sensor sends pulses that travel through air and are reflected back by the surface of the measured substance. The instrument is calibrated to calculate the difference between the distance from the sensor to the tank floor and that from the sensor to the level of the measured substance (Figure 6.27).

Sonic and ultrasonic wave velocity depends on temperature, pressure (somewhat), and humidity (to a minor extent). If changing conditions are anticipated, automatic temperature compensation can be provided (compensation is unnecessary for the other two factors).

Sonic and ultrasonic level sensors can measure distances from 152 mm to 61 m (6 in. to 200 ft), depending on the sensor. The sensors are available in frequencies ranging from about 9 kHz (sonic) to 20 kHz or more (ultrasonic). The pulse generator

FIGURE 6.27 Typical ultrasonic level transducer mounting. (Courtesy of the City of Roseville, California.)

is available in various shapes (e.g., cone, parabola, or a threaded-pipe configuration for direct connection to a stilling well pipe).

The signal strength is proportional to the square of the distance between the sensor and the measured substance. The signal can be attenuated via absorption in the vapor space, reflection away from receiver's sensing area, and absorption by foam on top of the measured substance or by the substance itself. The cone shape minimizes the attenuation related to reflection. Propagation distance and wave frequency affect attenuation by absorption. Sonic waves attenuate less than ultrasonic. Foam on liquid surfaces or carbon-dioxide-rich vapor spaces (e.g., anaerobic digesters) may completely absorb ultrasonic waves.

Persistent, dense foam is typically problematic for sonic and ultrasonic devices.

Corrosion and freezing can also be problematic. Sensor heaters are available to prevent freezing, and proper material selection can reduce corrosion.

Construction Materials. Transducer probes are available in various materials, so they can be used in many processes. Typically, the transducers have polyvinylidene fluoride (PVDF) facings or Teflon (PTFE) flange facings to resist corrosion. PVDF facings can be used in temperatures ranging from -40 to 149 °C (-40 to 300 °F). If the sensor will be monitoring an aggressive chemical or abrasive process, ethylene tetrafluoroethylene (ETFE) facings are available.

Most transducers are designed to work in vented tanks, but some can operate in vessels with pressures more than 345 kPa (50 psi).

Accuracy and Precision. Sonic and ultrasonic level monitors typically are accurate within ± 0.25% and precise within ± 0.1% of span. Airspace conditions, turbulence, foam, and interferences (e.g., mixers or proximity to fill stream) can reduce accuracy and precision. Manufacturers can help calculate the total attenuation attributable to interferences and process variations.

Installation. The transducer's mounting site is determined based on manufacturer recommendations. Typically, the sensor is mounted at least 152 mm (6 in.) above the maximum level to be measured; this additional distance is called the *blanking distance*. Refer to the manufacturer's installation requirements for an individual transducer's blanking distance.

The transducer should be mounted far enough from the tank walls to prevent false echoes. The distance depends on the angle of the transducer beam. The transducer also should be mounted away from mixers, agitators, physical obstructions, or the process fill stream (Figure 6.28).

A stilling well can be used to dampen liquid-level turbulence, reduce foam, increase signal strength, eliminate noise from stray echoes, or reduced condensate problems. The well is cut from a piece of pipe that is at least 102 mm (4 in.) in diameter. The bottom is cut at a 45-degree angle. Air holes should be drilled near the top of the stilling well, where the transducer is mounted.

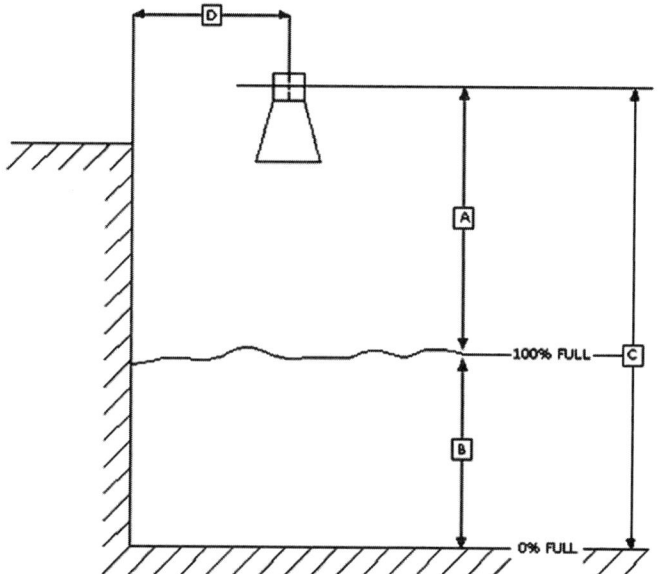

FIGURE 6.28 Typical ultrasonic level transducer installation. (Courtesy of the City of Roseville, California.)

Maintenance Requirements. The transducer facing may need periodic cleaning, depending on how fast a coating accumulates on the transducer surface. Staff also should check the sensor calibration annually, unless it is being used for compliance or custody transfer, in which case the calibration checks should be semiannual or quarterly. A depth gauge (staff gauge) is used to verify calibration, usually when the transducer is used as a secondary element in a flow-measuring system.

Microwave (Radar). While sonic, ultrasonic, and microwave level sensors may seem identical, microwaves measure the distance differently.

Operating Principles. There are two types of microwave level sensors: frequency-modulated and guided-wave. The frequency-modulated level sensor involves a frequency-modulating, continuous wave transmitter and antenna assembly. The device is mounted on top of the process vessel. The transmitter sends radar signals into the tank, and the antenna detects both those and the signals that reflect off the measured substance. The transmitted signal has a different frequency than the reflected one, and this difference is proportional to the level of the measured substance.

Microwave sensors typically operate at frequencies between 6 and 24 GHz; the higher-frequency units receive more reflected signals. The antennas are 51- to 254-mm (2- to 10-in.) cones or parabolas. Engineers should choose the largest antenna possible because this will yield the strongest, most concentrated signal and, therefore, the most reliable level measurement. Most substances measured at wastewater treatment facilities have dielectric constants larger than 10, so they are ideal for radar installations. Radar instruments are also suitable for high carbon dioxide environments (e.g., anaerobic digesters).

The guided-wave level sensor consists of a transmitter and a guide probe, which is immersed in the measured substance. The probe is sized according to the level being measured. Low-power microwave pulses are transmitted to the guide probe. When the pulse hits the surface of the measured substance, which has a different dielectric constant than air, part of the energy is reflected back to the transmitter. The instrument notes how much time it took the pulse to be reflected back and uses this information to calculate the level of the measured substance. The intensity of the reflected wave depends on the dielectric constant of the substance that the wave touched, so the transmitter can differentiate among various substances (e.g., wastewater and sludge) in the same tank. So, users typically can track the level of one or more substances simultaneously.

Guided-wave radar devices are available with single or dual lead probes, and both are available in fixed or flexible coaxial versions. Single probes are preferred when bridging could occur across the twin probes. The flexible probes typically are used with a tie-down or a weight affixed to the bottom of the probe. Pressure, temperature, density, or small variations in the dielectric constant have minimal effect on the sensor. They also are relatively unaffected by vapor, dust, physical obstacles, or turbulence.

Construction Materials. The antennae typically are made of 316L stainless steel but are available in other materials. Also, PTFE isolation windows are available that separate the antenna from the process. Some windows resist coating and condensation buildup.

The guided-wave system's fixed and flexible probes typically are made of 316L stainless steel.

Accuracy and Precision. Accuracies typically range from ± 0.1% to ± 0.5% of measured distance, and precision typically is within ± 0.025% of the measured distance. Temperature typically is a minor factor (± 0.01% per degree Celsius) when its fluctuations are small. When large temperature variations are expected, automatic temperature compensation is available from most manufacturers.

The frequency-modulated sensor's largest errors are due to condensation, foam, or coatings on the antenna. These conditions can compound errors up to ± 10% of the measured distance.

Installation. The installation guidelines for frequency-modulated radar sensors are similar to those for sonic and ultrasonic sensors. The antenna's mounting site is determined based on manufacturer recommendations. Typically, the antenna is mounted at least 152 mm (6 in.) above the maximum level to be measured; this distance is called the *blanking distance*. Actual blanking distances depend on the antenna and manufacturer selected. A flanged PTFE window typically is installed between the antenna and measured substance to reduce the potential effects of condensation, corrosion, and coatings.

The antenna should be mounted far enough from tank walls to prevent false echoes (Figure 6.29). The distance depends on the antenna's beam angle. It also should be mounted away from mixers, agitators, physical obstructions, or the process fill stream.

A stilling well can be used with frequency-modulated sensors to increase return signal strength and reduce the effects of turbulence, foam, condensate, and noise from stray echoes. The stilling well is cut from one piece of pipe that is at least 76 mm (3 in.) in diameter. The bottom is cut at a 45-degree angle. Air holes should be drilled along the length and near the top of the stilling well, where the antenna is mounted.

If the sensor transmits frequencies that are within a communications bandwidth, it must be installed in enclosed or vented metal tanks.

162 Automation of Wastewater Treatment Facilities

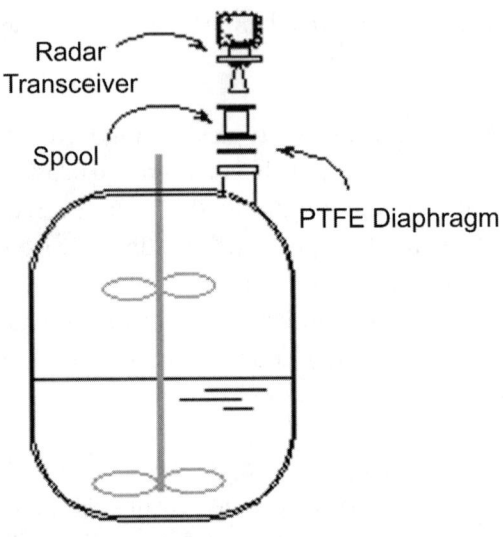

FIGURE 6.29 Microwave level sensor installation. (Courtesy of the City of Roseville, California.)

Guided-wave radar probes are hung along the entire distance of desired level readings (Figure 6.30). For best performance, keep them away from process fill streams, agitators, or other objects that may contact the probe. If the probe could come within 0.3 m (1 ft) of any object, then it should be secured to the bottom of the process vessel.

Maintenance Requirements. The antenna or PTFE window of a frequency-modulating sensor may need periodic cleaning, depending on how fast coatings accumulate. Staff should check the instrument's calibration annually, unless it is used

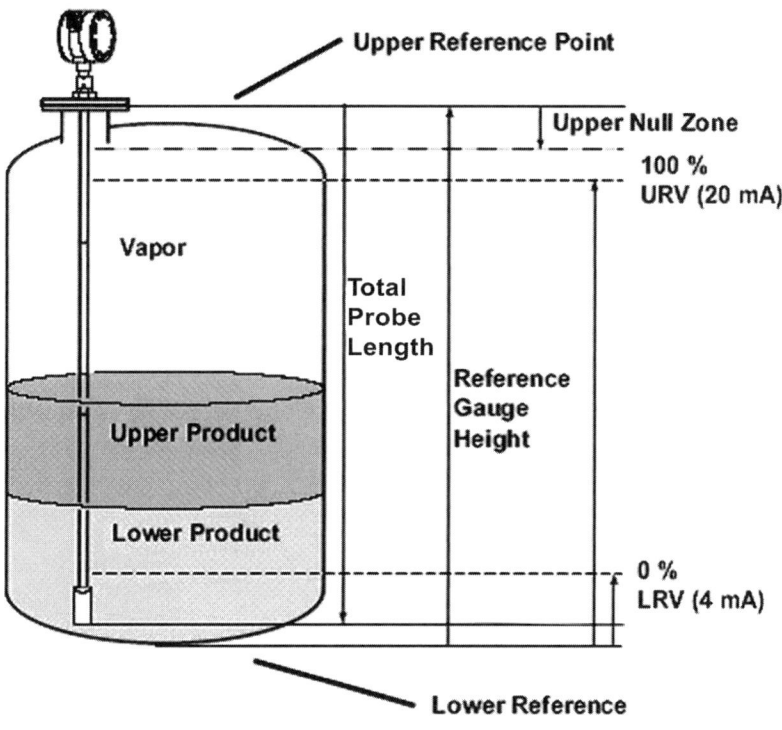

FIGURE 6.30 An example of a guided-wave radar probe. (Courtesy of the City of Roseville, California.)

for compliance or custody transfer, in which case calibration checks should be semi-annual or quarterly.

Guided-wave radar probes should be inspected and cleaned based on the rate of accumulation of coatings on the probe. Generally, annual calibration checks of the instruments are sufficient, with more frequent checks when used for compliance or custody transfer.

POINT. There are several types of point level monitors, including ultrasonic gap, float, frequency-shift tuning fork, thermal dispersion, hydrostatic, conductivity, capacitance, and inductance (Table 6.7).

Operating Principles. Although there are many types of point level monitors, they basically can be divided into two groups: direct-acting switch devices and inference-based level switches.

Direct-switch devices (e.g., float and some hydrostatic) directly operate discrete switches via interaction with the process medium. For example, as the process level increases to the point of contact with a ball float, the ball float's buoyancy will cause it to change its angle from vertical to horizontal as the process level continues to rise. When the float achieves a specific angle, it will open or close a mercury switch (depending on the ball float configuration).

TABLE 6.7 Point level instrument application guidelines (+ = excellent, 0 = good, - = poor).

Level measurement method	Interface measurement	Light foam	Sludge >5%	Corrosive liquids[a]	Conductivity <300 mho
Ultrasonic gap	-	0	0	0	+
Float	-	0	-	0	+
Tuning fork	0	-	-	0	+
Thermal dispersion	+	+	0	0	+
Hydrostatic	-	+	+	0	+
Conductivity	0	-	-	0	-
Capacitance	-	-	-	0	0
Inductance	-	+	-	0	0

[a]Selection depends on the characteristics of the liquid being measured.

Inference-based level switches use electronic circuitry to sense the changes when the sensing element touches the measured substance. An example of a thermal dispersion switch is described in the following paragraph. Most other inference-based level switches function similarly.

Thermal dispersion technology measures the heat loss (cooling effect) of a fluid flowing across a heated cylinder. A typical flow element configuration uses two RTDs, sheathed in thermowells, separated by a gap. Heat is applied internally to one RTD, creating a differential temperature between the two. This differential temperature is greatest at no-flow conditions and decreases as flow increases, cooling the heated RTD. Immersing the flow element into a liquid directly affects the extent to which heat is dissipated and, in turn, the magnitude of the temperature differential between the RTDs. This differential is electronically converted into an electrical signal that can be used to trip a relay in level or interface-level switch applications

Construction Materials. Point level monitors are constructed of materials compatible with all wastewater-related substances. Engineers should consider temperature, pressure, and chemical compatibility when choosing a point level sensor.

Installation. Point level switches are mounted in a variety of orientations with the sensing element mounted on a flexible or rigid guide assembly. Flexible point level switches should be attached to a rigid guide assembly or have a weight affixed next to the switch to keep it from moving in the process vessel with flow or process movement. Insertion-style point level switches can be used for pressurized vessel applications. If the point level switch is likely to be cleaned frequently, then the switch should be mounted in a hot-tap configuration for ease of maintenance without having to drain the process vessel level below the switch depth. The sensing element should be mounted away from mixers, agitators, physical obstructions, or the process fill stream. Many manufacturers offer multipoint sensors on one guide assembly, thereby simplifying installation of multiple switches in a single process vessel.

Point level instruments can also be installed via threaded or flanged penetrations in the wall of tanks or vessels. These installations should only be used when there is no access via the top of the tank, because the sensing element will be difficult to maintain.

Maintenance Requirements. The sensing element of a point level instrument typically needs to be cleaned or flushed. Cleaning frequency should be determined by maintenance staff based on the fouling rate.

PRESSURE

Pressure sensors are frequently used to monitor compressed-air distribution systems, pump suction and discharges, and pressure-vessel pressures. Differential-pressure transmitters are used to measure gauge pressure and liquid level—or to measure flows when used with primary elements. With diaphragm seal systems, pressure sensors can be used in any wastewater treatment process. Diaphragm seals can be connected to capillary tubing so the instrument can be installed at a convenient location for maintenance.

OPERATING PRINCIPLES. Bellows and diaphragms typically measure the actual pressure in a process. Bellows are deeply corrugated metal cylinders that are closed at one end. When process pressure or differential pressure is applied to the bellows, they expand. Restoring springs increase the instrument's maximizing operating range and reduce wear on the bellows assembly by reducing the assembly's range of operation.

The diaphragms are flat or concentrically corrugated metal or ceramic disks that compress in response to pressure increases. As the diaphragm compresses, an electrical component (e.g., capacitor, strain gauge, resistor, or inductor) connected to it produces electrical signals proportional to the differential pressure, as reflected by the degree of diaphragm deflection. This changes the associated electrical property (e.g., the distance between the plates in a capacitor, piezoelectric response, or loop reluctance).

Diaphragm seal systems consist of a pressure sensor, a diaphragm seal, fill fluid, and direct-mount or capillary tubing. During operation, the thin, flexible diaphragm and fill fluid separate the measured substance from the pressure sensor. When process pressure is applied, the diaphragm is displaced, transferring the measured pressure through the fluid filled system (capillary tubing) to the sensor. This displacement is proportional to the process pressure, less losses in the capillary tubing.

When a pressure transmitter is installed in a system with pressure perturbations, surges, or fluid hammers, a snubber should also be installed to protect the sensing element from damage. A snubber works like a shock absorber. It operates by driving a piston up against an orifice during rapid pressure fluctuations, choking the flow and thereby absorbing the pressure perturbations. By changing the piston diameter, the snubber can be sized to accommodate various viscosities and magnitudes of pressure surges.

CONSTRUCTION MATERIALS. Process-sensing elements are typically made of stainless steel or ceramic. When the process being measured is chemically incompatible with these materials, then a diaphragm seal should be used to provide a chemical barrier while still transmitting the pressure to the sensing element via a fill fluid. Typical fill fluids include glycerin, silicone, and Halocarbon.

ACCURACY AND PRECISION. A directly connected pressure transducer can be accurate within ± 0.04% of span, but typical accuracies are in the range of ± 0.1% of span.

Differential-pressure instruments using diaphragm seal systems must add the inaccuracies associated with the seal system, including diaphragm responsiveness and the coefficient of thermal expansion for the fill fluid. For seal systems using capillary tubing, headloss also must be added to the inaccuracy calculation.

INSTALLATION. The sensing element should be installed as close as practical to the transmitter to minimize the response time delays that can occur in capillary tubing. The installation should provide easy access for maintenance, unless doing so would compromise process-control requirements. However, a diaphragm seal system can solve this dilemma.

Differential-pressure transmitters typically are installed with valves that aid calibration efforts. Three-valve manifolds should be installed when calibration verifications are infrequently required. Five-valve manifolds should be installed when the instrument is used for billing or custody-transfer applications where frequent calibration verifications are required.

For applications involving solids-bearing liquids, flushing provisions or diaphragm isolation may be needed. Diaphragms for solids applications are available with nearly every conceivable process connection; the connections should be at least 25 mm (1 in.) in diameter. Diaphragm flanges are available with integral flushing connections or diaphragm extensions to extend the sensing diaphragm into the process.

For steam applications, precautions should be taken to prevent overheating the manifold and transmitter. A condensate pot should be installed to provide a water seal between the steam environment and the sensing element. Side-mounted pressure taps allow steam into the tap while permitting excess condensate to drain back into the process. To avoid overheating the sensing element, blowdown should not be done on differential-pressure instrument piping.

MAINTENANCE REQUIREMENTS. When used in a process that can coat the surfaces, the pressure sensor's sensing taps or flange faces need periodic manual or automatic flushing. If flushing connections cannot be installed, then the sensing taps or flange faces must be cleaned periodically. Staff should check the pressure sensor's calibration annually, unless it is used for compliance or custody transfer, in which case calibration checks should be semiannual or quarterly.

TEMPERATURE

Three types of temperature sensors typically are used at wastewater treatment plants: thermocouples and thermal bulbs, resistance temperature detectors (RTDs), and thermistors (see Table 6.8). Thermocouples and thermal bulbs are rugged and inexpensive. Resistance temperature detectors are used when accuracy is important. Applications involving temperature compensation of flow and level meters often merit use of thermistors because they have a large response to temperature changes and do not need a reference point.

THERMOCOUPLE. Thermocouples operate on the Seebeck effect—the flow of current caused by heating a circuit junction composed of two wires of dissimilar metals. The open-circuit voltage is a nonlinear function of the junction's temperature and the wires' compositions.

All junctions of dissimilar wires produce the Seebeck effect, but some combinations produce more voltage per degree than others (Table 6.9). So, junctions other than those at the point of interest must be prevented or compensated for. For example, thermocouple extension wires may be made of the same metal as the thermocouple, or they may be maintained at a known temperature so unwanted voltage can be taken into account. All thermocouple circuits have at least two junctions: one at the point of measurement and the other to obtain two copper leads to connect to a voltmeter, which measures the voltage caused by the junction of interest.

Two compensation methods are used for the secondary junctions: softwire and hardwire. In softwire compensation, the thermocouple wires are terminated at an isothermal block, which keeps the two terminations at the same temperature (Figure 6.31). This temperature is measured by an RTD, thermistor, or integrated circuit. The sensor's resistance is measured, converted to the reference temperature, and then converted to a reference junction voltage. This voltage is subtracted from that mea-

TABLE 6.8 Temperature measurement application guidelines.

Measurement type	Temperature measurement application	Application comments
Thermocouple	All	
Resistance temperature detector	0 to 300 °C	Stable; accurate; more expensive than thermocouple
Thermistor	0 to 300 °C	Limited range; fragile
Integrated circuit	0 to 200 °C	Limited range; fragile
Thermal bulb	0 to 500 °C	Limited range; repair usually not feasible

TABLE 6.9 Typical thermocouple materials.

Type	Positive metal	Negative metal	Standard color code Positive	Standard color code Negative	Application range
E	Chromel[a]	Constantan[b]	V	R	-100 to 1000 °C
S	Iron	Constantan	W	R	0 to 760 °C
K	Chromel Platinum	Alumel[c]	Y	R	0 to 1370 °C
R	13% rhodium Platinum	Platinum	—	—	0 to 1000 °C
S	10% rhodium	Platinum	—	—	0 to 1750 °C
T	Copper	Constantan	B	R	-160 to 400 °C

[a]Chromel is nickel—10% chromium.
[b]Constantan is copper—nickel of composition specific for each thermocouple type.
[c]Alumel is nickel—5% aluminum silicon.

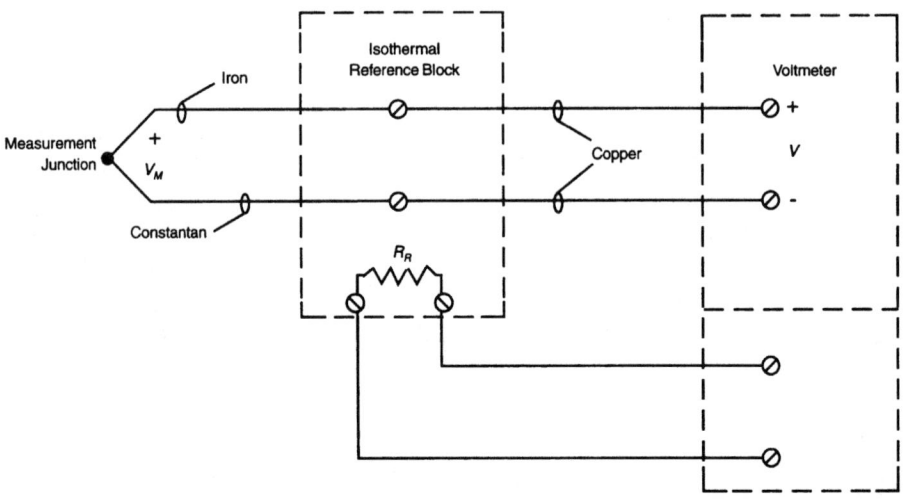

FIGURE 6.31 Thermocouple circuit with external reference (R_R = resistance at reference temperature, V_M = voltage at measurement junction, and V = voltage measured).

sured by the voltmeter to obtain the thermocouple voltage, which is converted to the measured temperature. Multiple thermocouples at the same reference temperature can be added to the measuring unit to reduce the cost of conversion per thermocouple.

In hardwire compensation (electronic ice point compensation), a voltage is generated to drive the reference junction voltage to equal 0 °C (32 °F; Figure 6.32). Electronic compensation may be applied to more than one thermocouple connected to the measuring unit, but the electronic compensation is set for a particular type of thermocouple.

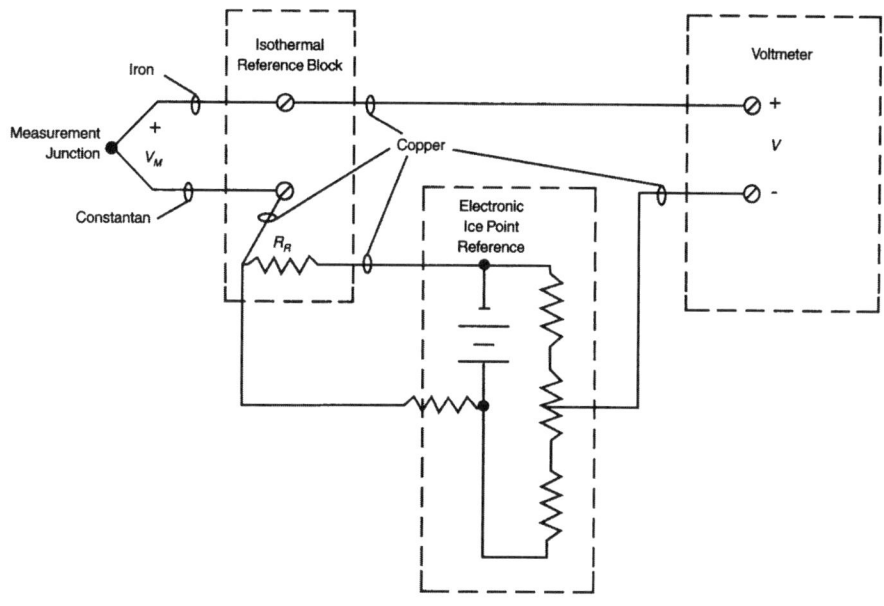

FIGURE 6.32 Representative thermocouple circuit with electronic ice point compensation (R_R = resistance at reference temperature, V_M = voltage at measurement junction, and V = voltage measured).

RESISTANCE. Resistance temperature detectors operate on the principle that the resistivity of a metal changes with its temperature. By measuring the resistance at a point of interest, a temperature can be determined.

Resistance temperature detectors consist of a thin wire or film supported by a substrate with matching thermal-expansion characteristics. The thin metal element is necessary to obtain adequately high resistance [typically, about 100 ohms at 0 °C (32 °F);]. It is supported so the RTD will be rugged enough for process measurement. Laboratory RTDs can be less rugged; they may use a wire helix that is largely unsupported. The temperature-detecting elements may be placed in glass or ceramic bobbins, embedded in glass, or mounted on ceramic substrates of various shapes.

To measure the resistance change at the point of interest, lead resistance from the instrument wires must first be eliminated. One method for doing this involves using a constant current source and measuring the voltage change across the element. This technique requires a four-wire lead or a three-wire lead and a wheatstone bridge, which will introduce some error unless the bridge is balanced for the temperature being measured. A self-heating error is also introduced by passing a current through the resistive element. The heat dissipated by the element will raise its temperature.

THERMISTOR. Thermistors are semiconductors whose resistivity changes in response to temperature changes. Most thermistors' resistances decrease as temperature increases. Their temperature coefficients are higher than those for RTDs; a typical thermistor element has a resistance of 5000 ohms at 25 °C (77 °F). So, lead resistance errors are not appreciable, and three-wire leads are typically used. Self-heating errors can be significant because of the thermistor's small mass.

Integrated circuit sensors are semiconductors whose resistivity increases as the temperature increases; otherwise, they are similar to thermistors. These sensors have not been widely used for process measurement.

THERMAL BULB. Thermal bulbs are sealed capillary-bulb systems that operate on the principle that the absolute pressure of a sealed fluid is proportional to its absolute temperature. Because the capillary lead's volume is smaller than that of the bulb, the bulb's temperature causes the greatest change in pressure. The pressure of the capillary-bulb system is measured by a Bourdon tube or other pressure-measurement device and read in temperature units.

The measured fluid may be liquid, vapor, or gas. Mercury is available as a fill fluid but is only used when accidental breakage will not cause a problem. Gas-filled thermal bulbs and capillaries are the most common thermal bulbs at wastewater treatment facilities because they have simple installation requirements, wide ranges, and wide overrange capacities.

ACCURACY AND PRECISION. The accuracy and precision of temperature-measurement devices depend on how well the installation minimizes sources of inherent error (Table 6.10). Thermocouple accuracy also depends on the type and size of element used and the fit of the curve used to convert millivolts to temperature.

TABLE 6.10 Accuracy and repeatability of temperature-measurement devices.

Device	Accuracy	Relative repeatability
Thermocouple	±1% (±0.1 °C)	Poor
Resistance temperature detector (4-wire)	±0.5% (±0.05 °C)	Good
Thermistor	(±0.1 °C)	Medium
Integrated circuit	(±0.1 °C)	Good
Thermal bulb	±1%	Medium

INSTALLATION. Except for thermal bulbs, all temperature probes inserted into process streams need protection. Thermal bulbs are usually inserted directly into the process to obtain sufficient thermal contact with the process fluid. Thermocouples, RTDs, and thermistors typically are inserted into a thermowell mounted in contact with the process fluid. A thermowell is a thin-walled metal tube capped at the process end, so it can be removed without disturbing the process (Figure 6.33).

In some applications (e.g., those involving high pressure), penetrating the process vessel or piping is not desirable. So, the temperature probe is mounted externally without a thermowell, but higher measurement errors and lack of response need to be considered, especially if the temperature must be controlled.

Temperature probes are directly imbedded in motor windings and used without thermowells in low-pressure air-handling systems.

In all applications, the probe should be protected from corrosive or erosive environments, because physical changes in the sensor will cause it to lose calibration. Thermocouple sensors also can lose calibration if the wires become annealed (cold-worked) at high temperatures. Annealing happens when a large temperature change suddenly occurs (e.g., removing the unit from a hot process to a cold ambient temperature). It can also occur because of vibration or if wires have been installed under tension.

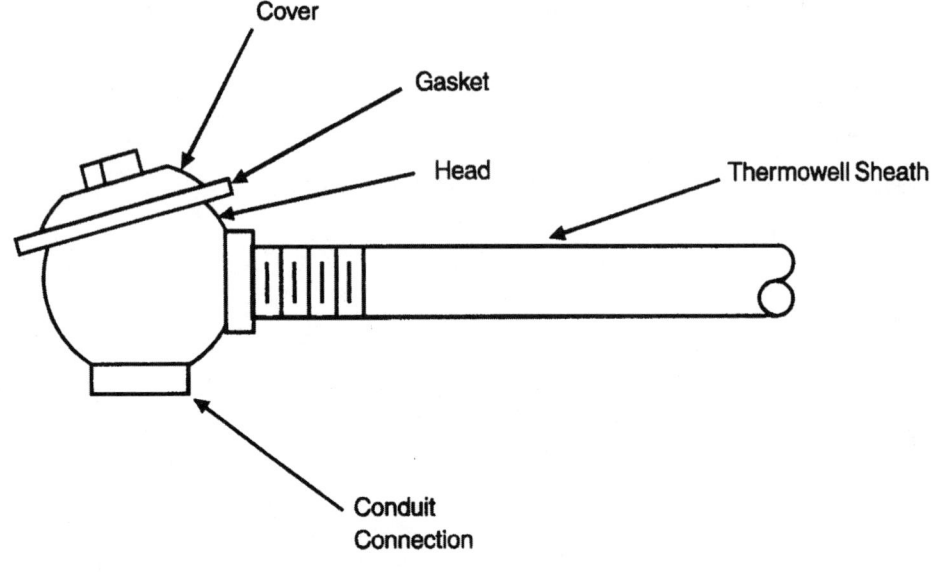

FIGURE 6.33 Thermowell assembly.

ON-LINE ANALYZERS

On-line analyzers are instruments that, when well calibrated and well installed, allow continuous measurement of a parameter that can otherwise only be measured by a time-consuming grab sample and laboratory testreturn an analytical result from a process stream. Most of the time instrument output is calibrated to a laboratory result. This section includes dissolved oxygen, chlorine residual, oxidation–reduction potential, streaming current, ammonia and ammonium, nitrate and nitrite, pH, phosphorous and orthophosphate, solids analyzers, and sludge blanket level. Recent improvements in these instruments allow operators to continuously monitor wastewater treatment plant chemicals and constituents and improve process control.

DISSOLVED OXYGEN. *Dissolved oxygen*, a parameter used to measure water quality, is the amount of free oxygen dissolved in water. It typically is expressed in parts per million (ppm) or milligrams per liter (mg/L). The amount of dissolved oxygen in water is in a state of dynamic equilibrium because oxygen is water-soluble and its maximum solubility is proportional to the water's temperature and pressure.

Dissolved oxygen typically is monitored in biological wastewater treatment processes (e.g., aeration basins), which break down wastewater biochemically. These processes typically use aerobic bacteria to convert the wastewater's organic materials into inorganic byproducts (e.g., carbon dioxide, water, and more microorganisms). Maintaining the proper dissolved oxygen concentration in the aeration basin is necessary to keep the aerobic microorganisms alive. When the dissolved oxygen level drops below a desired concentration, the process-control system typically adds air automatically to ensure that the bacteria have enough to thrive. *Aeration*—the process of adding air or pure oxygen to a biological wastewater treatment process—is often a treatment facility's largest operating expense. Adding too much is wasteful and inefficient. Adding too little can allow troublesome microorganisms to proliferate, causing poor process performance, foaming, or sludge-settling problems. So, knowing the dissolved oxygen level is critical.

Typical Membrane Sensor Design. The dissolved oxygen sensors used for continuous process monitoring typically consist of the following three elements:

- Electrodes, which provide the necessary reaction site for reducing oxygen molecules and generating electrons;

- A gas-permeable membrane, which is designed to keep the electrolyte around the electrodes and only permit dissolved oxygen to diffuse into the measurement cell; and

- A electrolyte, which facilitates dissolved oxygen migration, provides an electrical path to complete the current loop, and removes metal oxides (a reaction byproduct) from the electrodes so their metal surfaces remain clean enough to react.

The electrolyte must be replenished periodically to ensure that the electrodes remain clean.

The sensor basically works by absorbing oxygen from the surrounding wastewater. Oxygen in the wastewater diffuses through the membrane into the electrolyte. The concentration of gases always tends to equalize on both sides of the membrane.

When the concentration is unequal, gas molecules migrate to the side with the lower concentration. So, when the membrane is functioning, the electrolyte's dissolved oxygen concentration is about equal to the wastewater's dissolved oxygen concentration. The diffusion process is critical. The dissolved oxygen must be allowed to migrate freely through the membrane for the sensor to function properly.

Most continuous-measurement dissolved oxygen sensors use galvanic (spontaneous voltage), polarographic, or amperometric electrolytic (applied voltage) measuring cells. All measure an electric current between two electrodes that is proportional to the dissolved oxygen concentration.

Conventional Galvanic Measuring Cell. A conventional galvanic cell operates like a battery (Figure 6.34). Two electrodes made of dissimilar metals are immersed in an electrolyte, and an electrochemical reaction occurs when oxygen in the electrolyte touches the electrodes. In this reaction, the cathode reduces the oxygen into hydroxide, thus releasing four electrons for each oxygen molecule. The electrons cause a current to flow through the electrolyte, the magnitude of which is proportional to the electrolyte's dissolved oxygen concentration.

The electrodes typically are made of gold, silver, copper, or lead, and the electrolyte is typically potassium hydroxide (KOH). The cathode must be a noble metal (silver or gold) for the cathode potential to reduce molecular oxygen when the cell circuit is closed. The anode should be a base metal (iron, lead, cadmium, copper, zinc, or silver) with good stability and no tendency toward passivation. The electrolyte is selected so it will not rapidly dissolve the anode when the cell is open. The oxidation–reduction reactions for an iron anode and silver cathode are:

Reaction at cathode:

$$O_2 + 2H_2O + 4e^- \Rightarrow 4OH^- \tag{6.6}$$

Reaction at anode:

$$2Fe \Rightarrow 2Fe^{2+} + 4e^- \tag{6.7}$$

This cell has some inherent disadvantages. It depends on oxygen reduction to generate a measurement voltage, making it susceptible to electrode and electrolyte contamination. If a contaminant permeates the membrane, it will cause the cell potential to shift, and the shift will be falsely interpreted as a change in dissolved oxygen concentration. Also, because a galvanic cell's output is linearly proportional to the amount of dissolved oxygen present, the potential for errors at low dissolved oxygen levels increases because the signal-to-noise ratio is low. Finally, the electro-

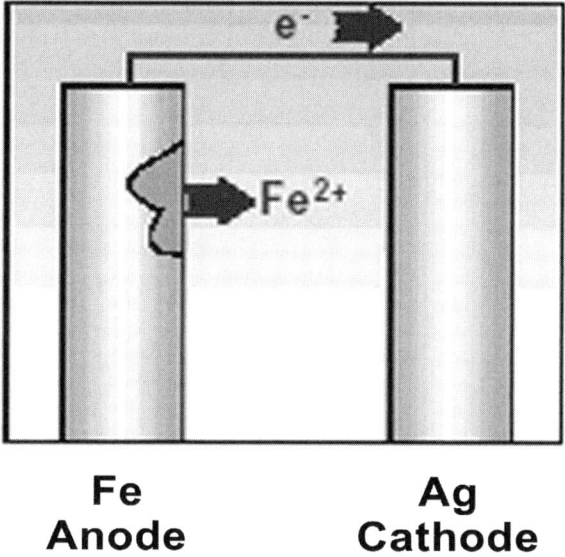

FIGURE 6.34 Conventional galvanic measuring cell. (Courtesy of Hach Company, Loveland, Colorado.)

chemical reaction consumes the anode electrode, so the cell will have to be replaced periodically.

Clark Polarographic Measuring Cell. In a Clark polarographic cell (Figure 6.35), the dissolved oxygen diffuses through the membrane and into the electrolyte, which is typically an aqueous potassium chloride (KCL) solution. When a constant polarizing voltage is applied across the gold and silver electrodes, oxygen will be reduced at the cathode. The resulting current flow is directly proportional to the electrolyte's dissolved oxygen content.

The oxidation-reduction reactions for a Clark cell are:
Reaction at cathode:

$$O_2 + 2H_2O + 4e^- \Rightarrow 4OH \tag{6.8}$$

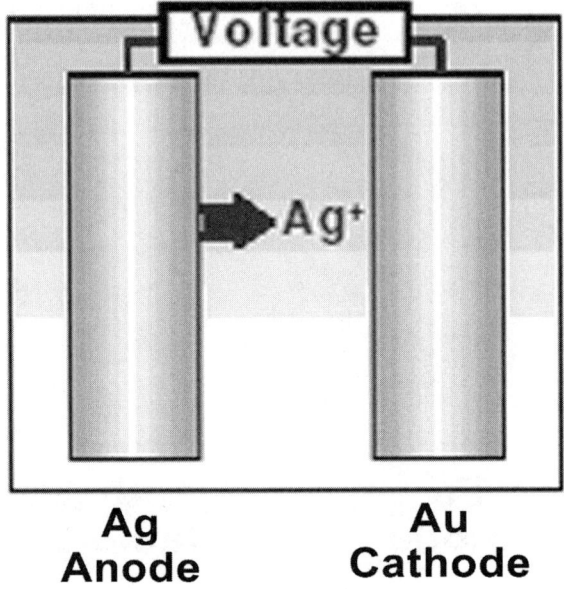

FIGURE 6.35 Clark polarographic measuring cell. (Courtesy of Hach Company, Loveland, Colorado.)

Reaction at anode:

$$4Ag + 4Cl^- \Rightarrow 4AgCl + 4e^- \tag{6.9}$$

This cell addresses some of the problems associated with a conventional galvanic cell. The polarographic cell requires a polarization voltage to be applied to the electrodes. As long as this voltage is constant, the cell is less susceptible to electrode and electrolyte contamination than a galvanic cell. If a contaminant permeates the membrane, the cell potential will not shift.

A polarographic cell measures the current flow due to oxygen reduction, and because this current flow is linearly proportional to the amount of dissolved oxygen present, the potential for errors at low dissolved oxygen levels is reduced, if the cell is properly calibrated. Under low dissolved oxygen conditions, cell contamination makes offset calibration a frequent activity to prevent zero drift. *Zero drift* is the gradual adjustment of the instrument's zero point.

Finally, the anode electrode will not be consumed by the electrochemical reaction.

Ross Polarographic Measuring Cell. In a Ross polarographic cell (Figure 6.36), the electrochemical reaction causes the anode to generate an amount of oxygen equal to the amount consumed by the cathode. The current flow is generated by this steady-state diffusion of oxygen from the anode to the cathode. This reaction continues until equilibrium exists across the boundary between the membrane and its surroundings. When the wastewater's dissolved oxygen concentration changes, oxygen will pass into or out of the sensor until equilibrium is re-established. The new steady-state dif-

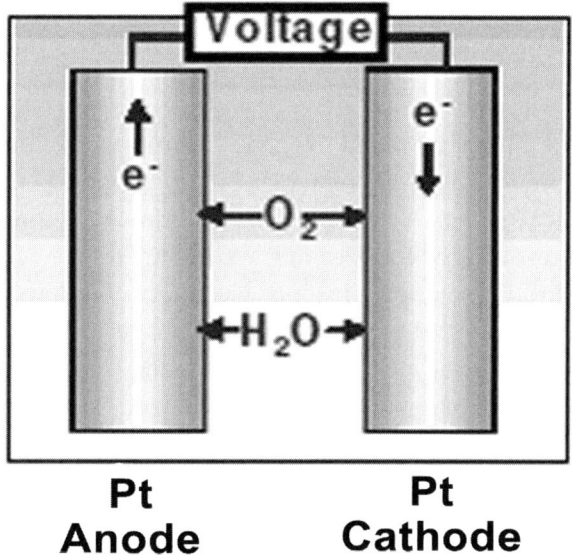

FIGURE 6.36 Ross polarographic measuring cell.
(Courtesy of Hach Company, Loveland, Colorado.)

fusion rate will result in a new current proportional to the dissolved oxygen concentration.

The oxidation–reduction reactions for a Ross cell are:

Reaction at anode:

$$2H_2O \Rightarrow O_2 + 4H^+ + 4e^- \tag{6.10}$$

Reaction at cathode:

$$O_2 + 4H^+ + 4e^- \Rightarrow 2H_2O \tag{6.11}$$

Net reaction = Zero (balanced)

By operating at equilibrium, the Ross cell helps minimize sensor maintenance and extends electrolyte life. The other measurement cells consume oxygen at the cathode without replacing it. When their membranes become coated, the oxygen consumed at the cathode will exceed the rate of diffusion through the membrane, causing the sensor output to trend toward zero. This cannot happen in a Ross cell because it is producing oxygen at the same rate as it is being consumed, so it is not affected by a reduction in the oxygen diffusion rate through the membrane. Because measurement accuracy is not affected by partial membrane fouling, the sensor will provide accurate readings as long as the membrane is not completely fouled. Measurement at equilibrium also eliminates electrolyte depletion and anode electrode consumption.

Zullig Sensor. The Zullig dissolved oxygen sensor is a galvanic sensor without a membrane; its electrodes are two independently spring-loaded concentric rings that are insulated from each other. A rotating diamond grindstone continuously polishes the electrode surfaces. This reduces cleaning and eliminates the need for replacement membranes or electrolyte solution. The open electrodes are protected from exposure to air bubbles and suspended solids in the wastewater by an oscillating sample chamber that provides wastewater to the electrodes and ensures that the wastewater is sufficiently sampled.

Luminescent Dissolved Oxygen Sensor. The newest sensor used to measure dissolved oxygen levels is based on luminescent technology. The sensor is coated with a luminescent material. A blue or green light from a light-emitting diode (LED) strikes the sensor's luminescent chemical, which instantly becomes excited. As the excited chemical relaxes, it releases a red light, which is detected by a photo diode. The higher the dissolved oxygen concentration, the less red light the sensor gives off.

Accuracy and Precision. The sensor's accuracy and precision depends on the manufacturer and type of cell. Oxygen-permeable membrane probes typically are accurate within 0.10% of span and precise within 0.05% of span.

The Zullig dissolved oxygen sensor typically is accurate within ± 0.2 ppm (mg/L) for solutions with 0 to 5 ppm of dissolved oxygen traveling less than 0.46 m/s (1.5 ft/s). Its accuracy drops to ± 0.3 ppm (mg/L) for solutions with more than 5 ppm of dissolved oxygen traveling less than 0.46 m/s (1.5 ft/s).

The luminescent dissolved oxygen sensor typically is accurate within ± 0.1 ppm for solutions with less than 1 mg/L of dissolved oxygen. It accuracy drops to ± 0.2 ppm for solutions with more than 1 mg/L of dissolved oxygen. The sensor's precision typically is 0.05 ppm, with a typical resolution of 0.01 ppm, or 0.01% of saturation. Sensitivity should be ± 0.05% of the span.

Installation. Dissolved oxygen sensors typically are placed at strategic points in an aeration basin—one-quarter, halfway, and three-quarters through the process—to adequately control the process. Some facilities install dissolved oxygen sensors before the aeration tank to monitor for anoxic or anaerobic conditions and help combat dissolved oxygen sag. The sensor installed at the three-quarter point (after aeration) is used to determine whether the process is overaerated and wasting money.

The sensing point typically is several feet from the basin wall and about 0.46 m (18 in.) below the wastewater surface. The sensors should have convenient mountings and adequate control wire so they can be removed easily and often for maintenance. Handrail-mounted instrument booms also can place a floating buoy several feet from the wall and retrieve it for regular maintenance.

Maintenance Requirements. The sensor must be kept reasonably clean to maintain measurement accuracy. During process operations, a portable meter typically is used to check the *in situ* instruments and ensure that the entire process has adequate dissolved oxygen. Because of the diurnal nature of the aeration process, this monitoring should occur at the same time every day or week. When troubleshooting, however, examine the process multiple times a day to ensure that it is operating as intended throughout the day and there are no dead zones (areas with excessively low dissolved oxygen levels).

Depending on process conditions, some installed dissolved oxygen sensor units may need to be gently wiped and rinsed each week to control growth on the sensing head. The time between cleanings (days, weeks, etc.) depends on wastewater characteristics and can only be determined by experience. For example, a sensor operating in oily or greasy wastewater may need more frequent cleaning. Consult the manufacturer's O&M manual for a cleaning procedure that will not damage the oxygen-permeable membrane.

The sensor membranes should be replaced regularly—typically quarterly or more often, depending on process conditions. To replace the membrane, staff simply removes the sensor membrane or cartridge and installs a new one. Electrolyte is also replaced during this service. New polarographic dissolved oxygen probes should be placed in the water and allowed to operate for at least 12 hours to polarize the electrodes before calibration.

Besides cleaning, optical dissolved oxygen sensors require little maintenance. However, abrasive environments or excessive cleaning can remove the luminescent material from the sensing cap. External calibration is not required because the instrument internally calibrates itself to a red LED of known intensity. However, a calibration procedure is required when a new cap is installed.

CHLORINE RESIDUAL. *Chlorination* is the process of adding liquid or gaseous chlorine to water or wastewater to reduce or destroy microorganisms that could endanger human health. *Dechlorination* is the process of removing residual chlorine from disinfected wastewater before discharge to the environment.

Throughout the United States, treated wastewater is chlorinated to kill pathogens and then dechlorinated before discharge. This common practice resulted from several comprehensive studies that quantified the toxicity of chlorinated effluents on aquatic life. The amount of total residual chlorine in the final effluent is regulated by a National Pollutant Discharge Elimination System (NPDES) permit. Typical permit limits range from 0.002 to 0.050 mg/L of total residual chlorine in final effluent. For wastewater treatment plants, this effectively means 0 mg/L of total residual chlorine in the effluent.

Dechlorination by sulfur dioxide (SO_2) is the most common method for eliminating chlorine from water. Sodium bisulfite and sodium metabisulfite also have been used for dechlorination. In an SO_2 dechlorination process, sulfurous acid is formed first:

$$SO_2 \text{ (sulfur dioxide)} + H_2O \Rightarrow H_2SO_3 \text{ (sulfurous acid)} \qquad (6.12)$$

Sulfurous acid then reacts with the various chlorine residual species:

$$H_2SO_3 + HOCl \text{ (hypochlorous acid)} \Rightarrow HCl \text{ (hydrochloric acid)} + H_2SO_4 \text{ (sulfuric acid)} \quad (6.13)$$

$$H_2SO_3 + NH_2Cl \text{ (monochloramine)} + H_2O \Rightarrow NH_4Cl \text{ (magenta color)} + H_2SO_4 \quad (6.14)$$

$$2H_2SO_3 + NHCl_2 \text{ (dichloramine)} + 2H_2O \Rightarrow NH_4Cl + HCl + 2H_2SO_4 \quad (6.15)$$

$$3H_2SO_3 + NCl_3 \text{ (trichloramine)} + 3H_2O \Rightarrow NH_4Cl + 2HCl + 3H_2SO_4 \quad (6.16)$$

According to the above reactions, it takes less sulfurous acid to neutralize monochloramine. Monochloramine is an excellent disinfectant and more effective than trichloramine, which requires three times as much sulfurous acid to eliminate it.

It is common practice to overdose the sulfur dioxide to maintain up to 5 mg/L of SO_2 in the effluent. This ensures the reduction of all chlorine residual species and maintains compliance, with little extra chemical costs. When analyzing dechlorinated effluent with a chlorine residual analyzer, a biasing agent (e.g, hypochlorite or iodine) may be added to the water sample to ensure that the dechlorination agent is working. The value of the biasing agent is subtracted from the test result.

The following are several techniques for measuring chlorine in chlorination and dechlorination systems.

DPD Colorimetric Method Analyzer. The DPD (n,n-diethyl-p-phenylenediamine) method for residual chlorine was first introduced in 1957 and has since become the most widely used method for determining free and total chlorine in water and wastewater. Basically, the DPD colorimetric method analyzer injects a chemical reagent into a fixed sample and measures the color change to determine the chlorine residual.

The chemical basis for the DPD–chlorine reaction is depicted in Figure 6.37. The DPD amine is oxidized by chlorine to two oxidation products. At a near neutral pH, the primary oxidation product is a semi-quinoid cationic compound called a *Würster dye*. This relatively stable free-radical species accounts for the magenta color in the DPD colorimetric test (DPD can be further oxidized to a relatively unstable, colorless imine compound). When DPD reacts with small amounts of chlorine at a near neutral pH, the Würster dye is the principal oxidation product. At higher oxidant levels, the formation of the unstable colorless imine is favored—resulting in the apparent "fading" of the colored solution. Occasionally, operators performing this test will interpret a colorless sample as no measurable chlorine, when actually an

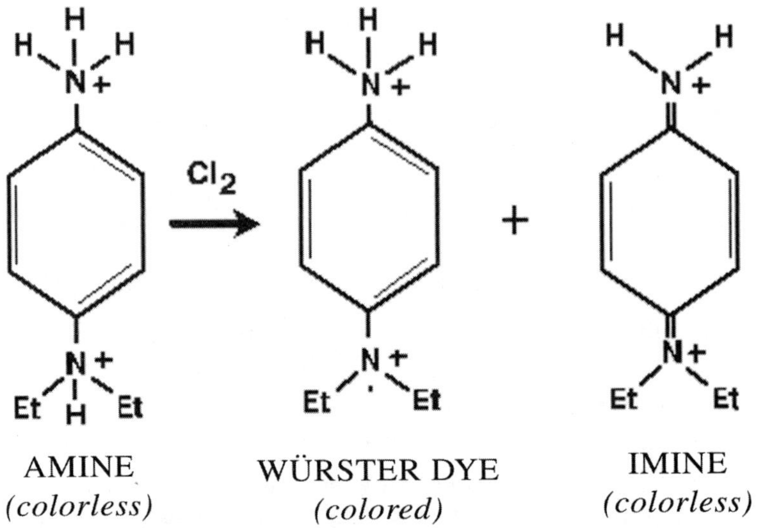

AMINE
(colorless)

WÜRSTER DYE
(colored)

IMINE
(colorless)

FIGURE 6.37 DPD–chlorine reaction products. (Courtesy of Hach Company, Loveland, Colorado.)

excess amount of chlorine can result in a colorless or yellow test sample. If this is a problem, adding extra powder packets will provide enough DPD reagent turn the sample magenta, but doing so will prevent the instrument from accurately reading chlorine residual. For proper measurement of high residuals, the sample must be diluted.

The DPD–Würster dye color has been measured photometrically at wavelengths ranging from 490 to 555 nanometers (nm). The absorption spectrum shows a doublet peak with maxima (the highest points) at 512 and 553 nm. For maximum sensitivity, absorption measurements should be made between 510 and 515 nm.

The process analyzers are used to measure free or total chlorine residuals in drinking water, to prevent biofouling in cooling water treatment, to protect membranes in reverse osmosis systems, and to ensure regulatory compliance in wastewater treatment.

A typical analyzer is equipped with a two-reagent system based on the DPD chemistry. The DPD indicator is prepared by adding powdered DPD reagent salt to the acidic indicator solution. The powder readily dissolves in the solution, and the mixed solution is stable for at least two months. The DPD reagents are carefully manufactured to ensure that the DPD indicator solution does not contain impurities that could oxidize ionic DPD. After dissolution, the DPD indicator reagent solution is free of insolubles that could exhibit a reagent blank or plug reagent tubing.

The buffer solution for the total chlorine analyzer is a citrate that also contains iodide. The free chlorine buffer reagent is a maleate buffer.

The buffer reagent and its complementary indicator reagent are added in equal volumes to a captured portion of sample. Premixing the two reagents before introduction to the sample is important. This is done by putting a "T" union in the reagent feed lines before the colorimeter block.

An on-line DPD instrument typically operates as follows:

1. The sample inlet line of the pump or valve module is opened, allowing pressurized sample to flush the sample tubing and the colorimeter sample cell.
2. The sample inlet is closed, leaving fresh sample in the cell. The cell volume is controlled by an overflow weir.
3. As the sample inlet line closes, the reagent lines open, allowing the buffer and indicator solutions to fill the tubing in the pump or valve module.
4. The untreated sample (sample blank) is measured for reference before the reagent is added. This compensates for sample color and turbidity
5. The reagent outlet block opens, allowing precise volumes of buffer and indicator to blend and enter the colorimeter cell to mix with the sample.
6. After about a 1-minute delay so the color can fully develop, the magenta color is measured at 510 nm (the sample measurement).
7. The concentration (as X mg/L of chlorine) is calculated (the log of the ratio of reference measurement to sample measurement) and then displayed.

The above sequence is repeated every 2.5 minutes.

Varying the test acidity and increasing the iodide concentration in the sample may shorten the color development time required. When using a portable analyzer, users should refer to the instruction manual to determine the time required for color development.

For example, let's consider the reaction chemistry of dichloramine:

$$NHCl_2 + 3I^- + H_2O + 2H+ \Rightarrow NH_4OH + 2Cl^- + I_3^- \qquad (6.17)$$

$$I_3^- + DPD \Rightarrow 3I^- + DPD\ (oxid.) \tag{6.18}$$

The rate of the first reaction is much slower than the rate of the second. Increasing the iodide addition or adjusting the acidity could increase the speed of the dichloramine–iodide reaction, but in this case, acidity cannot be significantly increased because of increased nitrite interference at a lower reaction pH.

Users should consult the equipment manufacturer if they suspect limited reaction times are causing errors in the instrument output. With different chemical formulations, it may be possible to quantitatively measure 5 mg/L dichloramine (as Cl_2) at cold sample temperatures without nitrite interference and within the 1-minute color-development time.

Amperometric Bare-Electrode Analyzer. The amperometric bare-electrode analyzer measures chlorine residual using the small current generated by hypochlorous acid (or iodine for total chlorine residual measurement) in a water sample. Amperometric measurements require that two dissimilar metals be held in a solution (electrolyte). A voltage is applied to the two metals, which act as electrodes, thus generating a current. The amount of current is proportional to the amount of chlorine in the solution.

A basic amperometric chlorine residual analyzer consists of an inlet sample tank, a flow regulator, reagent solution with metering pump, a measurement cell, and an electronic signal converter. The metered sample stream acts as the buffer solution for the pumped reagent [may use buffer or potassium iodine (KI) solutions] as they flow through the measurement cell. The current generated is proportional to the amount of chlorine in the solution.

The measurement cell is sensitive to temperature; the reading can change as much as 3% per 1 °C of temperature change. To compensate, a temperature sensor in the measurement cell provides data to the electronics converter that is used to correct the indicator and output signals to nullify the temperature effects. A pH measurement can be used instead of a buffer to compensate for the change in dissociation when the pH of the water is higher than 4.5 and less than 9.

Amperometric Membrane-Covered-Electrode Analyzer. The amperometric membrane-covered-electrode analyzer measures chlorine residual via the same method as the bare-electrode analyzer. It can measure free, total, or combined chlorine. The electrochemical cell contains two electrodes and an electrolyte. A polarizing voltage is applied to the electrodes. A microporous polymer membrane, permeable to chlorine species, separates the cell from the process stream. Chlorine enters the cell through the membrane, reacts with the electrolyte, and is electrochemically reduced at the measuring electrode surface. An electrical current is generated in proportion to the chlorine concentration.

Some amperometric sensors can be installed directly in the contact tank, eliminating the need for a sampling line.

Oxidation–Reduction Potential Probes. Oxidation–reduction potential (ORP) analyzers measure the water's potential to oxidize (or reduce) organic matter or compounds in the water. The potential is generated by the chlorine or sulfite in the sample. With pH correction, the ORP output in millivolts (mV) is proportional to the chlorine residual.

Oxidation–reduction potential is a potentiometric measurement in which the potential (tendency) of a medium to transfer electrons is sensed by an inert metal electrode and read relative to a reference electrode immersed in the same medium, according to the 20th Edition of the *Standard Methods for the Examination of Water and Wastewater* (Section 2580 B.). It also is called a "redox" (reduction–oxidation) measurement. In most monitoring systems, the inert metal electrode is a metal tip or band made of platinum (sometimes gold) and the reference electrode is the same one associated with the pH sensor (usually silver or silver chloride). The sensor's readout is a voltage (relative to the reference electrode), with positive values indicating an oxidizing environment (ability to receive electrons) and negative values (e.g., -300 mV) indicating a reducing environment (ability to lose electrons).

Knowing ORP is useful in waters with a relatively high concentration of reducing or oxidizing substances [e.g., chlorine (a strong oxidant) and sulfite (a strong reductant)]. However, ORP is a nonspecific measurement (the measurement notes the effect of all dissolved species in the medium) that is affected by multiple parameters (e.g., time and temperature in the sewer, nitrification, and denitrification).

Also, the results must be carefully interpreted. Results depend on the condition of the sensor and the makeup of the water being tested. The most common problem reported is that readings from various instruments (sometimes with exactly the same sensor type and electronics) differ by a significant margin (50 to 100 mV) even though the sensors are in the same container of water. To make the problem more perplexing, all of the sensors show identical readings in an ORP standard.

The ORP reading must be related to some known value (e.g., a known oxidant residual). Once the known values are established, this instrument can shift from total to free chlorine residual and back without recalibration. Use of the signal for chlorination and dechlorination also requires measuring the ORP value before chlorine is added. Oxidation–reduction potential analyzers work well when considered on a differential basis (e.g., the difference between unchlorinated and chlorinated water). In dechlorination processes, the ORP readings must also be related to the actual residual level, but there are fewer interferences at low chlorine or sulfite levels.

Currently, there is continuing interest and research in using ORP to control nutrient removal processes at wastewater treatment facilities. Although the use of ORP is well-documented in sequencing batch reactors (SBRs), it is much less widely used in conventional applications. In wastewater treatment plants, these analyzers typically are used to analyze the disinfection (chlorination) and dechlorination processes.

Accuracy and Precision. Chlorine residual analyzers typically are accurate within ± 3% of full-scale. Standard full-scale spans vary from 1 to 20 mg/L. The measuring error also varies from 0.03 to 0.6 mg/L, depending on the range selected. The analyzers' precision typically is within ± 1% of full-scale. With automatic temperature compensation, accuracy and precision should hold over a sample temperature range of 0 to 50 °C (32 to 120 °F).

Installation. Chlorination samples typically are pumped from the contact tank at the beginning and end of the process. Dechlorination samples typically are pumped from the contact tank at the end of the process. Chlorine residual analyzers typically are installed in a sample building near the contact tank(s) to reduce the sample piping length. A valved cross-connection between the two chlorination sample lines allows operators to switch the sample being pumped to an instrument if a mechanical problem occurs. The sample point should be somewhere that does not contribute unnecessary dead time to the chlorine residual analysis. The sample point also should be clean, thoroughly mixed, and representative of the monitored stream. Sample inlet lines should not be mounted next to walls or in corners. As with all compliance constituents, some level of redundancy is needed. If amperometric sensors will be installed in the contact tank, engineers should refer to the installation and maintenance guidelines provided for dissolved oxygen sensors.

The initial chlorine residual typically is not required for compliance, but assures operators that chlorine levels were adequate and allows for automatic residual control. The sample should be analyzed 5 to 10 minutes downstream of the chlorine addition point to allow time for side reactions that do not contribute to disinfection.

The effluent should be analyzed after enough contact time has elapsed for disinfection to occur. The total contact time is the time in the contact tank plus the time to deliver the sample to the analyzer (because reactions continue in the sampling line). If the required disinfection time is 30 minutes, samples should arrive at the analyzer 30 minutes after chlorine is added.

Oxidation–reduction potential probes are installed in flow-through reservoirs connected directly to the sample line.

Maintenance Requirements. Some amperometric systems include balls or beads to routinely clean the electrode surfaces. A bare electrode system may need manual cleaning. After cleaning, the sensor must be given time to reform its oxide layer before the measurement can be used.

In membrane systems, the membranes typically need to be replaced every 3 to 6 months. More cleaning may be required, depending on water conditions and any clogging that might prevent the membrane from operating.

Both amperometric and membrane systems need to have the probes replaced every one to two years, depending on system conditions.

Oxidation–reduction potential probes used to analyze chlorinated water typically need weekly cleaning to remove tarnish. Oxidation–reduction potential probes used to analyze dechlorinated water may need daily cleaning to control biogrowth, which can isolate the probe tip from the water. Cleaning typically takes one minute with a toothbrush.

POLYMER. When the correct polymer dose is added to a treatment process, the zeta potential of the water changes, destabilizing colloids and allowing flocculation so suspended solids can be removed from a treatment process more easily. It is impossible to measure residual polymer in wastewater so operators can optimize a treatment process. However, a streaming current monitor (SCM) can monitor the charge characteristics (zeta potential) of a process flow and provide an output that can be used to control the coagulant dose.

Streaming current monitors have been used in water treatment plants for years. At wastewater treatment facilities, tests have shown a strong correlation between polymer dose and centrate or filtrate charge characteristics (zeta potential). The monitors have been used in thickening and dewatering processes, and SCM could be used before tertiary filtration if more coagulant doses are required.

Operating Principles. Streaming current monitors measure the electrokinetic charge in the wastestream. They respond to changes in wastewater characteristics (TSS, percent solids, blends, pH, conductivity, etc.) and flow rates. A typical SCM consists of a sensor, a monitor, a sensor maintenance module, and a controller (Figure 6.38).

The sensor is a plastic reciprocating piston in a plastic body with metal electrodes. The reciprocating piston shears the water near the metal electrodes so its charge characteristics can be measured. The signal is sent to the monitor, where it is amplified. The monitor may have a PID controller that can adjust the coagulant chemical feed rate.

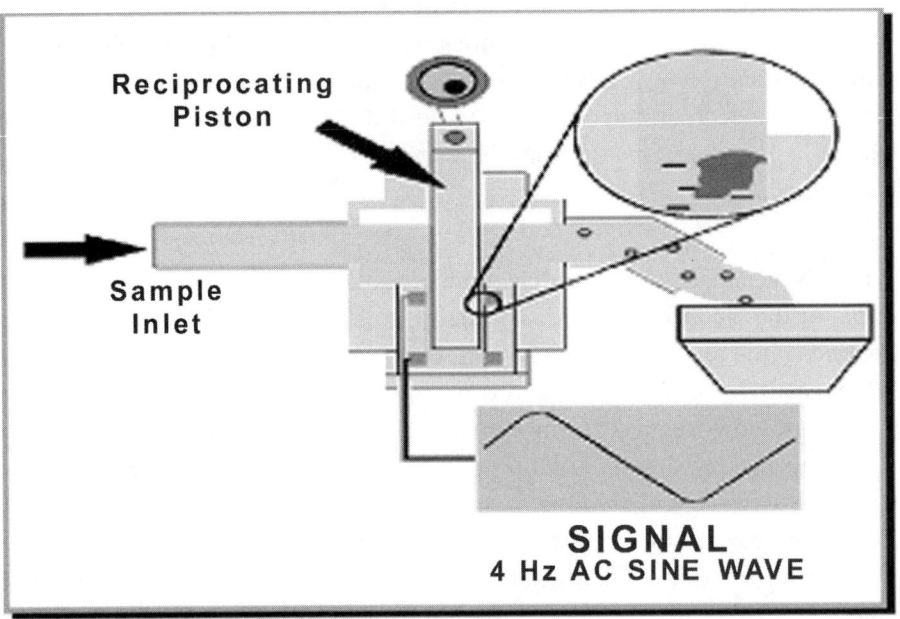

FIGURE 6.38 A diagram of a streaming current monitor unit. (Courtesy of Chemtrac Systems, Inc., Norcross, Georgia.)

The sensor maintenance module operates the cleaning water solenoids, which are connected to clean plant water or potable water. The solenoids flush both the instrument and the sample supply line, when required. Flushing them every 15 minutes helps prevent the accumulation of debris that could lead to blockages. Cleaning agents, acids, or bases may be added to the flush water to prevent precipitate or grease from accumulating on the metal electrodes.

In areas where pH or conductivity changes are known to occur or sludge is fed from multiple sources, SCM field trials should be performed before plans for a permanent installation are made. Streaming current monitors with PIDs require an appropriate setpoint that the control algorithm can track. If a sludge source changes, then the PID setpoint must also be adjusted. The unit also must be adjusted if the instrument becomes fouled over several days of operation or if the setpoint is no longer appropriate.

Construction Materials. The sensor is a plastic reciprocating piston in a plastic body with metal electrodes. A 25-mm (1-in.) sample line prevents large debris from entering the instrument and allows for a sampling rate of 11.4 to 38 L/min (3 to 10 gpm).

The sensor mechanics (excluding the body of the motor) and all of the sensor's wetted surfaces are outside the NEMA 4X enclosure that houses the drive motor and electrical components. A stainless steel cover protects the sensor mechanics.

The signal monitor box contains all circuitry and signal processing cards to display readings, set up parameters, and produce a 4- to 20-mA signal. The system uses a microprocessor PID controller with isolated inputs, capable of single or multiple (cascading) PID control, autotuning, and communications. The signal monitor also includes:

- A meter zero adjustment (full-scale on all ranges);
- A signal gain switch (adjustable 1 X to 20 X);
- Internal, continuous adjustments for higher gain;
- Alarm indication for sensor operation or failure;
- Four auxiliary inputs and three outputs for monitoring or controlling other process variables (e.g., turbidity and sludge flow);
- Programmable high and low alarm settings (and relay contracts for both);
- A programmable ± 1000 span for the electrokinetic value or setpoint;

- Manual and automatic modes; and
- Programmable high and low limits for control output.

Installation. The instrument is typically installed below the dewatering equipment or filter feed piping so there is at least 1.5 m (5 ft) of head on the sample line, which should deliver between 11.4 to 38 L/min (3 and 10 gpm) to the instrument. The sample line should be sloped to the instrument, with no sharp bends. For easier maintenance, a screen should be provided to prevent large debris in the sample from entering the sensor.

Maintenance Requirements. The sensor typically is maintained via the sensor maintenance module. During a cleaning cycle, the monitor holds the 4- to 20-mA output signal. Sample flow through the sensor is then stopped, and flush water is injected into the bottom of the sensor. A pump is then activated, and cleaning solution enters the probe. The sensor is then flushed again, and the sample is brought back on-line. After a stabilization period, the sample hold is disengaged, and the unit is brought back on-line.

If there are problems with blockages in the instrument, the unit is easily disassembled by unscrewing the outer plastic body. The unit may occasionally be clogged by gradually accumulating hairballs, but these are easily removed when found during weekly routine maintenance. In some installations, gradual scale buildup may require weekly chemical cleaning.

NITROGEN. *Ammonia and Ammonium.* Most wastewater treatment plants must comply with a permit limit typically expressed in terms of ammonia as nitrogen (NH_3-N). However, ammonia is a gas; when dissolved in wastewater, it exists in equilibrium with the hydrated ion, ammonium (NH_4^+).

The relationship between dissolved ammonia and ammonium is pH-dependent. As pH increases, the equilibrium is driven toward dissolved ammonia. Above pH 11, all of the ammonium is converted to dissolved ammonia. Under normal circumstances, however, almost all of the dissolved ammonia will escape from the liquid, leaving only the ammonium ion. So, in wastewater treatment facilities, the ammonium concentration is critical.

Depending on the measuring principle used, manufacturers may express results as either ammonia or ammonium. An ion-selective electrode (ISE)-based instrument, for example, measures ammonium. A gas-sensitive electrode (GSE)-based instrument, however, actually measures ammonia gas (after raising the sample's pH to convert all the ammonium to ammonia). The best way to avoid confusion is to express the results in terms of nitrogen (e.g., NH_3-N or NH_4-N), and most manufacturers do that.

Ammonia is typically monitored in the following places:

- In the plant influent to allow load balancing and feed-forward control of aeration;
- In the aeration tanks to monitor nitrification and optimize aeration control;
- In the plant effluent to monitor compliance with permit requirements;
- In intermittent aeration systems to control aeration time based on mixed-liquor ammonia levels; and
- In the supernatant of dewatered biosolids to monitor the high ammonia levels returning to the head of the plant.

Operating Principles. There are three principal methods for measuring ammonia with an online instrument: ISE, GSE, and colorimetry. Ammonium ISEs are similar in principle to a pH electrode, can measure over a wide range, and have a fast response time. However, they are subject to interferences and operate over a limited pH range.

Gas-sensitive electrodes are more selective than standard ammonium ISEs and more commonly used in wastewater treatment applications. An ammonia GSE is effectively a pH electrode behind a gas-permeable membrane. Sodium hydroxide is added to the sample to raise the pH above 11 and drive all the ammonium to dissolved ammonia. The ammonia gas is allowed to pass across the GSE membrane and redissolve in an electrolyte (typically NH_4Cl) surrounding a pH electrode. The ammonia gas decreases the electrolyte's pH. The analyzer then converts this change in pH to an ammonia reading.

There are several colorimetric methods for measuring ammonia, including

- The indophenol blue method; and
- The monochloramine-F method, in which hypochlorite is added to the sample to convert the ammonia to monochloramine. A second reagent containing an indicator is then added to the sample, and the resulting color development is measured via a colorimeter.

The method is similar to the GSE. Sodium hydroxide is added to the sample to raise the pH above 11. The ammonia gas is then stripped from the sample and redissolved in a liquid pH indicator solution. The change in the indicator's color is measured via a colorimeter and converted to an ammonia reading. Some instruments measure at multiple wavelengths, providing compensation for light scattering by particles in the sample and variations in the sample color.

Construction Materials. The exterior structure of ammonia analyzers is typically made of a rigid ABS plastic or stainless steel. Internally, the key components that touch the sample are the tubing (typically made of Tygon®, Norprene®, or a similar material) and the mixing and holding chambers (typically made of glass).

Accuracy and Precision. Manufacturers claim an accuracy of between 3 and 5% of reading, which is more than adequate for most wastewater applications. Most manufacturers claim a lower detection limit of 0.1 mg/L NH_3-N or NH_4-N. The instrument's range depends on the measurement method. Ion-selective electrode and GSE analyzers typically have the widest range, measuring as high as 1000 mg/L, which makes them valuable for some industrial wastewater applications. Colorimetric analyzers can also measure high levels but require different reagents to extend the range. In practice, this is not a great disadvantage because ammonia levels typically do not vary widely in a given application. The range is also defined by the photocell's geometry and physical characteristics.

Installation. Ammonia analyzers are available as freestanding or submersible units. The submersible analyzers can be installed directly in a channel or aeration tank, thereby providing faster response time and eliminating the need to pump the sample to the analyzer. Their chief disadvantages are the inconvenience of having to remove the analyzer for reagent replacement and maintenance, and its weight (a hoist may be needed to remove it for maintenance).

Freestanding ammonia analyzers typically are wall-mounted and require shelter from the elements. Samples must be pumped to them. They also need some form of filtration: either a filtration system suspended in the basin or a free-standing filtration system mounted close to the analyzer. (Manufacturers typically offer filtration systems as an option.) Also, a sample-preparation system (chosen based on the application) is the key to a successful installation.

Maintenance Requirements. Most manufacturers provide a variety of features to reduce maintenance. These features include automatic cleaning, automatic calibration, peristaltic pumps, and pinch valves (so the sample stays inside the tubing and does not touch moving parts). Nevertheless, ammonia analyzers are sophisticated instruments that require careful routine maintenance. Most require the following maintenance:

- Routine replacement of reagents (typically monthly or quarterly, but can vary according to the measurement, calibration, and cleaning intervals chosen);
- Cleaning or replacement of ISE or GSE electrodes (monthly);
- Replacement of tubing (every 3 to 6 months); and
- Regular inspection and cleaning, if necessary, of the sample-filtration system.

Nitrate and Nitrite. Nitrate (NO_3) is typically measured in biological nutrient removal (BNR) systems to monitor and control the denitrification stage. It can also be used to monitor the nitrate concentration in effluent. Nitrite (NO_2) is also present in BNR plants, but it is relatively unstable and quickly oxidized to nitrate or reduced to nitrogen. So, it is of less interest than nitrate. Nevertheless, nitrite monitors are available.

On-line nitrate instruments typically are used to control the sludge recirculation rate in biological treatment systems, to control aeration in intermittent aeration systems, and to monitor the nitrate concentration in effluent. Monitoring nitrate also can reduce aeration costs, avoid floating sludge in settling tanks because of unintended denitrification, and optimize both the RAS rate and the anoxic zone size.

Operating Principles. Ultraviolet (UV) absorbance probes typically are used to monitor nitrate. In water, nitrate absorbs UV light at 215 nm (up to 240 nm). On-line UV absorbance probes direct UV light through a wastewater sample and measure the amount of light that passes through to a detector. The absorbance is calculated and converted to a nitrate value (measured as X mg/L of NO_x-N, where $NO_x = NO_3 + NO_2$). Light with a different wavelength is also passed through the sample to compensate for interference due to turbidity.

The advantages of the UV absorbance method are the absence of reagents and the ability to directly measure nitrate in the process. However, a high, variable concentration of organics with adsorption at 254 nm may interfere with nitrate measurement.

Another nitrate monitoring method involves ISE technology, as described in the section on ammonia analyzers. However, nitrate ISEs suffer from interferences and are rarely used in wastewater treatment because the UV absorbance method is more convenient.

In practice, UV absorbance instruments also measure nitrite because the two compounds have similar UV absorbance characteristics. Nitrite can also be measured colorimetrically using the on-line Azo dye method. However, nitrite levels typically are insignificant at wastewater treatment plants, and the NO_x value provided by UV absorbance probes is widely accepted for process control purposes.

Construction Materials. Ultraviolet absorbance probes are typically made of stainless steel with a quartz glass optical window. They may have a mechanical cleaning system with a silicon rubber wiper.

Accuracy and Precision. Ultraviolet absorbance probes typically have a lower detection limit of 0.1 to 0.2 mg/L of NO_x-N. The upper detection limit may be as high as 50 mg/L of NO_x-N, although this may require a longer path. Manufacturers claim that the probes' accuracy varies from 2 to 5% of reading, and their precision is in the same range.

Installation. Ultraviolet absorbance probes typically weigh less than 4 kg (10 lb) and can be mounted directly in an aeration basin. Wall-mount units are also available. If direct immersion is impossible (e.g., a closed pipe), the probe can be installed in a bypass or flow-through system. Engineers should carefully plan the sample pumping and piping to keep the piping as short as possible and eliminate problems due to debris blockages.

Maintenance Requirements. Ultraviolet absorbance probes are relatively easy to maintain. The mechanical wiper system keeps the optics clean, and the units typically do not need calibration. So, the only maintenance required is annual replacement of the wipers and mechanical seals.

PH. In wastewater treatment facilities, pH sensors are used to monitor plant conditions, track biological treatment process conditions, and control acid–base additions for pH adjustment. Regulators require the pH of treatment plant influent and effluent to be measured to indicate overall conditions. Specific industrial discharges also may require pH monitoring to provide advance warning of possible toxic conditions.

While some biological processes (e.g., activated sludge systems) can tolerate a pH variance of 5 to 9, others are pH-sensitive. The anaerobic digestion process, for example, requires pH to be between 6.6 and 7.6; the process fails when pH is less than 6.2. So, the pH of anaerobic digester liquor should be monitored. On-line monitoring of digester pH is not recommended, though, because the solids will continually foul the sensor. Periodic sampling is preferred.

An on-line pH monitor also can provide feedback in other processes requiring pH adjustment. For example, pH adjustment may be required to neutralize low-pH industrial wastes, enhance phosphorus removal by alum addition, or optimize nitrification or denitrification processes.

Operating Principles. Typical pH Sensor. Figure 6.39 shows a typical pH sensor arrangement. The heart of the sensor is a glass membrane, across which an electrical potential is generated. This potential varies as pH changes. The difference between this potential and a reference electrode is measured and amplified by an electronic signal conditioner.

The complete electric circuit includes the glass electrode wire, glass membrane, process fluid, reference electrode fill solution, and reference electrode wire (Figure 6.40). Voltage at the input of the amplifier is defined as

$$E_i = E_g - E_r \qquad (6.19)$$

Where

E_i = amplifier input (mV);
E_g = glass electrode potential (mV); and
E_r = reference electrode potential (mV).

The glass electrode has the approximate characteristic

$$E_g = K_1 + K_2 \,(\text{pH}) \qquad (6.20)$$

So, voltage at the amplifier input is

$$E_i = K_1 + K_2 \,(\text{pH}) - E_r \qquad (6.21)$$

Where

K_1 = asymmetric potential (mV); and
K_2 = electrode gain (mV/pH).

The reference electrode is designed so potential E_r is constant with the process fluid's pH and other chemical characteristics. The asymmetric potential K_1 varies

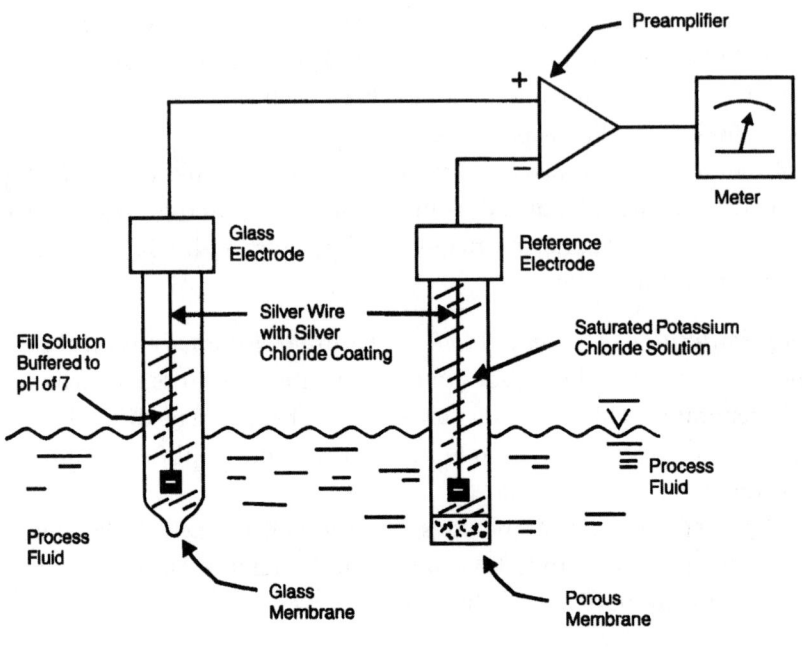

FIGURE 6.39 Typical pH sensor arrangement.

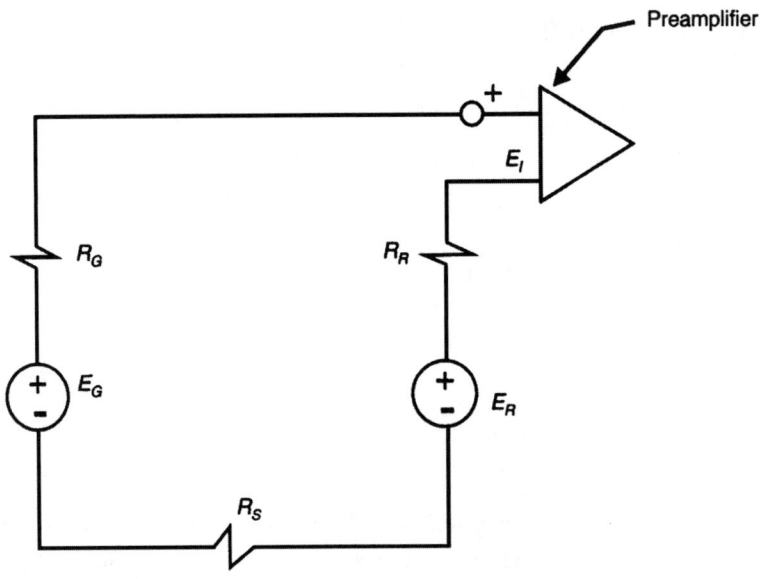

FIGURE 6.40 Equivalent electric circuit of pH sensor.

from sensor to sensor according to design preferences among manufacturers. It also changes as the sensor ages. For this reason, pH sensors must be periodically standardized against buffer solutions of known pH.

The electrode gain K_2 is a function of temperature. So, most commercial pH sensors include automatic temperature compensation. A temperature sensor in the process fluid adjusts amplifier gain to compensate for changes in electrode gain that are caused by temperature. Figure 6.41 illustrates the effects of temperature on asymmetric potential.

Ion-Sensitive Field Effect Transistor. The ion-sensitive field effect transistor (IsFET) sensor measures pH via an IsFET chip embedded in a sensor body made of PEEK, a

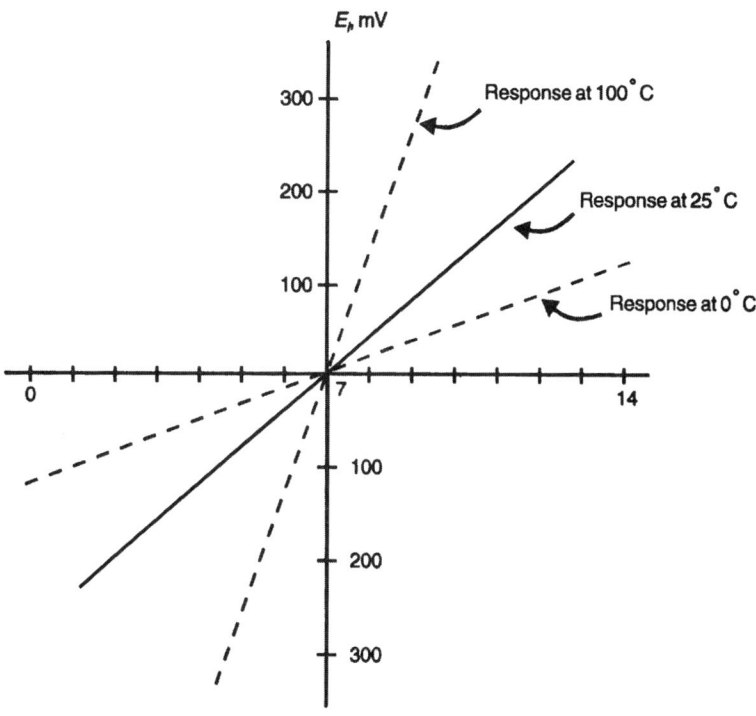

FIGURE 6.41 Effects of temperature on asymmetrical potential.

very stable material. The sensor has the same dimensions as a traditional 120-mm-long glass pH electrode, so it can be used with all the commercially available assemblies (Figure 6.42).

The chip's gate is the part of the chip touching the medium, where a pH-dependent Nernst voltage forms. This voltage generates mirror charges inside the chip that modifies the resistance between its source and drain (Figure 6.43). The modified resistance is proportional to the pH level. Measurement then takes place via two circuits. The voltage circuit to the medium (via the gate) has a high impedance (more than 10 to 12 ohms), while the electric circuit via source and drain has a low imped-

FIGURE 6.42 An example of an IsFET sensor. (Courtesy of Hach Company, Loveland, Colorado.)

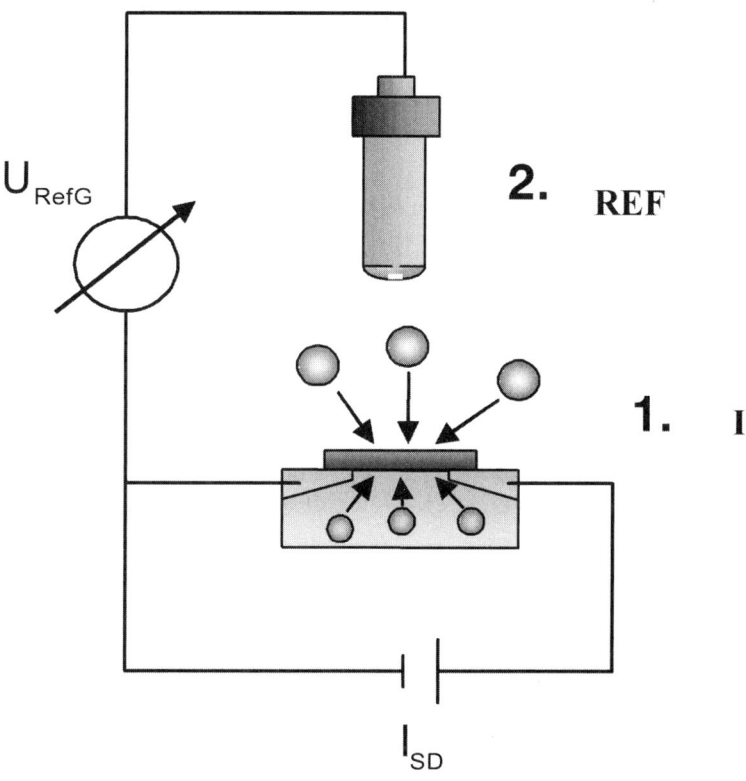

FIGURE 6.43 Operating principle of an IsFET sensor. (Courtesy of Hach Company, Loveland, Colorado.)

ance (1 to 3 ohms). So, the measuring method is fundamentally different than that used with glass electrodes.

This sensor has several advantages. The Nernst layer on the gate does not need to undergo hydrolysis (i.e., there is no gel layer). It is not made of glass, and there is no internal electrolyte. However, a reference electrode is still necessary, because IsFET technology requires a known voltage reference point.

Accuracy and Precision. Methods of reporting performance specifications vary among manufacturers. When these are adjusted to equal units of measure, the accuracy and stability claimed by different manufacturers varies greatly. Manufacturer claims for pH meter accuracy range from ± 0.02 to ± 0.2 pH units for both the electrodes and the signal conditioner or transmitter. Without temperature compensation, an additional error of 0.002 pH units per 1 °C difference from the calibration temperature can be expected.

The precision of pH meter measurements varies by manufacturer from 0.02 to 0.04 pH units.

Stability (drift) is an important performance parameter that indicates how often meters must be recalibrated. Manufacturer claims for stability vary from 0.002 to 0.2 pH units per week. In flow-through probes, the sample velocity can cause a shift (0.2 to 0.3 pH) in measured values.

Typically, good pH meters achieve an accuracy of ± 0.1 pH units, a precision of ± 0.03 pH units, and a stability of ± 0.02 pH units per week.

Installation. When pH control is not the goal, a pH analyzer can be installed with other on-line analytical instruments in a sample system. This makes maintenance easier because the buffer solutions needed for standardization can be conveniently stored with other analytical instrument reagents.

Flow-through pH sensors typically are used in sample system installations. The flow-through sensor should be designed so electrode tips are flush with the tube wall and do not obstruct flow. To make maintenance easier, bypass and shutoff valves should be installed to provide for instrument removal and service. Sensors with electrodes that are easily removed from the flow-through housing allow for easier cleaning and replacement. Putting the pH analyzer near the probe mounting assembly makes for easier standardization. A sample valve can be installed next to the sensor to help with conformance checks. A work surface for the buffer solution containers is also helpful.

Submersed electrode assemblies with integral pre-amplifiers typically are used in tank and open-channel installations (Figure 6.44). The electrode assembly is attached to a polyvinyl chloride (PVC) pipe with a bracket typically mounted on a guardrail. The bracket is designed so the pipe and electrode assembly can be removed for maintenance without using tools. All fastening devices are secured to prevent them from being dropped into the tank or channel. The signal conditioner or transmitter is mounted next to the electrode assembly mounting bracket. Enough spare cable is needed to allow the sensor and pipe assembly to be lifted clear of the tank.

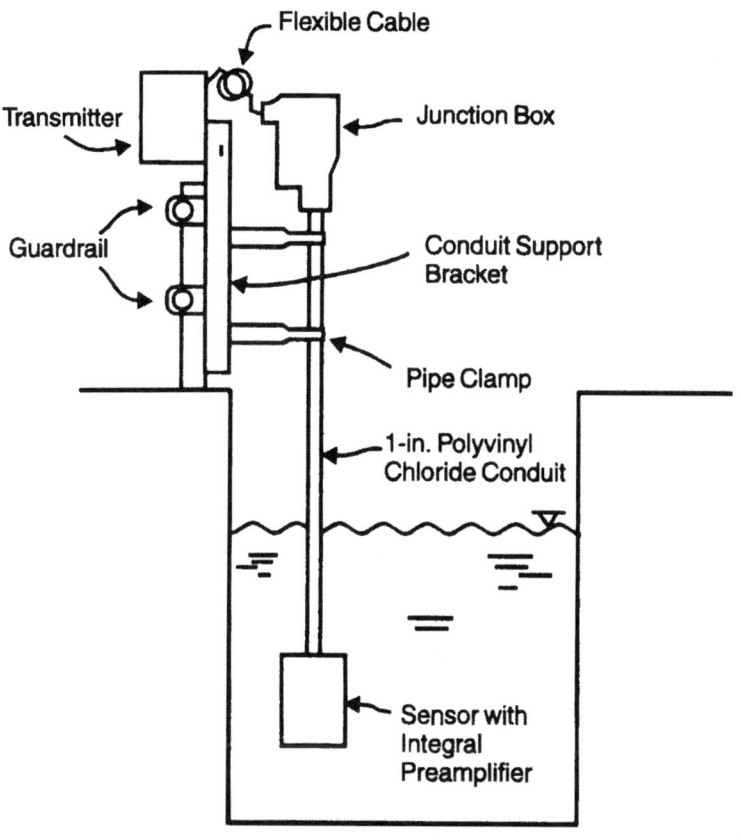

FIGURE 6.44 Submersion-type pH electrode with integral pre-amp (in. × 0.025 4 = m).

A submersion probe installed in a well-mixed zone will provide a representative sample of the process. If the probe is installed in an open channel, it should be in a free-flowing zone. The electrode assembly and support pipe installation should be designed to inhibit collection of debris.

PHOSPHORUS AND ORTHOPHOSPHATE. Most NPDES permits note phosphorus limits in terms of total phosphorus. On-line total phosphorus monitors are available but are relatively expensive and require a lot of maintenance. So, orthophosphate (PO_4) monitors are more widely accepted. An analyzer measuring 3 mg of PO_4 is measuring the equivalent of roughly 1 mg of phosphorus, so an orthophosphate analyzer measuring 0.3 mg/L PO_4 is equivalent to a total phosphorus analyzer measuring 0.1 mg/L of phosphorus. Expressing orthophosphate values as PO_4-P avoids confusion.

Orthophosphate. On-line orthophosphate instruments are used to monitor plant influent for feed-forward control of chemical dosing and flow balancing, aeration basin effluent to ensure that phosphorus uptake was adequate, and final plant effluent as a surrogate for total phosphorus. Orthophosphate instruments also are used to monitor anaerobic effluent to ensure adequate release of phosphorus from phosphorous accumulating organisms (PAOs). In addition, they can be used to monitor and control phosphate release by controlling mixed-liquor nitrate recirculation back to the anaerobic zone.

Operating Principles. All on-line orthophosphate instruments use colorimetry. The two colorimetric methods typically used are the molybdovanadate method (also called the yellow method) and the ascorbic acid method (also called the blue method or the molybdenum blue method). The yellow method is most widely used because of the relative simplicity of the instrument and related reagents. When very low detection levels (less than 0.1 mg/L of PO_4-P) are required, the blue method is used.

Construction Materials. The materials used in orthophosphate analyzers are similar to those in ammonia analyzers.

Accuracy and Precision. Most orthophosphate analyzers using the yellow method can measure a range from 0.1 up to 50 mg/L of PO_4-P. Accuracy and precision typically are reported to be on the order of 2%.

Installation. Both conventional wall-mounted and immersible orthophosphate analyzers are available. Their installation requirements are the same as those for ammonia analyzers.

Maintenance Requirements. The maintenance requirements of orthophosphate analyzers are similar to those for colorimetric ammonia analyzers.

Total Phosphorus. Because of their inherent complexity (and corresponding expense), online total phosphorus analyzers typically are only used to monitor the final effluent at wastewater treatment facilities. Many treatment plants prefer to use orthophosphate analyzers as a surrogate parameter in final effluent.

Operating Principles. On-line total phosphorus instruments first convert polyphosphate and organic phosphorus compounds to orthophosphate. They then measure orthophosphate colorimetrically (typically using the molybdenum blue method). The conversion step requires high pressures and temperatures, which is why on-line total phosphorus instruments are relatively expensive and maintenance-intensive.

Construction Materials. The materials used in total phosphorus instruments are similar to those in orthophospate instruments.

Accuracy and Precision. Total phosphorus analyzers are available with measuring ranges from as low as 0.01 mg/L to as high as 5 mg/L of total phosphorus. Their accuracy and precision are typically 2% or less.

Installation. The installation requirements of a total phosphorus analyzer are similar to those of a conventional ammonia analyzer, except the sample must not be filtered. Wastewater solids contain organic phosphorus, so they must be delivered to the analyzer for a true total phosphorus reading to be obtained.

Maintenance Requirements. Total phosphorus analyzers require a relatively high degree of maintenance because of the oxidation stage. Reagents must be replaced as necessary, depending on measurement interval.

ON-LINE SOLIDS MEASUREMENT. On-line solids analyzers may be installed in most wastewater treatment processes, as long as engineers consider the nature of the solids (e.g., concentration, color, and variability) and the site (e.g., basin, channel, or pipe) before specifying an instrument (Table 6.11). Currently, most solids instruments use near-infrared technologies, although instruments based on microwave, nuclear, or ultrasonic technologies are also available.

On-line solids analyzers work by developing an electronic correlation between a sludge's characteristics and actual solids concentrations. This correlation is established during calibration, which compensates for some variation in solids characteristics. However, if the solids characteristics vary significantly, so do the correlation constants, and recalibration typically will be required. Most optic-based instruments,

TABLE 6.11 Treatment processes and the range of solids concentrations that are observable.

Process	Solids concentration	Possible interference	Purpose[a,b]
Plant influent	200 to 1000 mg/L	Grease, rags, color change	Monitor solids load to plant
Primary sludge	10 000 to 100 000 mg/L	Grease, rags, color change	Optimize sludge pumping, monitor digester loading
Mixed-liquor suspended solids	400 to 8000 mg/L	Attached growth	Monitor pounds under aeration, F:M, MCRT
RAS/WAS	400 to 30 000 mg/L	Attached growth	Optimize sludge pumping, F:M, MCRT, control secondary solids
Secondary effluent	1 to 1000 mg/L	Attached growth	Monitor compliance
Thickening/dewatering feed	400 to 30 000 mg/L or 10 000 to 100 000 mg/L	Grease, rags, color change	Control and optimize loading or polymer dose
Thickening/dewatering liquid discharge (underflow, filtrate, centrate)	50 to 10 000 mg/L	Grease, rags, color change, air	Optimize polymer dose, monitor recycle streams
Thickening/dewatering solids discharge (underflow, filtrate, centrate)	3 to 30%	High solids, abrasions	Optimize polymer dose, monitor loading or moisture content to next process

F:M = food-to-microorganism ratio.
MCRT = mean cell residence time.

for example, are sensitive to size distribution and major color changes. Other instruments may be sensitive to air bubbles, conductivity, or precipitate buildup.

This chapter describes near-infrared and microwave analyzers. Nuclear and ultrasonic solids analyzers have been used at wastewater treatment plants but are rare and not discussed here.

Operating Principles. *Near-Infrared Analyzers.* Optical suspended solids analyzers typically consist of a controller or interface device that communicates with a sensing head that is either inserted into a pipe or suspended in a process tank. There are two types of optical solids analyzers developed for use in the wastewater treatment industry: transmittance (two-beam and four-beam alternating light) and reflectance (single- and multiple-angle backscatter). Transmittance analyzers are suitable for light colored (brown) and light concentrated sludge (typically RAS, WAS, or less). The transmittance meter measures the relative loss of light and converts the signal to a solids concentration. The four-beam version uses two emitters and two detectors, creating an array of light paths that produce a matrix of results. A ratiometric comparison of these signals produces a reading that effectively eliminates interferences, compensates for fouling, and allows measurement in harsh environments. In addition, the analyzers may have an integral purge port that allows solenoid-activated air or water pulses to clean deposits from the sensing head at user-programmed intervals.

Reflectance analyzers can accurately measure a wider range and up to 10% solids concentration depending on sludge type. The multiple-angle version consists of a light source and two detectors positioned with their optical axes at 90 degrees and 140 degrees from each other (Figure 6.45 and Figure 6.46). The emitted light is backscattered by solids particles and received by the detectors. The detectors have filters to reduce interferences from ambient light sources and compensate for color changes. The analyzers typically use a mechanical device (e.g., a wiper) to keep the light source and detector windows clean. Multiple outputs from the detectors are used by the instrument processor (transmitter) to determine solids concentration.

All optical instruments are sensitive to environmental conditions that may interfere with output over time. Possible problems include interference from entrained air, grease or fiber accumulation, changes in water or sludge color, and changes in sludge particle size. Most optical instruments have internal software to compensate for sensor head fouling, or changes in sludge color or size, but some manufacturers are better at it than others.

Microwave Analyzers. On-line microwave analyzers typically consist of a 400-mm-long spool piece with 68 kg (150-lb) ANSI flanges. They measure total solids based on the measured phase difference of microwaves passing through the sludge. They operate on the principle that when microwaves go through a substance, the physical density can be determined by measuring the phase lag of the waves (Figure 6.47). The phase difference correlates to density, which can be directly related to the total

208 Automation of Wastewater Treatment Facilities

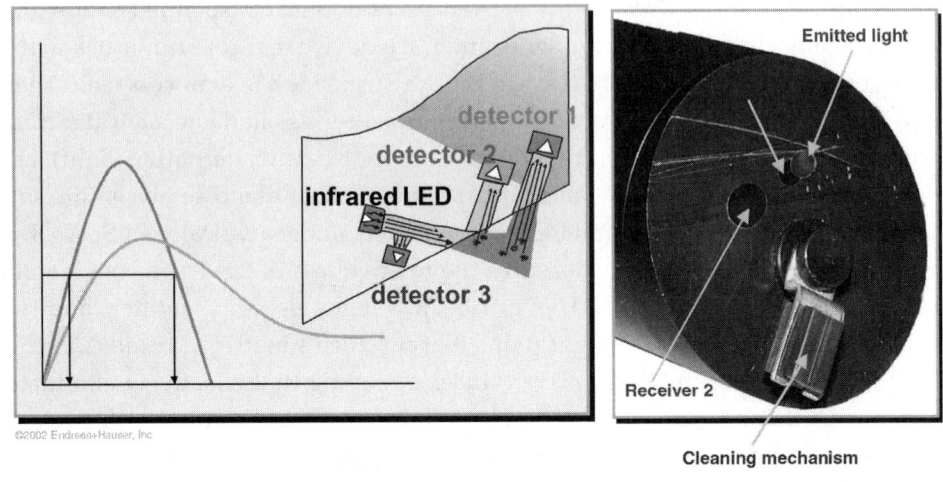

FIGURE 6.45 Illustration of how backscattered light principle instrument works. (Courtesy of Endress + Hauser, Greenwood, Indiana.)

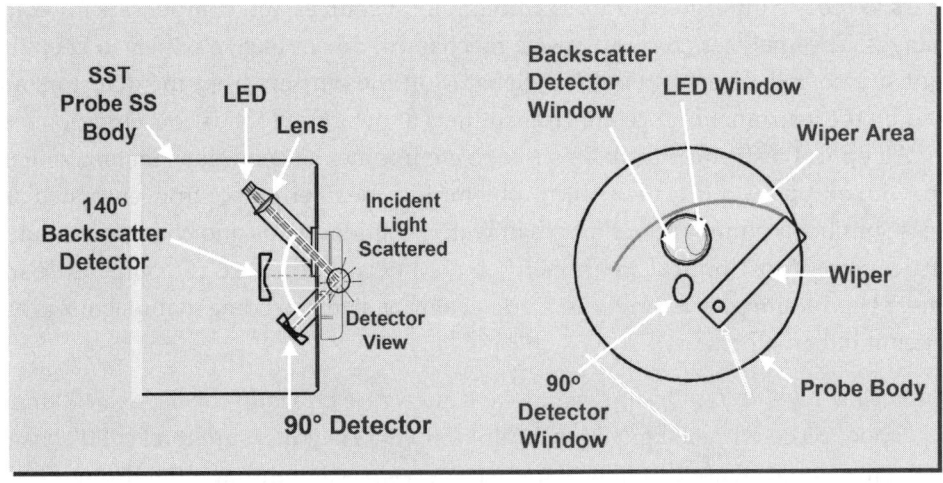

FIGURE 6.46 Illustration of how backscattered light principle instrument works. (Courtesy of Hach Company, Loveland, Colorado.)

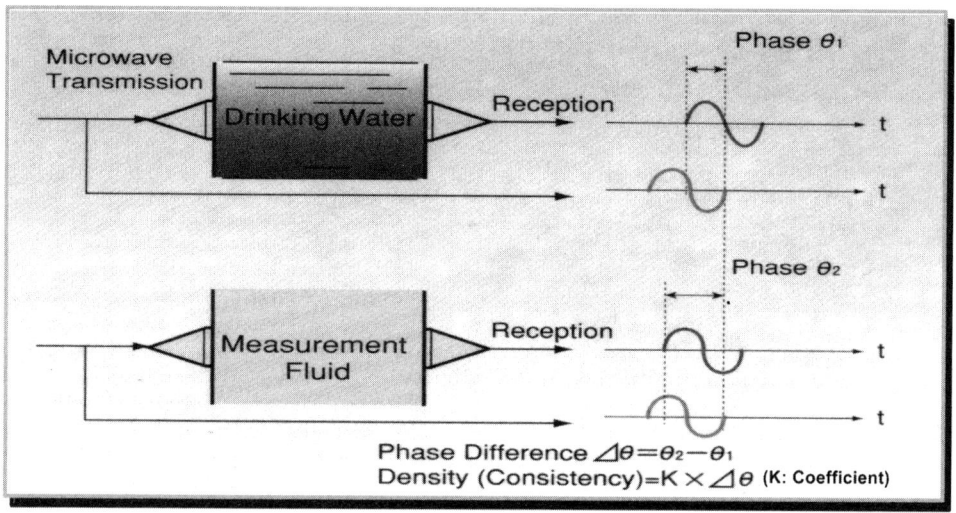

FIGURE 6.47 An example of how microwave phase difference is measured. (Courtesy of Toshiba International Corporation, Houston, Texas.)

solids concentration after compensation for ambient conditions. The transmitter is a microprocessor-controlled unit that displays the total solids reading in terms of percent. It provides both digital and analog outputs.

On-line microwave analyzers are very linear and easy to calibrate. Correction factors for span, conductivity, and the slope and intercept of the solids density are used to calibrate the instrument, using grab-sample percent solids results. However, on-line microwave analyzers work well even without a formal calibration.

The analyzers do not seem to be sensitive to upstream or downstream fittings. Color and color changes do not affect the reading. However, the analyzers are sensitive to conductivity, changes in conductivity, changes in temperature, air entrainment, and partially filled pipes. To prevent gas bubbles from interfering with the measurement, the sensing element must be full and under pressure at all times. On-board instrumentation can compensate for many of these parameters, but engineers should advise the manufacturer of process characteristics before installation.

Construction Materials. The instrument head of an optical analyzer is typically made of plastic or stainless steel. The lens may be constructed from epoxy or a scratch-resistant glass (e.g., quartz or sapphire) and should resist grease accumulation.

On-line microwave analyzers are available in various pipe sizes. The instrument is made of 316 stainless steel and can operate in up to 10 bar (150 psi). Teflon (PFA) coatings are available for applications with potential fouling problems.

Accuracy and Precision. The accuracy of an optical analyzer is typically 2% of reading (± 0.2 g/L), and several ranges of operation are available. Its precision is typically ± 1% of reading or ± 0.1 g/L, whichever is greater.

On-line microwave analyzers can measure 1 to 50% solids. Precision is ± 0.01% of the current reading above 1% total solids.

Installation. Optical suspended solids analyzers may be installed as submersible probes. Various devices are available to attach them to concrete walls or handrails. Most manufacturers also have an insertion assembly so the probe can be installed in a pressurized pipe. The insertion assembly allows the meter to be installed and removed from service for maintenance while the process pipe is in service and pressurized. However, the meters should not be installed in pipes with more than 276 kPa (40 psi) because of the force required to reinstall the meter after normal maintenance.

The meters should be installed near a work platform for easy access. Sample taps (to collect samples to verify instrument readings) should be installed near the instrument.

The on-line microwave analyzer is a flow-through spool piece. It should be installed so it is full at all times. However, if full pipes cannot be guaranteed, the instrument can be externally synchronized to process equipment. The instrument is not sensitive to upstream or downstream fittings, and because it is fully ported, it can be installed next to a magnetic flow meter.

Gradual precipitation or other buildup in the meter may degrade accuracy, so a bypass around the meter (similar to a magnetic flow meter) should be installed to allow for infrequent cleanings. The analyzer does not come with a cleaning mechanism.

Maintenance Requirements. All optical instruments will need periodic cleaning and occasional recalibration to maintain an accurate output. The actual cleaning and recalibration interval depends on site-specific conditions. Typically, a sensing head only needs to be wiped with a rag once a week or less. Cleaning may include removing the probe from the process and washing it with a mild detergent or possibly a mild acid or caustic to remove solids buildup. Daily or weekly samples may be collected to verify instrument output.

Suspended solids analyzers with effective cleaning devices require little operator intervention. Analyzers with air purging systems effectively remove debris (e.g., rags or plastic wraps) that may block the sensing elements. Operators should assess the fouling characteristics at their plant and implement appropriate maintenance schedules.

Calibration requirements vary by manufacturer and installation conditions. Analyzers may only need calibration after 6 or 12 months of operation if their measurements are independent of solids color, size, and particle shape, and they have integrated cleaning devices. Those with minimal color compensation, however, need to be calibrated every time the particles change. Typically, optical meters need a seasonal recalibration, so the instrument can continue to produce an accurate output. A three-point calibration is typically recommended.

On-line microwave analyzers require little maintenance. In grease or precipitate-depositing environments, bypass piping will allow the unit to be removed from service and wiped out, when necessary.

INTERFACE/SLUDGE BLANKET LEVEL ANALYZERS. Traditionally, operators measured sludge blanket depths manually by inserting a core sampler into the clarifier. The core sampler consists of several interchangeable 1.5-m (5-ft) sections of clear PVC tube with markings 0.3 m (1 ft) apart and a bottom 1.5-m (5-ft) section with a foot valve (ball valve) that is seated by the water's static head (Figure 6.48). When assembled, the core sampler should be 0.3 m (1 ft) longer than the clarifier depth at the measurement point. An operator gently lowers the sampler vertically down to the clarifier floor, allowing the displaced water column (including the sludge blanket) to fill the sampler through the unseated valve. The top of the sampler must not be submerged when its bottom is resting on the clarifier floor or the core sampler will underestimate the sludge depth. Once the sampler touches the bottom, the operator lifts it back out vertically, and the static head of water in the sampler seats the ball in the valve. The operator then "reads" the blanket height by looking for the point on the tube between the sludge and the clearer water above it. The height of the sludge blanket typically is estimated to the closest 152 m (6 in.).

An interface/sludge blanket level (ISBL) analyzer (sludge blanket detector) automatically identifies and reports the depth of the sludge blanket in a clarifier. It can replace the manual core sampler, and more importantly, eliminate the variability of the manual readings.

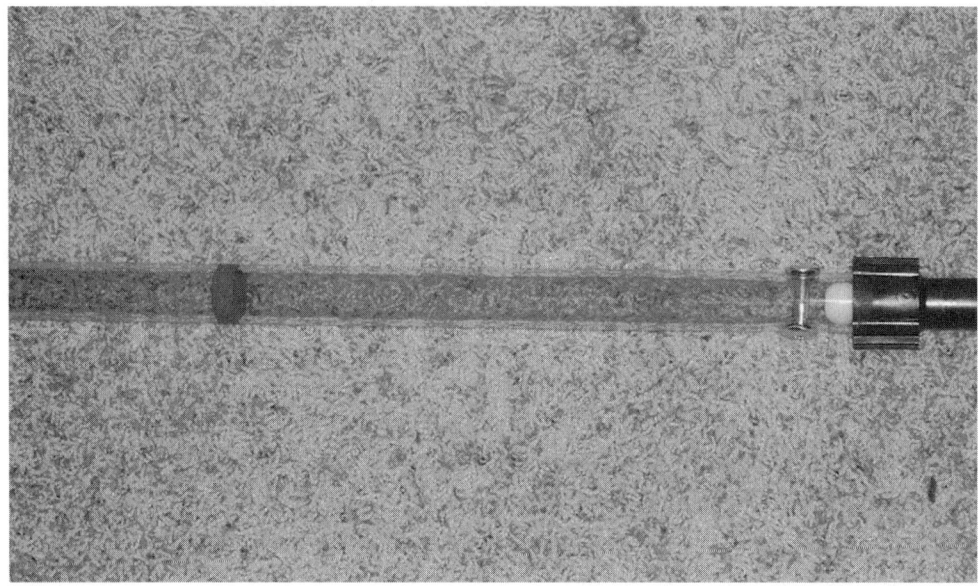

FIGURE 6.48 Sludge core sampler.
(Photo courtesy of Carollo Engineers, Sacramento, California.)

Operating Principles. Currently, there are two types of automatic ISBL analyzers available: ultrasonic and optical.

Ultrasonic ISBL Analyzers. Ultrasonic ISBL analyzers are similar to the level sensors used in tanks or channels. In other applications, the sound waves travel through air and bounce off the top of the liquid. In clarifiers, the sound waves travel through water and bounce off the top of the sludge blanket.

The electronics typically are rail-mounted over the point where the blanket needs to be measured. The sensor is mounted on a rigid or hinged mount with its face submerged, and the distance between the sensor face and the tank floor is measured immediately after installation and keyed into the analyzer. The blanket height is calculated as the difference between this distance and the echo depth to the top of the blanket, similar to a level transmitter.

Optical ISBL Analyzers. Optical ISBL analyzers are similar to turbidity-suspended solids sensors. They basically function by transmitting two or more beams of light across a gap and calculating how much light was absorbed by the media in the gap. In a clarifier, the sensor typically samples the medium continuously while traveling past the sludge blanket interface. The interface is the transition point in the absorbance values.

The electronics typically are placed in a rail-mounted cabinet over the point where the blanket needs to be measured. Ideally, this housing should be large enough to accommodate both the electronics to analyze the data and a retractable motorized reel to lower and raise the sensor through the sludge. If site constraints do not allow this, then the analyzer electronics must be housed elsewhere.

The sensor is mounted on the end of a retractable cable. It samples the wastewater and sludge continuously as it is lowered to a point just above the tank floor. Once it reaches the bottom, the cable and sensor are retracted. The distance between the fully retracted sensor face and the tank floor is measured immediately after installation and keyed into the analyzer. The blanket height is calculated as the difference between this distance and the depth at which the light absorbance changes.

Applications. An interface/sludge blanket level analyzer typically is used to measure the height of a variable sludge blanket, develop a suspended solids profile in addition to measuring blanket height, and sound an alarm when a sludge blanket reaches a predetermined depth. The choice of technology depends on the use and the plant's control strategy (Table 6.12).

Variable Blanket Height. This is the typical application at wastewater treatment plants. It allows process decisions to be based on actual blanket depths and whether the blanket is rising or falling. The choice of technology will depend on sampling frequency. If continuous feedback is important, an optical device that has to be lowered into the water is not suitable. It is suitable, however, if the sampling frequency is once every 30 minutes.

Variable Blanket Height and Suspended Solids Profile. When both the sludge blanket height and the suspended solids profile are desired, an optical unit is the only suitable technology. Sometimes, history shows that a clear interface will not occur at the

TABLE 6.12 Which ISBL analyzer to use based on installation classification and application.

Installation classifications	Application classifications		
	Height of variable blanket	Height of variable blanket and suspended solids profile	Alarm on reaching fixed blanket height
In-line with skimmer/flights	Ultrasonic Optical	Optical	Ultrasonic Optical
Off-line from skimmer/flights	Ultrasonic Optical	Optical	Ultrasonic Optical
Fixed submerged	Not Possible	Not Possible	Optical

top of the blanket; instead, a diffused layer is typical. If so, a solids profile throughout the entire depth of the clarifier is useful information. Even though the light absorbance can be calibrated to a suspended solids value, data available from relative light absorbance at different depths (without a suspended solids value) is equally helpful.

Blanket Height Alarm. This option is good when history shows that a plant's sludge blanket is always below a certain threshold, with minimal variability, and personnel are only interested in knowing when the threshold is exceeded. An optical analyzer is the only option available for this application. The analyzer is installed at a fixed location in the tank and is set to alarm when the absorbance indicates that the blanket interface is at that level. Changes in light absorbance can also indicate whether the blanket is rising or falling, but the actual blanket height cannot be determined except at that point.

A variation of this option is installing two or more sensors at different depths to bracket the blanket.

Sensors

Construction Materials. Ultrasonic ISBL analyzers consist of a probe with an ultrasonic transducer to transmit and receive signals, a probe mounting arm, and an electronics module to receive, transmit, and process the signal. Optical ISBL analyzers consist of a probe with a light source to transmit and receive signals; a retractable reel and cable; an electronics module to receive, transmit, and process the signal; and interlocks between the reel and clarifier mechanisms. The interlock prevents the reel from lowering the instrument head when the clarifier equipment is in the vicinity.

The construction materials typically available for various components are presented in Table 6.13. The choice of construction materials depends on the reactivity of the environment and the monitored substance. However, the construction materials matter less than the mounting assembly, according to Instrumentation Testing Association (ITA) research (ITA, 2002). Researchers found that the weak point was the design of the pivot for the probe mounting arm.

TABLE 6.13 Typical construction materials for ISBL analyzers.

Part	Material
ISBL probe body (wetted part)	• Chlorinated polyvinyl chloride (CPVC) • Polyvinyl chloride (PVC) • Polyoxymethylene (POM) • Ryton • 316L stainless steel
Electronics module enclosure	• Aluminum • 316L stainless steel • Polycarbonate
Probe mounting arm (ultrasonic)	• Aluminum • 316L stainless steel • Galvanized steel • Chlorinated polyvinyl chloride (CPVC) • Polyvinyl chloride (PVC)
Retractable cable (optical)	• 316L stainless steel

Accuracy and Precision. At best, manual core samplers are accurate to within 76 mm (3 in.), and for most plants, an accuracy of 152 mm (6 in.) is acceptable. Precision is the critical criteria.

On-line ISBL analyzers are reported to be accurate within 2.5 to 38 mm (0.1 to 1.5 in.). Manufacturers report the analyzers' precision to be in the same range as accuracy. However, according to the ITA (ITA, 2002), not all analyzers were consistently accurate and precise. Users are cautioned against relying solely on vendor information for these data and strongly urged to perform long-term tests or evaluate independent reports and prequalify or preselect manufacturers for their applications.

Also, when ultrasonic ISBL analyzers are installed in-line with scum skimmers and surface flights, they are hinge-mounted and designed to swing out of the way of the skimmers and flights. Because these are continuous reading devices, signal averaging is important to avoid the output being skewed by the readings taken when the sensor was being moved out of the way.

Installation. Installation of ISBL analyzers can be divided into three classifications: in the path of skimmer arm or surface flights; out of the path of skimmer arm or surface flights; and a fixed submerged point out of the path of skimmer arm, sludge collectors, and surface flights.

In-line with Skimmers or Flights. The on-line analyzer should be installed at the point where the core sampler traditionally was used. In a circular clarifier, this would be a point along the bridge that is easily accessible and typically in the path of a surface scum skimmer. In a rectangular clarifier, this would be a point on the bridge typically over the cross collector and in the clarifier flight's path. The probes are mounted on a two-piece mounting pole that is hinged in the middle so the unit can swing out of the way of the flights or skimmers.

The optical units that must be lowered into the water could become snagged in the skimmer arm or flights. To avoid this, manufacturers typically either

- Interlock the reel mechanism with the clarifier drive or chain-and-flight drive so the skimmer or flight cannot move while the probe is lowered and retracted; or

- If there is enough time to lower and raise the probe after the skimmer or flight has passed, a limit switch is tripped when the skimmer arm passes underneath the ISBL analyzer, triggering the reel on which the optical probe is mounted.

Off-Line with Skimmers or Flights. This mounting location is not a physical challenge to ISBL analyzer operations. In a circular clarifier, this location would be the annular ring between an in-board launder that is mounted away from the clarifier wall. In a rectangular clarifier, this location would be over the cross collector sump. While minimizing the skimmer or flight interference problem, neither location is ideal for measuring sludge blankets.

Fixed Submerged Location. Only optical units can be adapted for this installation. The sensor is wall-mounted below the water surface at a fixed elevation. It provides continuous feedback on light absorbance. This installation eliminates the reel mechanism, but requires the tank to be drained when personnel need to access the probe for cleaning and maintenance. An alternative is installing the probe at a fixed elevation in an off-line area and mounting it on a support that can be retrieved from the top without draining the tank.

No matter which installation is undertaken, the electronics module is typically housed in a NEMA 12 enclosure. The enclosure should be chosen based on the operating environment. If operators will take local readings, a sun shield would be helpful.

Maintenance Requirements. The sensing element of the ISBL analyzer may need periodic cleaning, depending on the plant and the wastewater involved. Rags and plastics will interfere with sensor functions, as will grease and scum. While some sensors come equipped with features to clean the sensor head, ITA researchers did not find a difference attributable to the cleaning feature (ITA, 2002). The clean sensors were attributed by the manufacturers to the constant pivoting motion of the sensor, according to the ITA test report. As each flight passed the sensor, the sensor pivoted out of the water and then fell back in after the flight had passed. This motion served to clean off the sensor face.

The probe mounting arm was the most maintenance-intensive piece of equipment on the ISBL analyzers evaluated by the ITA. In a pivoting mount, the lower arm attached to the sensor has to be fastened to the upper arm with a chain or another device. In at least half the installations tested, the lower arm fell off after repeated use over a 3-month period.

Retractable reels on an optical sensor need to be serviced periodically to ensure smooth operation. In an interlocked system, a jammed reel mechanism will keep the interlocks in place and prevent the drive from operating. In a limit-switch operated system, reel failure will result in the extended cable being wrapped around the skimmer and the sensor being damaged or lost.

VIBRATION. Vibration sensors are used to monitor the mechanical condition of pumps, motors, compressors, blowers, centrifuges, and other rotating and reciprocating devices. They can detect gradual or sudden mechanical failures, as well as abnormal process conditions that are stressing the machine.

Vibration sensors typically consist of two or three main components: a transducer (sensor), signal conditioners, and a computer and software (if necessary). The transducers are mounted on the case or inside the monitored machine, typically as close as possible to the anticipated source of vibration.

Signal conditioners convert transducer output to analog or digital signals. The conditioner could be

- a portable data collector that operators use to manually collect vibration samples at regular intervals,

- a permanently installed continuous-protection system that provides for unattended alarms and shutdowns, or

- a computer-based SCADA system with expert software that automatically detects and reports machine problems.

Whichever conditioner is used, its frequency sensitivity should match that of the transducer.

The choice of system depends on several factors, including personnel and plant safety, the monitored machine's repair or replacement cost, the severity or criticality of the machine's service, the plant's O&M philosophies, integration with other control-system components, available staff, and budget.

Operating Principles. There are three types of vibration transducers: velocity, acceleration, and displacement. Their outputs are oscillating voltage or current signals that are proportional to the vibration's velocity, acceleration, or displacement characteristics. The choice of transducer depends on the monitored machine's construction, the speed of its moving components, the mechanical conditions to be detected, and the conditions' vibration characteristics.

Velocity Transducers. A velocity transducer measures a machine case's absolute vibration (i.e., its movement relative to free space). It consists of a magnet and a bobbin of coiled wire. The magnet is affixed to the transducer case and is surrounded by the bobbin, which is suspended by springs inside the case. As the magnet vibrates inside the relatively fixed bobbin (an inertial reference), an electromotive force is generated in the coiled wire. The output voltage is linearly proportional to the velocity (within the transducer's amplitude limits).

The transducer's sensitivity is the ratio of millivolts (mV) to inches per second (in./s) or millimeters per second (mm/s). Velocity transducers respond well to low and mid-range frequencies (typically 1.5 to 1000 Hz), so they work well when monitoring low- and medium-speed (up to 1750 rpm) general-purpose machines. They are less sensitive at the lower and upper ends of their frequency range, and are susceptible to phase errors at very low frequencies.

A velocity transducer's advantages are external installation and a self-generating signal; its disadvantages are sensitivity to mounting angle and moving parts that can wear out. To deal with the moving parts problem, a piezoelectric version has been developed. Constructed as an accelerometer, it has additional internal electronics that convert the acceleration measurement into an equivalent velocity.

Acceleration Transducers. Like velocity transducers, acceleration transducers (accelerometers) measure casing vibration. It consists of a piezoelectric crystal sandwiched between a metal mass (inertial reference) and the transducer case by an internal preload bolt. The compression force on the crystal varies sympathetically with the machine vibration, and the crystal generates a picocoulomb charge proportional to the rate of change of this force.

The accelerometers typically used in water and wastewater treatment plants are "voltage-mode" type with built-in impedance-matching electronics that convert the charge to a voltage output that is linearly proportional to the acceleration (within the transducer's amplitude limits). This transducer's sensitivity is the ratio of millivolts (mV) to g's or millimeters per second squared (mm/s^2), and 100 mV/g is an industry standard.

Another type of accelerometer uses piezo-resistive technology. In this design, beams of piezo-resistive material suspend a mass in the transducer case. To keep the mass relatively fixed, the beams flex with varying degrees of resistance, and internal electronics convert that resistance into an acceleration output signal that corresponds to the vibration.

Acceleration transducers respond well to medium- to high-range frequencies (10 to 15 000 Hz). In fact, their sensitivity increases drastically above the upper end of their useful frequency range, ultimately peaking at the resonant frequency of the mass-and-crystal system. They typically are used on medium- and high-speed machines (3600 rpm and above), and those with ball bearings and gearboxes, because they emphasize high-frequency vibrations.

An acceleration transducer's advantages are external installation and a broad frequency response; its disadvantages are occasional sensitivity to electrical noise and spurious vibrations. Filtering may be needed to eliminate unwanted lower or higher frequencies. In some applications (e.g., high-temperature machinery), a separate "interface module" may be used to amplify the signal and filter it away from the transducer.

Displacement Transducers. Displacement transducers (proximity probes) measure the position of a metal surface—usually the shaft of the machine being monitored—relative to the machine's case. They are considered noncontact devices because the sensor measures the distance (gap) between its tip and the metal surface.

Displacement transducers typically consist of three components: the probe, which is mounted where it can observe the shaft; a coaxial extension cable; and an oscillator-demodulator (often called a driver). These components are designed and manufactured (within interchangeability tolerances) to operate together as a tuned system with a fixed electrical length (typically 5 or 9 m—the cable's length). The driver generates a high-frequency signal that is sent through the cable to the probe. A coil in the tip of the probe turns this signal into an electromagnetic field. When this field encounters the conductive material of the machine shaft, small electric currents (eddy currents) are generated on the shaft surface, with resulting energy losses. The driver senses the magnitude of these losses and produces a voltage output. Within the transducer's linear range, the output voltage is proportional to the instantaneous gap. the transducer's scale factor is the ratio of millivolts (mV) to the probe gap in mils or millimeters (mm), and the industry standard is 7.87 mV/μm (200 mV/mil).

Displacement transducers respond well to a wide range of frequencies (0 to 10 000 Hz); the industry standard is 3dB down point, but transducers with higher or lower frequency ranges are available. They typically are used to measure a shaft's radial vibration and its axial position on machines with fluid-film bearings. They can also be useful when a machine's vibration is significantly attenuated before it reaches the machine case or is masked by other sources of (internal and external) vibrations.

The displacement transducer's chief advantage is its lack of moving parts; its disadvantages are the extra installation work and its sensitivity to mechanical and electrical discontinuities on the observed surface.

Construction Materials. Velocity transducer cases are available in aluminum and stainless steel, depending on application. Piezoelectric transducers and accelerometers typically have hermetically sealed 304L or 316L stainless steel cases. The transducers typically include threaded or quarter-turn connectors with two or three conductors at the top of the device. Other connector options [e.g., integral cables (no connector), miniature compression-type terminal blocks, and threads for connecting directly to conduit] are available.

The cables typically have a two or three-pin MIL-C-5015 or other plug to ensure that the transducer–cable connection is reliable and weatherproof. If splashing water or other hazards are expected, a Fluorosilicone elastomer "boot" may be available to provide more environmental protection. The wiring itself is typically 18 to 22 AWG (American Wire Gauge) shielded multiconductor cable with an overall jacket, although flexible stainless steel armor is often available if the cable will not be run in conduit or will be susceptible to mechanical damage.

Displacement transducers typically have 303 or 304 stainless steel threaded or nonthreaded bodies, with polyphenylene sulfide (PPS) plastic tips that encase the coil. Some have an aluminum body for the oscillator-demodulator electronics that also contains a potting compound to provide moisture and condensation resistance.

All three types of transducers are available in versions that carry agency (UL, etc.) approvals for installation in hazardous or potentially explosive atmospheres.

Accuracy and Precision. Between 20 and 30 °C (68 and 86 °F), the sensitivity of velocity transducers and accelerometers typically is accurate within ± 5%. Their amplitude linearity error—the variance in sensitivity over the design amplitude range—is typically within ± 1%.

Velocity transducers also have phase errors at low frequencies. However, their *cross-axis (transverse) sensitivity*—sensitivity to vibrations perpendicular to the principal axis of sensitivity—should not exceed 5% of their principal axis sensitivity. The *broadband noise floor*—the intrinsic level of apparent vibration caused by the transducer's electronics—should be no more than 0.004 g rms.

Between 0 and 45 °C (32 and 110 °F), the accuracy of displacement transducers is determined by two parameters: incremental scale factor (ISF) error and deviation from straight line (DSL) error. *Incremental scale factor error* is the maximum amount that the scale factor (mV/μm) varies from the design scale factor when measured at specified increments throughout the transducer's range. It is associated with errors in radial vibration readings and should be within ± 5%. *Deviation from straight line error* is the maximum error (in mils) in the gap reading at a given voltage compared to a "best fit" line at the design scale factor. It is associated with errors in axial position or gap readings and should be within ± 1 mil.

Temperature can also affect transducer accuracy. The magnitude of temperature-related errors depends on the manufacturer and type of transducer. Typically, an accelerometer's temperature sensitivity is -14 to +7.5% over its operating temperature range. A displacement transducer's ISF typically varies ± 10%, and its DSL typically varies ± 3 mils over its operating temperature range.

Installation. First, engineers must choose the proper transducer for the application; it should have the appropriate sensitivity and amplitude limits, frequency range, and operating temperature. For example, when selecting a displacement transducer, engineers should consider the following:

- Where it will be installed (inside or outside the monitored machine);
- The vibration frequency range needed;
- The probe's minimum target area requirement, which must be smaller than the surface (shaft) to be observed (a smaller tip diameter does not necessarily correspond to a smaller target area);
- The calibration requirements, which depend on the material being monitored;
- The transducer's thread size, case style, and overall body length; and
- The system's electrical length (and, therefore, the physical length of the cables).

Vibration and accelerator transducers should be mounted as close as possible to and in the most direct path of the expected source of vibrations—not on the end of cantilevered mounting brackets, metal covers, or water jackets or the midspan of flexible members because this may attenuate the vibration or superimpose resonant vibrations on the measurement. They can be flush- or nonflush-mounted on the machine case, depending on whether the expected vibration frequency is higher or lower than 2 kHz, respectively.

Flush-Mount. First, installers drill and tap the mounting location to accommodate the full depth of the threaded mounting stud. The hole should be perpendicular to the mounting surface within ± 5 minutes of an arc. Then, they should finish the mounting surface according to the transducer manufacturer's specifications. The typical requirement is a roughness of 0.8 μm (32 μin.) Ra or better, and a surface flatness below 25 μm (1 mil) TIR. A rough or uneven surface under the transducer can attenuate higher-frequency vibrations or introduce false vibrations to the measurement.

Installers next use the threaded stud (¼-28 UNF-2A typical) supplied by the manufacturer to connect the base of the velocity transducer or accelerometer to the machine case. A light lubricant can be applied to the threads and base of the transducer before installation to ensure good coupling between the adapter and the mounting surface at high frequencies. When tightening the transducer to the mounting surface, installers should meet the manufacturer's torque requirements so the sensor is not damaged and will have the proper sensitivity. (Significant strain on the transducer base, either via improper installation or the mounting surface itself, will distort the measurements. The manufacturer may have a base strain specification.)

Other mounting designs are available, including a three-hole (triangular) pattern in which three screws secure the transducer against the machine case.

Nonflush-Mount. Nonflush-mount transducers do not require a smooth surface, and are a more expedient, versatile option when high-frequency sensitivity is not required. They have an integral ¼-18 NPT (National Pipe Thread) threaded stud on the base. Installers drill and tap a mounting hole with an appropriate NPT thread in the case, and then screw in the transducer according to the manufacturer's maximum torque requirements.

Mounting displacement transducers is a little more challenging. First, installers must inspect the surfaces to be monitored for scratches, roughness, out-of-roundness, plating, or other surface imperfections that could cause electrical and mechanical runout (glitches). The surface finish should be less than 1.0 μm (32 μin.) RMS (root mean square).

Then, installers must determine where to put the probes for optimal operations. Radial probes should be mounted perpendicular to the shaft axis and 90 degrees apart from each other in the same plane. Axial probes should be mounted in pairs perpendicular to the shaft's end or collar (parallel to the shaft centerline). There should be enough clearance around each probe tip and between the probes to ensure measurement accuracy. Probes range from approximately 25 to more than 250 mm long (1 to more than 10 in. long). Special "button" probes are available for tight installations; the probe tip is mounted to a small circular "button" that can be glued to an adjustable bracket. (This may require minor modifications to the machine or brackets.)

The probes can be mounted externally or internally. The preferred method is external mounting via holders that allow a probe to be removed, replaced, or adjusted while the machine is running. This basically involves threading the probe into a hole or bracket and securing it—via flats on the back end of the case—with a lock nut. (Thread sizes typically are ¼-28 UNF and 3/8-24 UNF.) Some probes have flats on the front end of the case (near the tip), so they can be "reverse-mounted" back end first into a threaded holder or tube that positions the probe near the shaft. If the probes must be internally mounted under machine covers or caps, engineers should consider installing "in place" spares as well.

However the probes are mounted, installers should be sure to secure probe leads with clips to prevent the cables from "whipping". They should protect the connectors both electrically and physically by wrapping them with self-fusing Teflon tape or using neoprene covers provided by the manufacturer. They should secure probes and probe holders to solid machine components that will not vibrate. They also should ensure that probe holders do not have natural frequencies that could be excited by machine-generated vibrations.

Maintenance Requirements. All vibration transducers should be physically inspected at regular intervals. Also, their calibration should be checked. The calibration of velocity and acceleration transducers can be checked via a shaker table that subjects the instrument to a known vibration frequency and amplitude.

The calibration of displacement transducers involves a specially designed test kit. The kit includes a fixture that holds the probe so it faces a metal target affixed to a micrometer. The transducer's electrical output signal then is measured throughout its range to check for acceptable ISF and DSL errors. Another fixture holds the probe over a rotating disk with a slanted surface. As the probe is moved across the rotating disk, it will observe a varying gap of known amplitude. Its output signal can then be compared to that amplitude.

REFERENCES

Endress and Hauser Consult AG (1998) *Document TIF 008 3.98*; Reinach, France.

Grant, D.; Dawson, B. (1997) *Isco Open Channel Flow Measurement Handbook*, 5th Ed.; Teledyne Isco Inc.: Lincoln, Nebraska.

Instrumentation Testing Association (2002) *Online Interface/Sludge Blanket Level Analyzers Performance Evaluation Report*; Instrument Testing Association: Henderson, Nevada.

Liptak, B. G. (1999) *Instrument Engineers Handbook*; CRC Press: Boca Raton, Florida.

Chapter 7

Final Control Elements

Control Valves	227	*Piston*	238
Characteristics	228	*Hydraulic*	238
Types	228	Positioners	239
Globe	229	Pumps, Aerators, and Mixers	243
Solenoid	230	Pumps	243
Ball	231	*Displacement Pumps*	243
Diaphragm	231	*Nondisplacement Pumps*	245
Plug	231	Fans, Blowers, and Compressors	245
Butterfly	231	Aerators and Mixers	245
Gate	231	Electric Motors	248
Other	231	Starters	250
Standards	232	Drives	250
Selection Process	232	*Variable-Frequency Drives*	250
Sizing	233	*Wire-Wound Motor Control*	251
Actuators	234	*Variable-Torque Speed Controls*	252
Solenoid	234	Control Signal Interface and Smart Pumps and Drives	253
Electric Motor	235	Other Final Control Devices	254
Pneumatic	237		
Diaphragm	237		

A *final control element* is any device (e.g., pump, valve, heater, or motor) that can change a process. The changes can be discrete (on–off) or variable. When combined with instruments, controllers, and a communication system, final control elements can maintain a process variable within a desired range.

A final control element typically consists of three or four components, and each performs a specific function (Figure 7.1). Traditionally, there were only three functional operations: a signal conditioner (e.g., a valve positioner), an actuator (e.g., hydraulic cylinder), and an end device (e.g., a valve). Digital control systems introduced a fourth operation: the digital signal translator (Figure 7.2). Typically, different manufacturers make all four components, and the failure of any one component can upset the process. So, engineers must understand these components and ensure that they work together seamlessly.

Two-way communication with the final control element is also necessary because the elements both receive commands and provide operational feedback. Traditional systems only provided simple position or speed data, but today's digital devices can provide real-time operating information about the process' "health".

Specifying or repairing today's final control elements requires knowledge of the mechanical, electrical, fluidics, electronics, and digital communications fields. To select the correct valve, pump, or motor for an application, for example, engineers must understand each device's operating characteristics and know how to size it properly. Many control problems can be traced to improper sizing. (This chapter only contains a brief introduction to sizing; detailed sizing characteristics should be obtained from the device's manufacturer.)

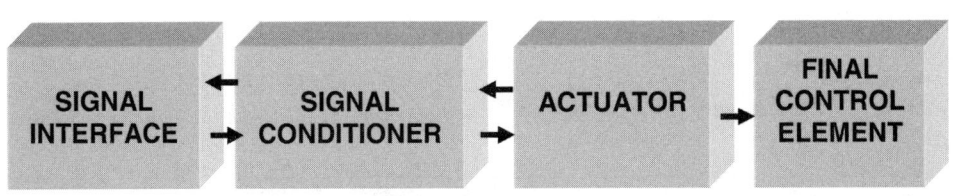

Figure 7.1 Components of a final control element.

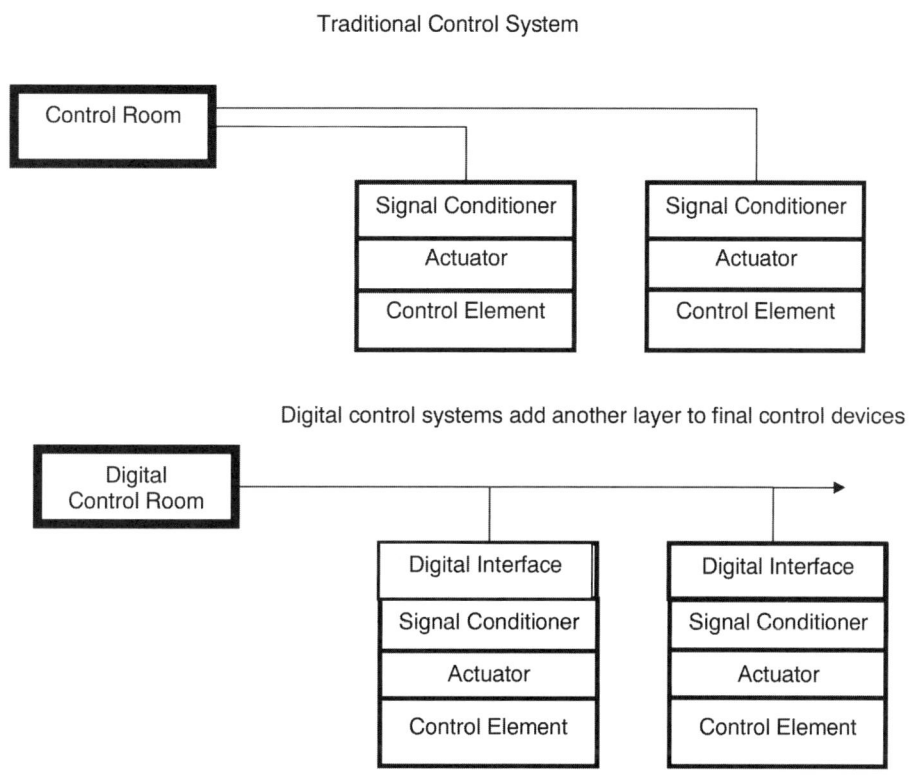

FIGURE 7.2   A comparison of traditional and digital controls.

CONTROL VALVES

A valve is a device that throttles or stops flow via the size of its opening. A control valve works with other components to regulate a process variable (e.g., pressure, level, flow, or temperature).

Valves basically consist of a body, a plug or disc, and a shaft or stem. The valve body holds the process fluid and connects the valve to the piping. It can be made of cast iron, steel, bronze, or polyvinyl chloride (PVC), depending on the fluid to be carried. If the fluid is aggressive or corrosive, a separate body lining made of Teflon (or a similar material) may be used.

Inside the valve body is a plug or disc whose position controls fluid flow. Its movement can be rotational or linear, depending on the type of valve. The plug or disc is moved by a valve shaft or stem, which is connected to the valve actuator. If the valve is designed to stop fluid flow or reduce it to a specific minimal value, both the valve body and the disc or plug will have seating surfaces made of metal or elastomer to ensure that they fit (sit) together tightly. Valves used at wastewater treatment facilities typically must have tight seals.

Valves also include packing or sealing components to prevent fluid from leaking out. In addition, valves typically require bearings so the shaft can move inside the body.

CHARACTERISTICS. When choosing a valve, engineers must consider the *valve characteristics*—the relationship between the valve opening and the flow through the valve (Figure 7.3). The change in flow volume as a valve opens and closes can vary greatly. When a *quick-opening valve* opens, for example, the flow increases rapidly. A *linear valve*, on the other hand, maintains a consistent level of flow. As an *equal-percentage valve* opens, the flow volume increases proportionally.

Most control valves are not designed to be operated near the ends of their operating range. Most valve sizing addresses the flow characteristics at between 20 and 80% of the device's operating range.

Control valves have two sets of characteristics: inherent characteristics and installed characteristics. *Inherent characteristics* describe the valve's theoretical performance at a constant pressure drop. The pressure-drop calculations are typically performed with specific upstream and downstream pipe lengths and connections. Engineers should use these characteristics as an initial basis of comparison among valve types and manufacturers.

Installed characteristics describe a valve's performance in a real application (because valves typically are not installed in the ideal conditions used to calculate inherent characteristics). Proper valve performance also depends on coordination between the choice of valve and the piping layout and design.

TYPES. Valves can reduce or stop flow in many ways. The valves described below are those typically used in the wastewater treatment industry (Figure 7.4). Wastewater applications are typically medium to low pressure [less than 1034 kPa (150

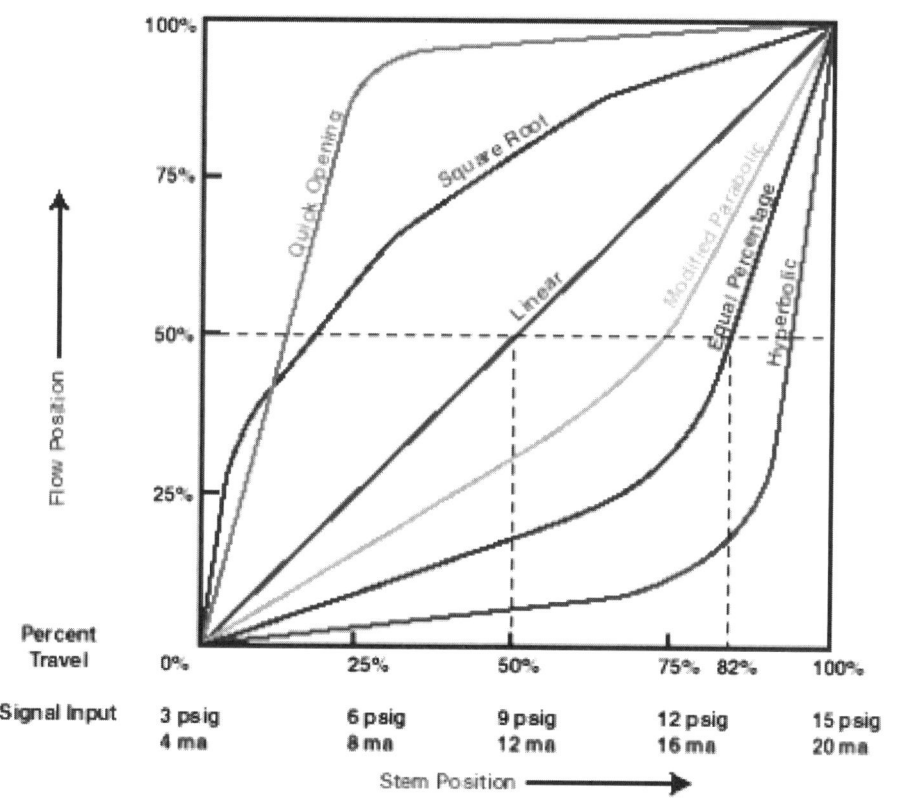

FIGURE 7.3 Control valve characteristics.

psi)], with flow rates less than 6 m/s (20 ft/s). The wastewater contains chemical and biological materials and a few abrasives. Its variables can swing widely and unexpectedly, and control valves must be able to handle such conditions.

Globe. Globe valves, which typically have large ball-shaped bodies with connecting flanges, consist of a plug connected to a vertical stem that is moved by either an external actuator or an internal piston via the process' pressure differential. The fluid

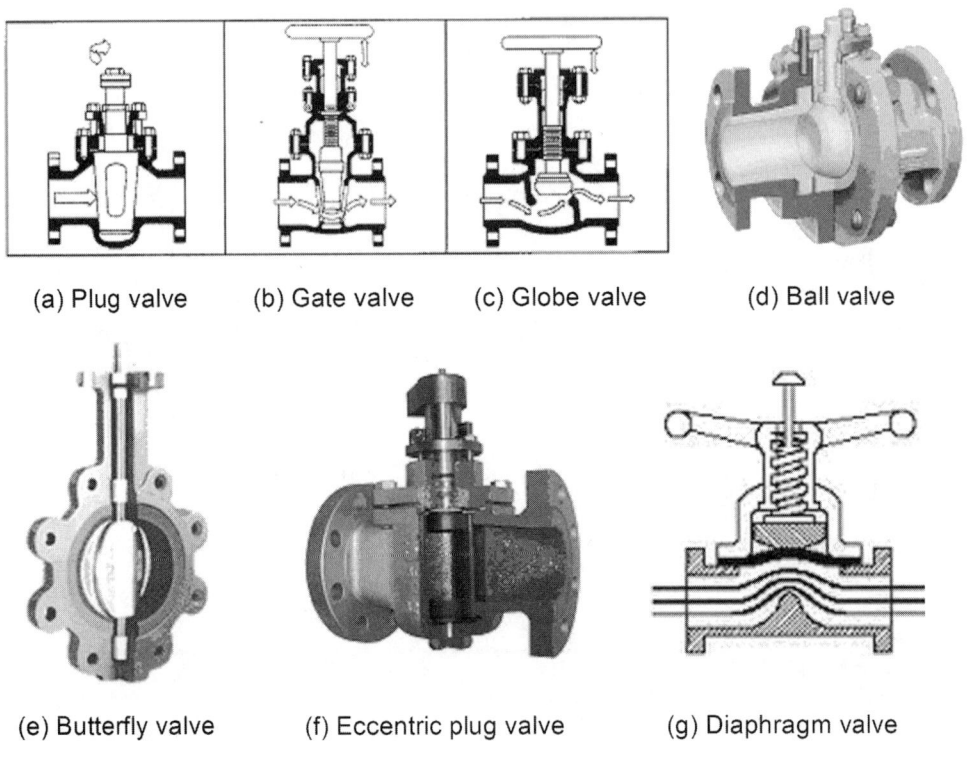

FIGURE 7.4 Types of valves.

flow through the valve depends on the clearance between the plug and its seat. This valve typically can provide tight shutoff and excellent control. It can also operate at high differential pressures. However, because the stem and plug arrangement are exposed, globe valves work best in relatively clean fluids. Also, larger valves typically use the hydraulic pressure difference to help move the plug, and this requires relatively clean fluids to operate properly.

Solenoid. Most solenoid valves are basically globe valves with an integral electrical solenoid designed to move the stem and plug. Smaller solenoid valves use electromagnetic force to close the plug and use a spring to open it. Larger valves also use the extra force provided by the differential pressure inside the valve to move the plug in combination with a solenoid.

Ball. Ball valves are quarter-turn valves. The valve body contains a shaft connected to a ball with an opening that typically is as large as the pipe, but it may be smaller. The shaft rotates the ball back and forth up to 90 degrees, as needed, to control flow through the tunnel.

Diaphragm. Diaphragm valves were designed to control corrosive fluids. Basically, the external valve stem and plug press against a flexible, corrosion-resistant membrane to restrict flow through the valve. These valves are limited by the temperature and pressure limits of their construction materials.

Plug. There are several types of plug valves. Lubricated plug valves, for example, were originally developed to handle gases. They use a replaceable internal lubricant to maintain a seal, so the seating surface can be "renewed" without removing the valve. However, they typically are not suitable for liquids.

Nonlubricating, eccentric plug valves are commonly used in wastewater treatment facilities. The plug rotates 90 degrees inside the valve body. It opens by swinging out of the flow path and can close tightly in various wastes because of the elastomer seating on its face. However, large operating forces are required to rotate the plug, so it may not be the best choice for modulating control applications.

Butterfly. One of the oldest styles, butterfly valves were originally developed as dampers. The body contains a shaft with a disc that typically is the exact size of the valve opening. The shaft rotates the disc 90 degrees to control flows. If used in low to moderate temperatures [typically less than 93 °C (200 °F)], the seating on the disc edge is an elastomer; if used in higher temperatures and pressures, the seating is metal. In higher-pressure applications [typically greater than 1724 kPa (250 psi)], the disc must be slightly offset to provide a better seal. Butterfly valves are used in many wastewater treatment facilities because they are small and relatively inexpensive, but they should not be used in raw wastewater or other fluids with stringy materials.

Gate. In gate valves, the rotating shaft lowers and raises a disc inside the valve body. The seating surfaces can be either elastomer or metal, depending on where they are used. Gate valves typically are used for clean water and on–off applications, not for modulating control applications.

Other. Many valves are variations of the types described above. Pressure-regulating and relief valves, for example, are globe valves that use internal pressure differences to regulate or release pressure. Knife-gate valves use a thin disc rather than the wider disc used in gate valves. Sluice-gate valves use rectangular plates instead of discs. Pinch valves are a variation of diaphragm valves. Needle valves are a variation of gate valves used in tiny lines.

STANDARDS. Currently, there is no design standard specifically for valves used in wastewater treatment. The American Water Works Association has issued standards for valves used in freshwater applications. The American National Standards Institute and the Instrumentation, Systems, and Automation Society have issued standards for control valves used in typical industrial applications. Before using a valve standard, engineers should make sure it is appropriate for the fluid to be controlled.

SELECTION PROCESS. Although all valves can change pressure and flow, some are better suited for control applications. Most of the information on control valves was developed for high-pressure industrial applications. However, most control applications in wastewater treatment facilities involve low pressure [less than 1034 kPa (150 psi)] and low flows [less than 6 m/s (20 ft/sec)]. Also, most wastewater includes abrasive and stringy materials that would be detrimental to any valve with small internal passages.

Before selecting a control valve, engineers should determine precisely where and how it will be used. Cost typically is also a consideration. The following are general guidelines:

1. Define the substance (e.g., liquid, gas, or suspended solids) to be controlled.
2. Define the substance's characteristics.
   a. Chemical composition (e.g., water, chemical, or wastewater), which will dictate the valve's construction materials.
   b. Minimum and maximum pressure (many control valves rely on pressure to operate and will perform poorly if the pressure is too low).
   c. Minimum and maximum flows.
   d. Minimum and maximum temperatures.
   e. Size of piping to be used (some valves are only available in certain sizes).
   f. Type of connections needed (e.g., flanged, screw end, or mechanical joint).

3. Define how the valve is expected to operate.
   a. Modulating (continuous movement over the complete range),
   b. Throttling (movement to several selected midpoints between open and closed), or
   c. Full open and full close only.
4. Define the required valve characteristics.
5. Determine which type of valve best meets these requirements.

SIZING. For the best pressure and flow control, engineers should select a valve size that will operate well in the range that the control system requires. Ideally, the valve should operate in the middle of its range. A valve that is too large may have to operate near its fully shut position, which may cause excessive valve wear and poor operations. A valve that is too small may have to operate near its fully open position, which eliminates much of its ability to control flows.

Valve sizing formulas are designed to match control-system requirements with valve characteristics. In the equations, valves may be characterized by a *flow coefficient* (C_V)—water flow rate through a fully open valve [L/min (gpm)] that will produce a pressure drop of 7 kPa (1 psi). Valve manufacturers provide C_V ratings using the maximum allowable pressure drop at a given angle of opening. Engineers should then select the valve with a C_V just below the calculated C_V.

The acceptable pressure drop across a valve depends on the type of valve chosen. Globe valves, for example, can be designed to handle the pressure drop of their full pressure rating. Butterfly valves typically can only handle a drop of about 10 to 25% of their rated pressure. Because of the cost difference between butterfly and globe valves, some systems use multiple butterfly valves to incrementally drop pressure.

The sizing equations below are for those fluids with viscosities close to that of water. Engineers should use different sizing equations for liquids whose viscosity, temperature, flow, and pressures are very different from those of water. Consult the valve manufacturer for the best control valve standard.

$$C_v = Q_L \sqrt{G_L / \Delta P} \qquad (7.1)$$

$$\Delta P = (G_L Q_L^2) / C_v^2 \qquad (7.2)$$

Where

C_v = flow coefficient for a particular valve normally in the full open position = flow of water through the valve in gpm when pressure drop through the valve is 1 psi;
Q_L = flow (gpm);
ΔP = pressure drop, $P_1 - P_2$ (psi): P_1 = inlet pressure (gauge psig), and P_2 = outlet pressure (gauge psig); and
G_L = specific gravity of liquid (water = 1)

$$C_v = \frac{Q_G}{963}\sqrt{\frac{G_G^T}{[P(Pa_1 + Pa_2)]}} \tag{7.3}$$

Where

C_v = flow coefficient for a particular valve normally in the full open position = flow of water through the valve in gpm when pressure drop through the valve is 1 psi;
Q_G = flow (scfh);
Δ_P = pressure drop, $P_1 - P_2$ or $Pa_1 - Pa_2$ (psi);
Pa_1 = inlet pressure, absolute (psia) = psig + 14.7;
Pa_2 = outlet pressure, absolute (psia; Pa_2 cannot be less than the critical pressure); the critical pressure is approximately equal to 0.5 times the inlet pressure;
T = temperature, absolute = °R = °F + 460; and
G_G = specific gravity of gas (air = 1 at 60 °F).

ACTUATORS. The valve actuator moves the disc, plug, or gate to the desired position via the shaft, which may move linearly, rotate a quarter turn, or rotate multiple turns (depending on valve type). Several types of actuators are available.

Solenoid. Typically used on pipes 102 mm (4 in.) in diameter and smaller, a solenoid actuator consists of a coil, a magnetic frame on which the coil is mounted, and a movable magnetic plunger (Figure 7.5). When current passes through the coil, the movable plunger is attracted to it. Moving the coil also moves the plunger and valve disc, opening or closing the valve.

Many solenoid valves rely on the plunger's movement to open an internal port on the plunger so differential pressure can help move the disc. This enables a small solenoid to direct the movement of a large disc. A spring returns the plunger and disc to its original position when the electrical field is removed.

Solenoid actuators typically are used with globe valves. They typically are two-position (open–close) devices. Special industrial solenoids can be modulated, but they typically are not used in municipal wastewater treatment applications.

Electric Motor. *Electric motors* are devices that convert electrical energy into mechanical energy. They rotate rapidly when energized. An electric motor may be single-phase or three-phase, depending on the size of the application. Single-phase motors, which operate via two wires and single-phase current, are typically used for smaller valves and pumps (fractional horsepower). Three-phase motors, which use the phase difference among three power circuits to determine direction, are typically used where motors 746 W (1 hp) or larger are required. To operate, a motor starter first applies voltage to the coils in the direction that the motor will turn. The electromagnetic field created by the induced voltage will cause the motor rotor to rotate inside the motor housing. The speed of rotation is a function of the rotor and stator design. Once operating, the motor will continue to rotate until the starter removes the

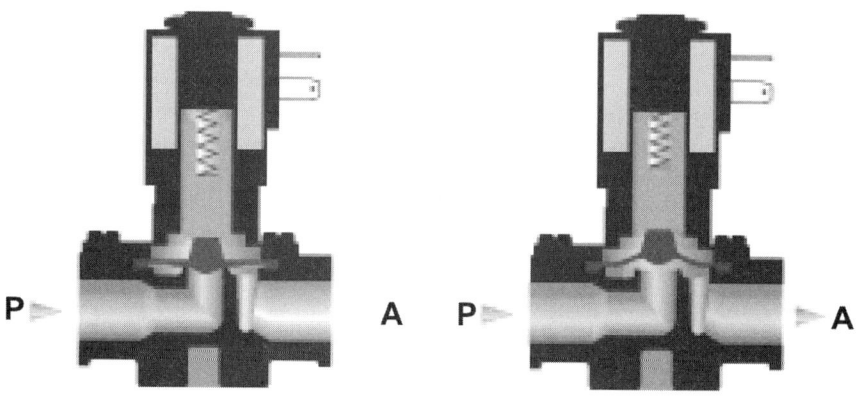

FIGURE 7.5 Solenoid valve.

voltage from the coils. Then, it will gradually slow and stop (stopping is not instantaneous because of the motor's momentum). Motor actuators must allow for the gradual startup and shutdown periods. Some actuators use motor brakes, signaling the motor to stop when the valve has nearly reached the desired position.

An electric-motor actuator typically consists of a motor connected to a manual gearbox, which is connected to the valve stem (Figure 7.6). (The connections must be tight enough to ensure proper valve movement.) The type of gearing depends on the application. Some electric actuators incorporate the gearing in their design. Others are connected to external gearboxes. The actuator usually includes a manual override so the valve can be manually operated during electrical power losses.

FIGURE 7.6 Multiple- and quarter-turn electric motor actuators.

The type of gearing also depends on whether the actuator is modulating or on–off. Gear units are rated for either multiple- or quarter-turn operation. Globe and gate valves require multiple-turn operations; butterfly, plug, and ball valves require quarter-turn operations.

The difference between on–off and modulating electric-motor actuators is related to the "rest time" between starts. Modulating actuators may have to operate continuously, thereby heating the coils to higher temperatures. Heat degrades motor winding insulation. Continuous-duty motors have special insulation that can tolerate higher operating temperatures. On–off actuators typically are designed to cool off for a certain period before being restarted, so the manufacturer can use lower-temperature, less-expensive coils. All electric-motor actuators should include an internal thermal-protection device to prevent heat damage.

Electric motor actuators are self-contained and can operate all types of valves, providing linear or rotary motion, as required. They only use energy when valve movement is required.

Pneumatic. Pneumatic actuators, which rely on air to function, have many advantages and disadvantages. The actuators themselves are typically simple, but their air compressors and related equipment may be complex. They are designed to always have air available, but this comes at an electrical cost. (Air sources based on stored energy can be provided for emergency power outages.) Manually operating a pneumatically actuated valve is more complicated than operating an electrically actuated one because a secondary gear must be added. The air pressure for pneumatic actuators typically ranges from 276 to 827 kPa (40 to 120 psi).

There are two types of pneumatic actuators: diaphragm and piston.

Diaphragm. The diaphragm in a diaphragm actuator is basically a large piston connected directly to the valve shaft (Figure 7.7). The piston's surface area is typically much larger than the valve opening. The piston's movement is controlled by air and a spring: one opens the piston; the other closes it. For example, if a diaphragm-actuated valve must "fail closed", air should be use to open the valve and the spring should be used to close it. If the valve must "fail open", then air should be used to close the valve and the spring should be used to open it. This arrangement can provide for very precise control.

Diaphragm actuators originally were developed to work directly off a low-pressure air signal [typically 21 to 103 kPa (3 to 15 psi)]. Because of the low-pressure air, a large piston surface was required. This limited the valve size on which the

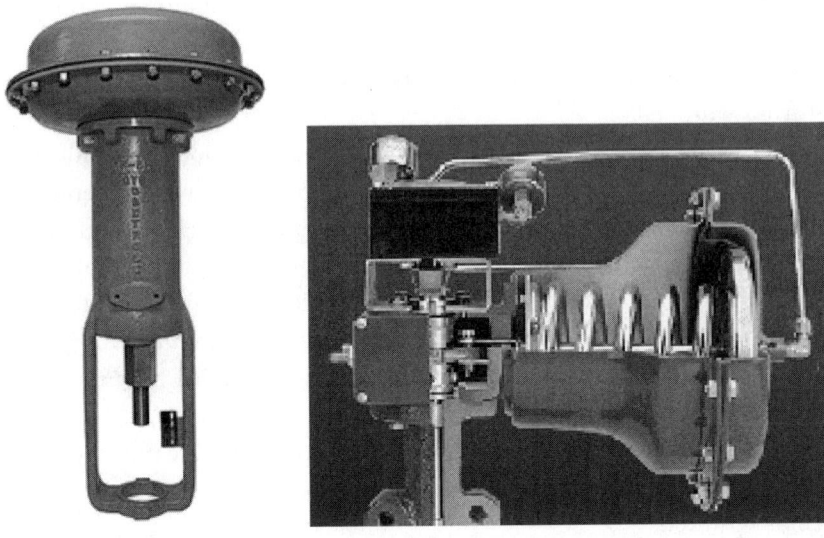

FIGURE 7.7 Diaphragm actuator alone and mounted on a quarter-turn valve.

diaphragm could be used. Most diaphragm actuators are used with globe valves in clean water and gas applications.

Piston. A piston actuator consists of a cylinder, piston, and a shaft (Figure 7.8). The force provided depends on the piston's diameter and the air supply available. The actuator stroke depends on the cylinder and shaft lengths.

Piston actuators function much like diaphragm actuators (including the springs), but they typically are smaller and rely on higher-pressure air [207 to 690 kPa (30 to 100 psi)]. Some pneumatic actuators incorporate multiple pistons and gearing in one enclosure.

Hydraulic. Hydraulic actuators are similar to pneumatic actuators, except they rely on the movement of hydraulic fluid to function (Figure 7.9). Also, a hydraulic actuator's pump, accumulator, and controls typically are more complex than a pneumatic actuator's compressor and controls.

Final Control Elements 239

FIGURE 7.8 Pneumatic actuator alone and mounted on a ball valve.

Hydraulic actuators typically are used when more force is required to operate larger valves. They can provide pressure ranging from several hundred to several thousand pounds per square inch. Operating at higher pressure requires specialty compressors and more safety considerations. Also, the more pressure available, the smaller the cylinder size must be.

Some hydraulic actuators incorporate the cylinder, pumps, and controls in one device. Some also include energy reserves for emergency valve closings.

POSITIONERS. Valve positioners tell the actuator where and when to move (Figure 7.10 and Figure 7.11). They can be simple open–close devices or complete loop controllers. Open–close devices (e.g., solenoid valves and motor starters) usually work with valve-position-feedback devices (e.g., limit switches). Some positioner systems combine open–close devices with an external loop controller to provide for

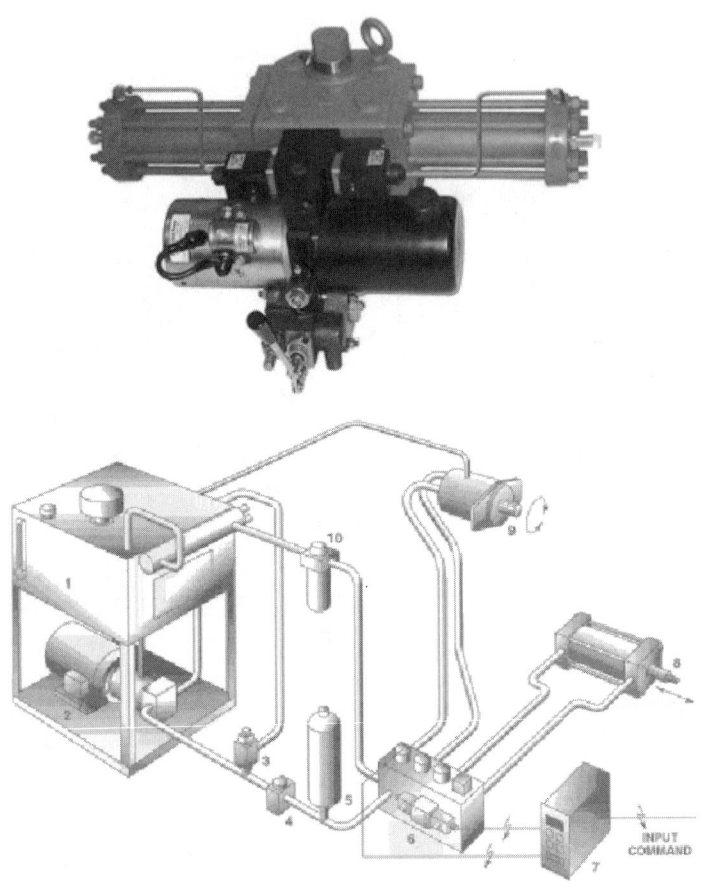

FIGURE 7.9 Hydraulic actuator and hydraulic control system.

intermediate valve positioning. This type of control is less expensive but less precise than direct control.

The positioners for diaphragm valves only push the piston or diaphragm in one direction. The return force is provided by a spring inside the diaphragm.

Final Control Elements   241

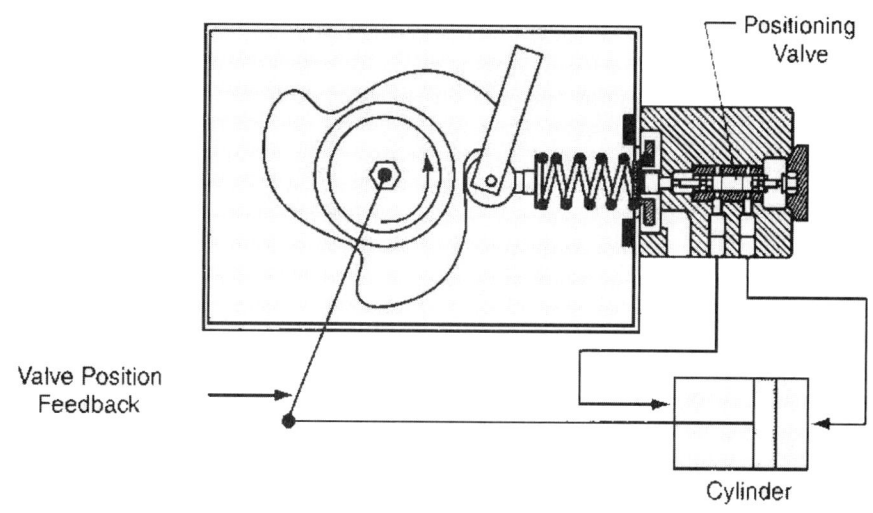

FIGURE 7.10 Pneumatic positioner.

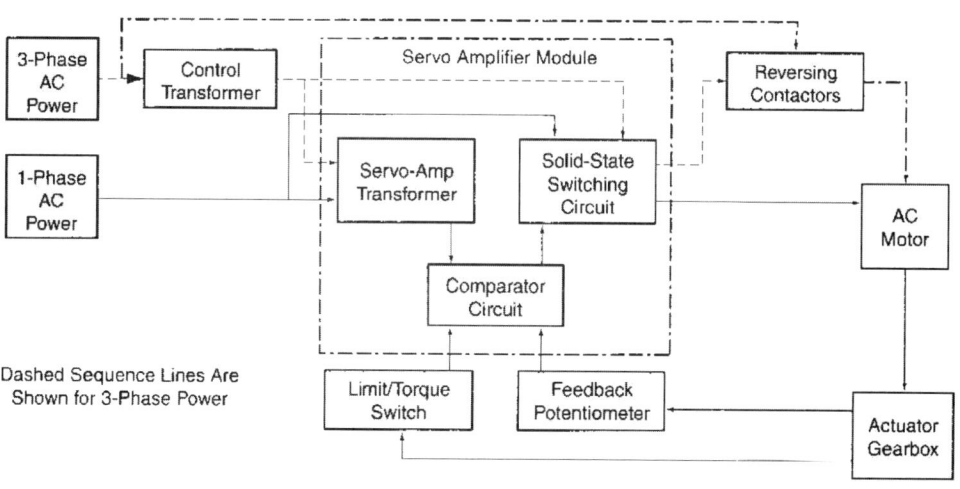

FIGURE 7.11 Electronic valve positioner.

A valve loop controller typically consists of three components: signal input, computation, and control force output. The controller will send a signal to the valve directing it to maintain a certain parameter (e.g., flow, level, or pressure setpoint) based on its own control requirements. Traditionally, this was a 4- to 20-mA signal for an electronic actuator and a 21 to 103 kPa (3 to 15 psi) air signal for a pneumatic actuator, although some pneumatic and hydraulic actuators directly converted 4- to 20-mA signals. The valve interpreted these signals as position commands. In today's digital positioners, this signal may be a specific position command (0 to 100%).

The valve positioner then compares the incoming signal with the valve's actual position via a potentiometer, another electronic measuring device, or direct mechanical linkage. The difference between the signaled and actual positions is the *position error*.

The valve positioner then sends an output signal to the appropriate movement mechanism (open or close) in the valve actuator. It continues to compare the signaled and actual positions until the position error equals zero. Then, it will stop all movement commands and maintain the new position.

Most positioners are designed to slow the valve's movement as it approaches the setpoint. They also have predefined failure modes if the control signal is lost. If the primary actuator power (e.g., air pressure, electricity, or hydraulics) is lost, electric actuators may be locked in place by their mechanical gearing (if properly sized). Pneumatic and hydraulic actuators, on the other hand, may be opened or closed by fluid flow forces. Engineers should consider such failures when designing the control system and provide appropriate backup mechanisms (e.g., springs, auxiliary hydraulic power, or auxiliary battery power).

Most valve positioners have internal settings (e.g., fully open, fully closed, and deadband) that must be calibrated in the field. The *deadband* is an adjustable range of input signal change that produces no output control change. Many valves tend to oscillate about their setpoints because of electrical noise in the control signal or environmental conditions affecting valve components. The deadband minimizes unwanted valve movement.

Digital control systems enable much more information to be transmitted to and from the valve. Microprocessors in final control elements enable them to store historical data and control their own operations in real time. This can improve control-system operations and provide for a level of preventive maintenance that is otherwise impossible.

PUMPS, AERATORS, AND MIXERS

Basically, pumps, aerators, and mixers add energy to a process. In wastewater treatment processes, pumps increase the pressure and flow of liquids and gases. Aerators and mixers condition fluids as they pass through a process.

Pumps, aerators, and mixers typically consist of a signal interface, a signal conditioner or positioner, an actuator (typically a motor), and the mechanical interface to the fluid or gas. Each device requires all four components to function properly, and the control system typically treats them as one unit. Each component, however, typically is made by different manufacturers.

Traditionally, final control elements were controlled by remote programmable logic controllers or large central control systems. Today, the final control elements often can control themselves locally via microelectronics.

PUMPS. A *pump* is any device designed to increase the flow or pressure of a liquid or gas. The choice of pump type depends on the fluid and size of the application (Figure 7.12). A pump that handles lower pressures and flows well may not be the best selection for higher pressures and flows.

There are basically two types of pumps: displacement and nondisplacement.

Displacement Pumps. Displacement pumps are designed to move a specific amount of material in each pumping cycle. The amount of material moved during each cycle depends on the pump's speed and configuration. It also may be variable or fixed, depending on the type of pump. There are three types of displacement pumps: rotary, piston, and diaphragm. The volume of displacement is fixed in rotary pumps but can be varied in piston and diaphragm pumps.

Rotary displacement pumps (e.g., gear, lobe, and screw pumps) typically are used to move highly viscous fluids. The pump cavity is designed to quickly and continuously accommodate fluids that would be problematic for other types of pumps. These pumps rely on continuous rotary movement rather than reciprocating action, so there is no internal adjustment to the amount of fluid moved during each pumping cycle. The pump's speed determines the total amount of fluid pumped.

Piston pumps typically are used in chemical feed systems because they have precise pumping rates. In smaller pumps, the piston (diameter of the cylinder or piston) can be changed, as can the stroke length. Piston size and stroke length determine the amount of fluid moved. This stroke length typically is set manually but can be done automatically.

244 Automation of Wastewater Treatment Facilities

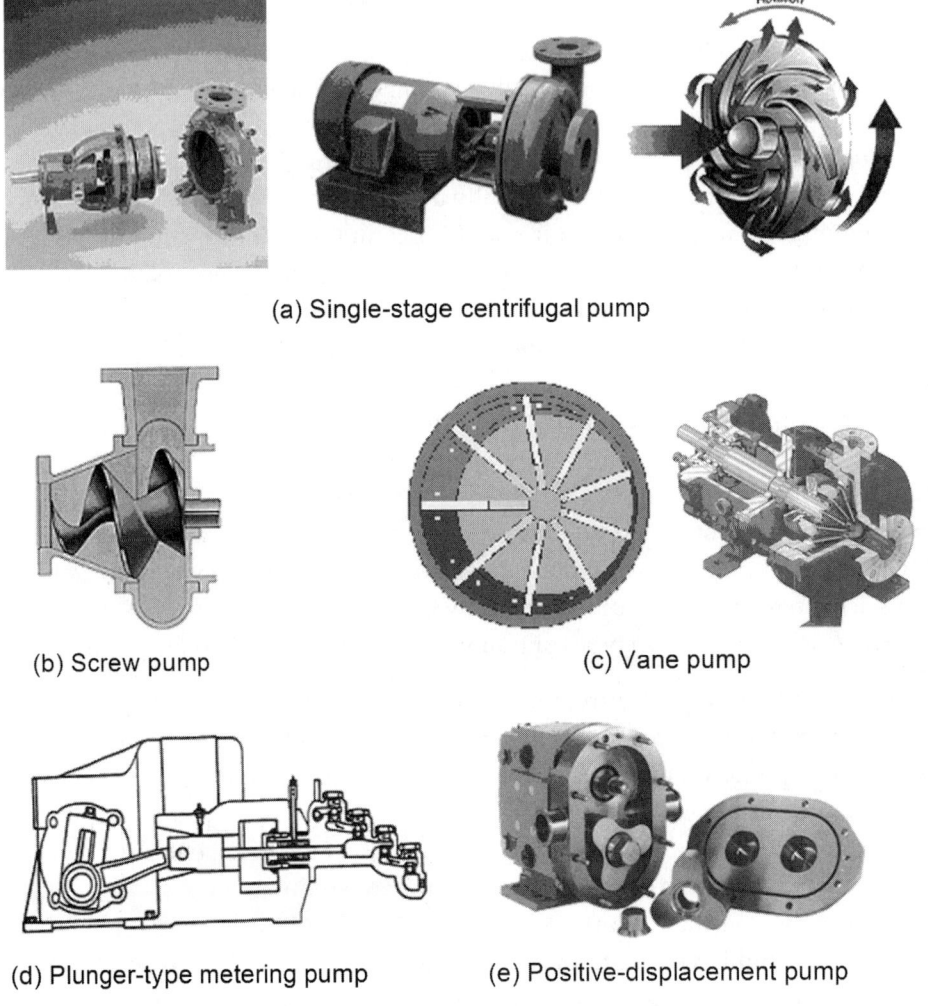

(a) Single-stage centrifugal pump

(b) Screw pump

(c) Vane pump

(d) Plunger-type metering pump

(e) Positive-displacement pump

FIGURE 7.12 Types of pumps.

Diaphragm pumps typically are used in the same manner as piston pumps to move specific volumes of fluid. They can change their volume displacement by varying the length of the pump stroke. The stroke adjustments are smaller than those of piston pumps but the diaphragm size is larger.

Nondisplacement Pumps. Nondisplacement pumps (e.g., centrifugal, propeller, and turbine) are not designed to move precise amounts of fluids with each cycle. These pumps' rotations move liquids or gases from the low-pressure side (suction) to the high-pressure side (discharge). They are more commonly used at wastewater treatment facilities, in applications involving large volumes of liquids or low-pressure gases. Drives can control these pumps' speed, but control valves can regulate their pressure and flow.

FANS, BLOWERS, AND COMPRESSORS. Air movement is a critical part of biological wastewater treatment, so fans, blowers, and compressors are common at wastewater treatment facilities. Fans are electromechanical devices for moving high volumes of low-pressure air [less than 6 kPa (25 in. of water)]. They typically are used in cooling operations. There are basically two types of fans: axial (propeller) and centrifugal. *Axial fans* (propeller) move air over their entire surface. *Centrifugal fans* pull air from a central collector via angled vents in their rapidly rotating outer ring housings. Another housing just outside the outer ring channels the exhaust to a single discharge opening. The axial fan's airflow is typically regulated via louvers, and the centrifugal fan's airflow is regulated by valves on the fan's exhaust or by changing motor speed.

Blowers move fixed volumes of low- to medium-pressure air [typically 7 to 109 kPa (1 to 15 psi)]. They typically are used in biological treatment tanks to provide oxygen for the microbes. There are two types of blowers: positive-displacement and centrifugal. Positive-displacement blowers provide a constant volume of air per revolution of their rotors or lobes (Figure 7.13). Centrifugal blowers work much like centrifugal fans. The blower's airflow typically is controlled via its speed.

Air compressors work like pumps taking atmospheric pressure air to higher pressures. Air compressors are primarily designed to increase air pressure rather than airflow. Because higher pressures are required, the clearance tolerances for compressors are much tighter than those for blowers and fans. Compressors are used to provide the operating power for valves and other pneumatic equipment inside a treatment plant. There are several types of air compressors, including reciprocating, lobe, centrifugal, sliding vane, and axial.

AERATORS AND MIXERS. Like blowers and diffusers, mechanical aerators introduce oxygen to wastewater. The mechanical aerator operates much like an axial fan.

Effective wastewater treatment often depends on the addition of oxygen and chemicals to the wastestream, as well as the controlled movement of the biological

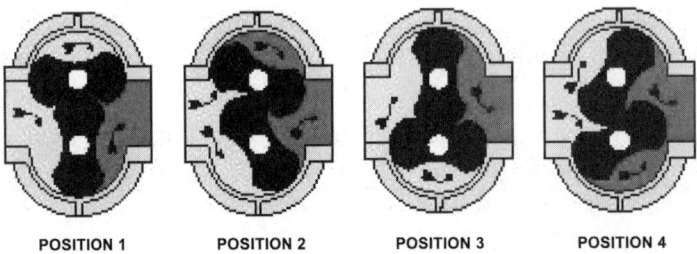

FIGURE 7.13 Positive-displacement blower.

mass. Mixers ensure that oxygen, chemicals, and the biomass are properly distributed. Mixers are commonly propeller elements inserted to selected tanks to increase the dispersion of inserted chemicals or prevent stagnation in a tank. These also operate much like axial (propeller) fans. (Figure 7.14 and Figure 7.15).

Final Control Elements 247

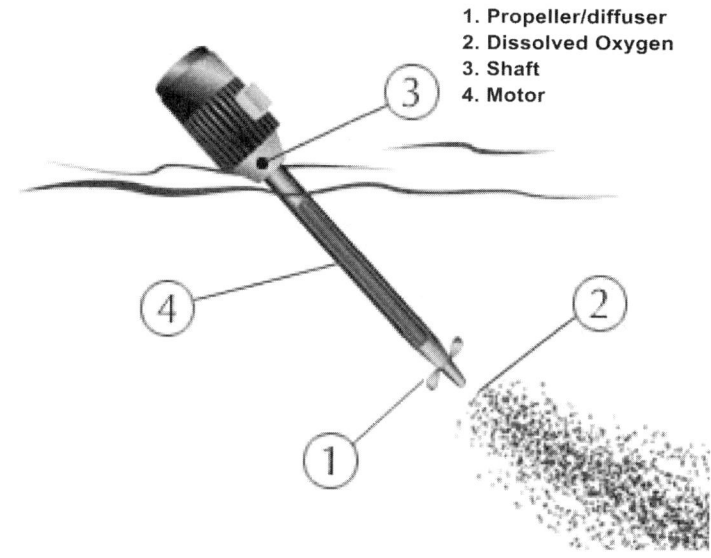

FIGURE 7.14 Horizontal aerator.

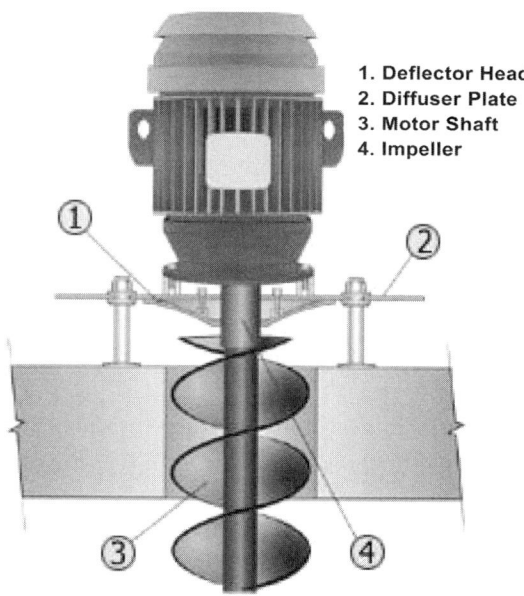

FIGURE 7.15 Surface aerator.

ELECTRIC MOTORS

Electric motors are electromechanical devices that convert electrical energy [either alternating current (AC) or direct current (DC)] into mechanical energy (Figure 7.16). This section will focus on the inductive AC motor because it is the type of motor most commonly used at wastewater treatment facilities.

An AC motor consists of two basic components: a stator and a rotor. The *stator* is the stationary component on the outside of the motor that has at least one long wire wound in it. Each wire in the stator is a *coil*. The *rotor* is a circular metal piece (typically made of multiple metal laminated pieces) that is firmly attached to a solid shaft. The shaft is connected to the outer housing via bearings that allow the rotor and shaft to freely rotate. The rotor is positioned inside the stator but does not touch the stator. When voltage flows though a coil (*current*), it produces an electromagnetic field. This induces a secondary current flow in the rotor with its own magnetic field. A secondary coil on single-phase motors (*starter winding*) determines the direction of rotation. The motor's torque is directly proportional to the current. The wire's length and path determine the degree of voltage drop. A motor's synchronous operating speed depends on the power's frequency and the number of coils in the stator. Some motors have multiple coils, so they can operate at several synchronous speeds.

The rotating electromagnetic field also generates heat via the wire's resistance and the load's effect on the motor. This heat must be controlled. The wiring inside the motor is protected from shorting by an insulation that is susceptible to attack by heat, moisture, or chemicals. If the insulation breaks down, the motor will fail. So, engineers should select motor insulation and thermal-overload protection devices based on the motor's expected operating temperature.

When a motor starts up, its rotating electromagnetic field must overcome internal friction, the inertia of the rotor mass, and the inertia of the mechanical load connected to the motor. So, a high current (10 to 100 times the operating current) is required during startup, and the motor's circuitry must be able to withstand this current. Once the motor reaches its typical operating speed, the current is dropped to operating levels. The time needed for the motor to reach this speed (*startup time*) depends on the motor's size and load, but is typically within a few seconds.

In single-phase motors, a separate coil (*start coil*) is used during startup. In three-phase motors, startup is a natural consequence of three-phase power.

Larger motors have longer coils and more resistance, so they typically need higher voltage. Single-phase motors operating at 120 V AC are common for small

Final Control Elements 249

(a) Electric motor cutaway

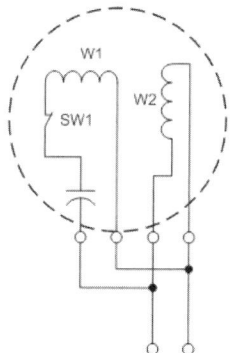

(b) Single-phase motor

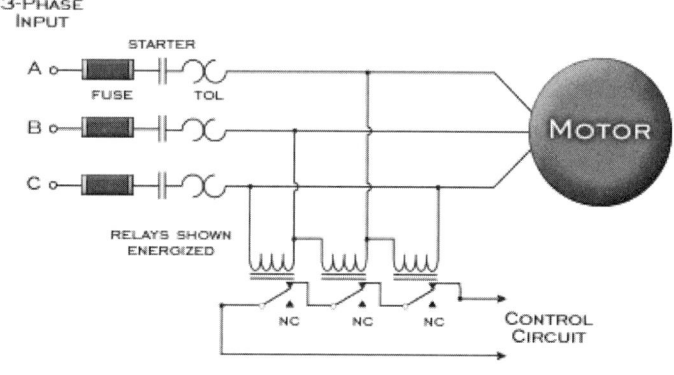

(c) Three-phase motor

FIGURE 7.16 Types of electric motors.

(fractional-horsepower) applications. Three-phase motors typically operate at 208, 230, and 460 V. Three-phase motors are a common choice at wastewater treatment facilities because they do not need the additional startup coil.

Motors have two current values: the locked rotor amperes (LRA) and the full load amperes (FLA). To determine a motor's LRA, manufacturers typically test how much current it can draw when its rotor cannot move. This is the motor's highest current draw. To determine FLA, manufacturers measure the motor's current draw while it is running at its full-rated mechanical load. Engineers should use these two values when designing the electrical circuits that will connect the motor to the electrical system.

STARTERS. A starter is an interposing relay between a high-voltage motor (typically any motor 115 V AC or higher) and the low-voltage (12 to 24 V DC) control circuits. These are used to increase safety and to minimize the motor's wiring requirement (Figure 7.17).

Starters are used on motors that only need to operate at one synchronous speed. They are specially designed switches or relays that allow for short-term operations at a high current level during motor startup. Once the motor reaches its synchronous running speed, the current drops to the motor's FLA. A starter also will shut down a motor if startup takes too long (exceeds its LRA) or the motor exceeds its FLA, because high current can cause a motor to overheat and fail. The starter also provides for low-voltage control of the higher voltage motor.

DRIVES. Starters are used to operate motors at a single set speed. Drives are used to operate motors at varying speeds. Because these devices are predominantly solid state, many of the original drives were severely limited by current and temperature restrictions. Advances in solid-state technology have overcome most of these issues.

Today, there are three types of variable-speed drives: variable-frequency drives (VFDs), wire-wound motor controls with secondary energy-recovery units, and variable-torque speed controls. The first two are the most frequently used in wastewater treatment facilities.

Variable-Frequency Drives. Variable-frequency drives (variable-frequency controllers) allow for changes in motor speed with or without the applied load (Figure 7.18). The drive basically changes a three-phase AC signal into a restructured DC voltage with controllable frequency and voltage. For example, if a motor's base frequency of operation is at 60 Hz, its harmonics will be at 120 Hz, 180 Hz, 280 Hz, etc.,

Final Control Elements 251

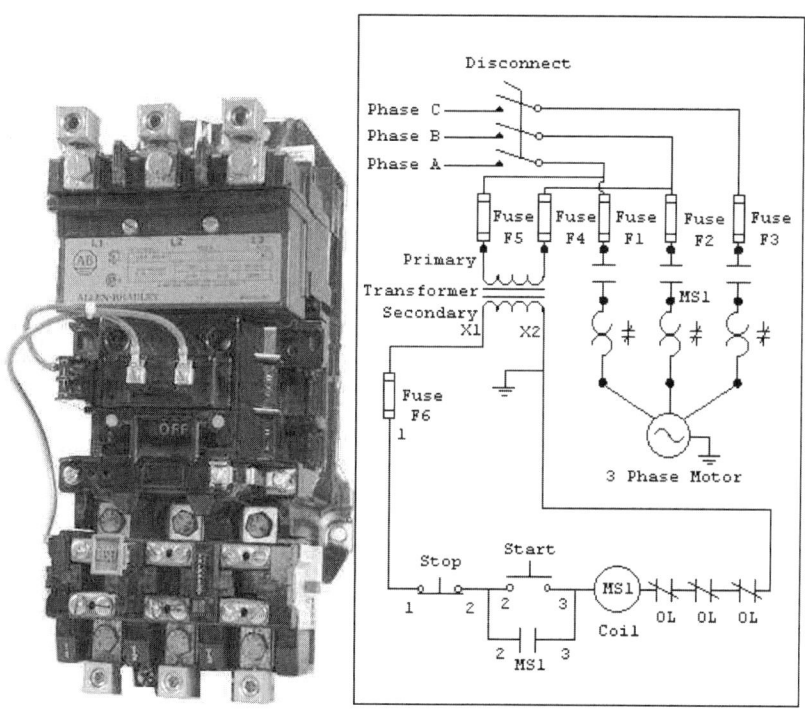

FIGURE 7.17 Three-phase starter.

and the VFD enables the motor to operate using these harmonic ranges. Although there is some power loss from full synchronous operation, VFDs have become very efficient.

Wire-Wound Motor Control. Wire-wound motors control speed by limiting the voltage to either the stator or rotor coils. The controls reduce current draw at startup and vary the impedance of the rotor coils via solid-state electronic rheostats. The solid-state electronic rheostats also enable most of the energy generated to be converted to fixed-frequency AC power and fed back into the plant's power grid. This

FIGURE 7.18 Variable-frequency drive.

control mechanism is typically used for larger motors [more than 75 kW (100 hp)] or on motors using more than 480 V AC. One drawback of this control is the amount of heat generated, which reduces the motor's efficiency. Nevertheless, the average efficiency of this method typically is between 85 and 90%.

Variable-Torque Speed Controls. While variable-speed drives directly limit motor speed, variable-torque speed controls allow motors to run at full speed but limit the amount of torque transmitted to the load (e.g., pump) by inserting a torque transmission device between the motor and the load. The two most common methods of torque control are eddy current clutches and liquid clutches.

Eddy current clutches use electromagnetic plates between the motor and the load. An electromagnetic clutch is connected to the output shaft (connected to the load) and a field coil is connected the input shaft (connected to the motor), which runs at a constant speed. A DC current in the field coil produces eddy currents. The resulting magnetic flux develops torque and tends to turn the load in the same direction. The output speed is proportional to the amount of DC applied to the magnetic field. The clutch's efficiency is typically 2 to 3% less than wire-wound motors with secondary resistance.

Liquid clutches are primarily used for very large pumping systems where electrical clutches are impractical. These clutches use the sheer properties of oil to transmit torque. They are typically more efficient at high rotational speeds.

Variable-torque speed controls typically are limited to large motors. Its primary disadvantage is that it generates large amounts of heat. It was a popular speed-control method before high-current, solid-state VFD controls were developed.

These controls are now available with an adjustable speed device, which is designed for use with larger motors. It consists of two components: a copper conductor assembly, which connects directly to the motor shaft, and a permanent magnet rotor assembly, which connects to the load shaft. The torque is transmitted across a thin air gap that can be continually adjusted to control the speed. Rare-earth, permanent magnets allow the adjustable speed device to maintain its magnetic properties throughout its lifespan.

CONTROL SIGNAL INTERFACE AND SMART PUMPS AND DRIVES

Traditional control systems required the final control element to be wired to the control panel. The signals were simple on–off commands (discrete output), "pump running" (discrete input), speed output control (analog output), and running speed (analog input). Many other variables (e.g., temperature, vibration, and current) were returned to the control panel via variable-specific wires. Speed controls relied on traditional 4- to 20-mA control signals.

Now that microprocessors are available in final control elements, this structure has changed. The first Fieldbus structures focused on moving traditional input/output (I/O) blocks from the control panel to the final control element. They eliminated the need for individually wired devices; instead, they rely on electronic addresses and one wire connected to all sensors and final control elements.

Also, once final control elements had microprocessors, engineers found that they could handle more than traditional I/O data. Each final control device could store and transmit current and historical operating data in real time. Moving the control connection point and increasing the availability of real-time data greatly improved pump and valve operations.

Final control elements are now composites of electronic, electrical, mechanical, and computer technology. So, the people who install and maintain them must be cross-trained in all these fields.

OTHER FINAL CONTROL DEVICES

Heaters and dry bulk conveyers can also be considered final control devices. Heaters increase the temperature in specific parts of a process and typically are used in areas where temperature control is critical. For example, heat may be required to keep wastewater treatment fluids from freezing or maintain the water above a certain temperature so specific biological actions will occur. The devices typically use resistive heating elements and can be thermostatically controlled to maintain a specific temperature or can operate over a defined temperature range.

Dry bulk conveyers are used in several locations in wastewater treatment plants to, for example, add dry bulk chemicals to tanks or remove dewatered sludge. The element being controlled is an electric motor, and the issues are similar to those for pump motors. Most conveyers use single-speed motors.

Chapter 8

Control Panels or Stations

Panel Devices	256	Local Control Panels and Stations	262
Analog Displays	257	Vendor-Supplied Panels	262
Digital Displays	257	Area Control Panels, Stations, and Consoles	262
Operator Interfaces	258		
Graphic Terminals	258	Main Control Panel, Station, or Console	263
PC-Based Terminals	258		
Portable Terminals	259	Graphic Panels	263
Annunciators	259	Alarm Stations	263
Horns	259	Termination Panels	263
Strobes	260	Fiber-Optic-Patch Panels	264
Pilot Lights	260	Control-Panel Specifications	264
Pushbuttons and Switches	261	CSI Format	264
Miscellaneous Panel Hardware	261	Standards, Codes, and Regulations	265
Programmable Logic Controllers	261	Submittal Requirements	265
Human-Machine Interface Software	261	General Construction Requirements	266
		Identification	267
Types of Control Panels and Stations	262	Environmental Requirements	267
		NEMA Rating	267

Thermal Management	268	Touchscreens	278
Electrical Systems	269	Resistive	278
Power Source	269	Capacitive	278
Safety Concerns	269	Infrared	278
Motor Control Centers	270	Acoustic Wave	279
Wiring	271	Near Field Imaging	281
Surge Protection	272	Dispersive Signal Technology	282
Factory Testing	272	Flat-Panel Displays	283
Installation	274	PC-Based Displays	284
Panel List	274	Monitors	285
Special Requirements	274	Plasma	285
UL 508: The Underwriters Laboratory Standard for Industrial Control Equipment	274	Digital Light Processing	285
		Wireless Operator Interface	287
		Digital Recorders	289
Options for Panels in Hazardous or Corrosive Areas	275	Data Loggers	289
		Selection Considerations for Recorders or Data Loggers	289
Intrinsically Safe Systems	275	Remote Annunciation	290
Purged Enclosures	275	References	291
New Developments or Trends	277	Suggested Readings	293
Operator Interfaces	277		

At water and wastewater treatment facilities, control panels include consoles, displays, operator interface devices, and the enclosures used to house the electrical controls. This chapter discusses control-panel selection criteria, as well as the many options available to designers and users.

PANEL DEVICES

Panel devices include all hardware mounted on or in a control panel. A panel's use typically dictates the type of devices it will have. Panels used as operator interfaces, for example, typically include a display of process information, while those housing electrical or electronic components [e.g., input–output (I/O) panels and termination panels] may not include displays.

ANALOG DISPLAYS. Analog displays are used when the total range of a process variable must be displayed. For example, an operator may need to know the instantaneous plant flowrate compared to maximum plant flow. Analog displays can be a hardwired panel meter or a computer screen image of an analog meter. Typically, such displays include a pointer on a graduated scale covering the process variable's range. Single-loop controllers (SLCs), which sometimes are used in standalone control applications, typically include a bar graph of the process variable, controller output, and setpoint. Today, manufacturers offer combination meters, which provide both analog and digital displays.

DIGITAL DISPLAYS. Digital displays are common in wastewater treatment plants because the information provided is concise and unambiguous. They present readouts in numerical form and typically are preferable because fewer mistakes are made reading a number than deciphering an indication on a scale.

Simple, inexpensive electrical counters are often used to display "run" hours of various motorized equipment. These counters are available for alternating current (AC) or direct current (DC) operation and should not include a zero reset feature because this would defeat the purpose of collecting accumulated equipment run time.

Today's electronic displays are microprocessor-based and can vary in the number of displayed digits, the digit height, the digit color, and the method used. The display span of a digital panel meter is rated by digit. By convention, a full digit can assume any value from 0 through 9, a 1/2 digit will display a 1 and overload at 2, a 3/4 digit will display digits up to 3 and overload at 4, etc. For example, a meter with a display span of +3999 counts is said to be a 3 3/4 digit meter. As a practical matter, four digits is the recommended minimum for process meters, and six digits offer more versatility (e.g., decimal point placement). Current offerings include a serial communications option, in which users can select RS-232, RS-485, and either ASCII, Modbus, Ethernet, or Internet from a pushbutton menu. Digital displays can connect directly to an Ethernet network and transmit data using standard transmission control protocol/Internet protocol (TCP/IP), the collection of communication protocols used by the Internet. It is now possible to monitor and control a process via a Web browser from anywhere on the Internet.

Light-emitting diode (LED) displays are the most common because they can be read in low ambient light. Available colors include red, green, and yellow. Red is the most common and brightest color.

Liquid crystal displays (LCDs) are also available and used where ambient light is adequate, although manufacturers now offer backlit versions.

Other displays offered by some manufacturers include:

- Vacuum fluorescent, consisting of a triode vacuum tube with phosphor-coated displays (used when various shaped segments form the display matrix); and

- Gas-discharge displays, in which shaped electrodes in the form of characters ionize surrounding neon gas. Such meters require a higher operating voltage (170 to 300 V DC) and are typically used for electrical metering.

OPERATOR INTERFACES. Operator interfaces are the operator's window into the process, providing monitoring, control, and display information graphically and intuitively. Operators can enter setpoint changes and timer values, start and stop equipment, and perform other control actions with relative ease. Color displays are relatively inexpensive and are replacing existing text-based, monochrome displays.

In nearly a decade, operator interfaces have evolved from simple panel displays to complex human–machine interface (HMI) technology using inexpensive hardware, thin clients, handheld computers, and cell phones that enable operators to access instrumentation and control (I&C) system data from anywhere in the plant (Liptak, 2002). There are basically three types of operator interfaces: graphic terminals, personal computer (PC)-based terminals, and portable terminals.

Graphic Terminals. Graphic terminals are inexpensive and vary in size from 165 to 381 mm (6.5 to 15 in.) or larger, depending on the manufacturer. They are available in thin film transistor (TFT) active matrix color displays with keypads, a touchscreen, or both. They are usually installed in the process area and are available with appropriate National Electrical Manufacturers Association (NEMA) ratings.

PC-Based Terminals. These terminals include both industrial grade and desktop versions. Industrial terminals with the appropriate NEMA rating can be installed in the process area. Typically, these PC-based terminals are loaded with Microsoft Windows-based HMI software (currently Windows 2000 and XP).

Portable Terminals. New standards in wireless technology [Institute of Electronics and Electrical Engineers (IEEE) 802.11b and IEEE 802.11g] have created a whole class of portable operator interfaces, according to a recent industry report (Shea, 2001). This report indicates that the growth of Web browser-based operator interface terminals in North America is expected to explode over the next 5 years, partially because of lower cost compared to other operator interface terminals.

Portable terminals with HMI software are another cost-efficient option. They enable plant personnel to access I&C system data and processes without being tied to one location. Current features include an on-screen keyboard, configurable navigation keys, a PC interface card, a touchscreen, and the Windows CE operating system.

ANNUNCIATORS. Annunciators use visual and aural mechanisms to alert operators to abnormal or dangerous conditions in the plant. Traditionally, annunciators were standalone, hardwired devices. They basically consisted of multiple alarm points [e.g., level switches, flow switches, programmable logic controller (PLC) outputs and analog trip modules], a logic module, and a visual display, as well as a dedicated power supply, a horn, and various pushbuttons. The horn and pushbuttons can be external or built in. The Instrumentation, Systems, and Automation (ISA) Society has issued standards for the annunciator's functions and sequence of actions (ISA S18.1, 1992). The pushbuttons allow operators to perform the "test", "acknowledge", and "reset" functions required in the annunciator sequence.

Although many annunciators are being replaced by PLC- and HMI-driven alarms, utilities prefer to use standalone annunciators for critical alarms (e.g., those associated with life, safety, or valuable assets). Annunciator manufacturers now offer systems with serial port connections as front ends to distributed control systems (DCSs). This hybrid links the visibility, reliability, and built-in redundancy of dedicated annunciators to the flexibility and record-keeping convenience of a DCS system (Liptak, 1995).

Horns. Aural alarms include vibrating horns, buzzers, chimes, sirens, and multiple tones. When specifying aural alarms, engineers typically should specify a sound level. Sound levels are measured in decibels (dB) and should include a distance to fully describe the sound level, because sound pressure fluctuations diminish with increasing distance from the sound source. For example, typical horn sound levels range from 97 to 103 dB at 3 m (10 ft) in indoor areas and 107 to 127 dB at 30 m (100 ft) in outdoor areas.

Vibrating horns are the most common audible alarms at water and wastewater treatment facilities. However, engineers should be judicious when adding horns to a control panel. Excessive use of horns is disruptive and results in the horns being disconnected from the alarm systems. In control rooms and other quiet areas, buzzers or chimes should be considered instead.

Strobes. Visual alarms include strobes and indicator lights with printed messages. Strobes are known for their exceptionally bright bursts of light. A strobe light basically functions by passing a brief, intense pulse of electric current through a gas (e.g., xenon or krypton), which then emits a brilliant burst of light. Strobe lights are vibration-resistant and will function for thousands of hours before needing replacement. They are available in various lens colors, although amber and red are typically used for alarms. Strobes typically are used in toxic gas alarm stations in hazardous areas, as mandated by the National Fire Protection Association (NFPA; NFPA 820).

PILOT LIGHTS. Pilot lights typically are used to convey equipment status information to operators. Although pilot lights are being replaced by other types of operator interfaces, they are still desirable in such applications as simple manual-control stations and supplemental equipment-status information (e.g., on, off, auto, opened, or closed).

Pilot lights can operate via either AC or DC voltage (e.g., 120 V AC/DC or 24 V AC/DC). Lamps are available with built-in transformers to operate at lower voltages (typically 6 V or less).

Three types of pilot lights are available: incandescent, neon, and LED. Incandescent lamps are bright and inexpensive but relatively short-lived. Neon lamps and low-voltage LEDs resist shock and vibration (Liptak, 1995). LEDs initially cost more but have a long life and are gaining popularity at water and wastewater treatment plants. They typically are the most economical option over the long term, because of the following advantages:

- Resistance to shock and vibration (solid-state LEDs are virtually impervious);

- Long life [LEDs typically have a service life of 100,000 hours (11 years), compared to 20,000 hours (28 months) for neon lamps and 2000 hours (3 months) for standard incandescent lamps]; and

- Reduced power consumption (today's LEDs use between 10 and 52% less power than an equivalent incandescent lamp).

Engineers should include some sort of lamp-test circuit in the panel. Light units with built-in push-to-test features are a popular option.

PUSHBUTTONS AND SWITCHES. Control panels typically include pushbuttons and switches for simple operator controls. Although the basic designs of these components have not changed significantly over the years, several enhancements (e.g., key-operated switches, padlock attachments, safety guards, and special contact blocks for low-voltage switching applications) have been made. One of the newest options is a switch or pushbutton with network connections that allows the device to be operated by a controller remotely in either manual or automatic (without the need for additional external hardware) via a built-in network connection using a standard protocol, such as the Open DeviceNet Vendors Association (ODVA) DeviceNet. The advantage of using a network connection, such as DeviceNet, is that a separate power supply connection is not required because the device is powered by the network. Other developments include zero force types. This type of button is designed to fit the shape of the hand and help alleviate the effects of carpal tunnel syndrome and other nerve disorders common to repetitive operator interfaces.

When selecting pushbuttons and switches, engineers should consider the application and the appropriate NEMA rating. Typically, heavy-duty industrial devices (oil-tight NEMA 13 as a minimum) are preferred at water and wastewater treatment facilities, although many installations may be better suited for NEMA 4 (watertight) designs.

MISCELLANEOUS PANEL HARDWARE. Control panels need various electrical components to be fully functional. Such components include control relays, DC power supplies, circuit breakers, wireways, electrical convenience outlets, and lighting fixtures.

PROGRAMMABLE LOGIC CONTROLLERS. For information on PLCs, see Chapter 12.

HUMAN–MACHINE INTERFACE SOFTWARE

When loaded into an operator interface terminal, HMI software allows operators to monitor and control a treatment process in real time. Users are no longer restricted to proprietary HMI software; today's software is based on open protocols and works with either a DCS- or PLC-based system. HMI vendors offer drivers for virtually all major PLCs and DCSs. The human–machine interface software typically is graphical, intuitive, and based on the Microsoft Windows operating system. (For more information on HMI software, see Chapter 11).

TYPES OF CONTROL PANELS AND STATIONS

LOCAL CONTROL PANELS AND STATIONS. Local control panels and stations typically provide manual and automatic controls for process equipment typically encountered in treatment facilities. Their components may include PLCs, remote I/Os, electromechanical relays, control switches and pushbuttons, alarms, pilot lights, and operator interface terminals. Engineers also should include manual controls for use in case of power outages and control-system failures. These panels typically are installed in the process area or an electrical equipment room.

Local control panels and stations also facilitate startups and testing by enabling equipment to be tested to satisfy contract requirements before the entire plant automation system comes on-line.

VENDOR-SUPPLIED PANELS. Vendor-supplied panels typically are part of an equipment package. They can include the same types of controls found in local control panels and stations. When specifying vendor panels, engineers should ensure that the contract specifications are closely coordinated with the rest of the I&C system controls so the entire system will be cohesive and, therefore, easier to maintain. This will also simplify staff training. Vendor-supplied panels should be able to share spare parts, use similar PLC program formats, and use the same HMI software.

AREA CONTROL PANELS, STATIONS, AND CONSOLES. Area control panels and stations are similar to local control panels, except they integrate the controls for a process area (e.g., solids handling) or segment of the plant. These panels typically are networked with the plant's main control panel or station and can be set up to allow control of other plant areas with the appropriate security measures. Traditionally, these panels would be installed in a controlled environment (e.g., electrical equipment room or auxiliary control room), but now they can be put in the process area with the appropriate environmental enclosures. Hazardous areas are no longer a problem because many manufacturers have found ways (e.g., EMI/RFI protection, purge systems, and washdown enclosures) to mitigate such hurdles. Electromagnetic interference (EMI) is produced in circuits or elements of a device by external electromagnetic fields encountered in treatment facilities caused by various large motors or other electrical equipment. Radio frequency interferences (RFIs) are caused by transmissions from devices, such as two-way radios, used by plant personnel. However, because operators only need access to the operator interface, the panels could simply be an industrial workstation, with the rest of the equipment (e.g., processors, I/O modules, and power supplies) in a separate enclosure.

MAIN CONTROL PANEL, STATION, OR CONSOLE. Main control panels or stations typically are installed in an area of the plant (e.g., a central control room) where operators can monitor and control the entire facility. They typically consist of control panels, PC-based workstations, log printers, alarm printers, process historian computer, engineering workstation, and large screen displays. Many also include process-area and building heating, ventilation, and air conditioning (HVAC) controls and power monitors.

The current trend is to use PC-based workstations and the latest PC technology because the hardware is not proprietary. The workstations use the same HMI software that the area control panels and stations do.

GRAPHIC PANELS. Graphic panels are mostly being replaced by PC-based graphic screen displays. Sometimes utilities prefer to use large mimic panels for certain applications, such as electrical power distribution, building security, and fire-alarm annunciator panels. The graphic panels typically provide a diagram of the system or process (e.g., power distribution, process graphics) with appropriate indicators and controls (e.g., breaker control switches, power metering process equipment, and valves).

ALARM STATIONS. Alarm stations (e.g., gas alarm stations and local process alarm stations) are simple alarm notification panels typically located in various process areas. Alarm station components may include horns, pilot lights, pushbuttons, and strobes.

TERMINATION PANELS. Termination panels are basic enclosures used for wiring terminations; they serve as junction boxes for collecting and distributing signals. The panels can be installed in process areas or electrical rooms. Their components include terminal strips, fuses, signal conditioners, and power supplies.

A recent development in wiring termination has been the introduction of insulation displacement connectors (IDCs). Insulation displacement connectors have been used for many years in the telecommunications and electronics industries, and have been gaining popularity in the automation industry. This technology has proven viable, passing the stringent testing required of existing screw-and-spring terminal blocks.

Insulation displacement connectors allow installers to terminate wires without stripping or ferruling, thereby saving up to 75% of installation time and an average of 60% of installation costs while maintaining or increasing reliability and functionality. Unlike screw clamps and spring cages, IDCs require no tools and eliminate much of the human error in connector installation. Basically, IDCs require installers to mechanically force an unstripped stranded or solid conductor into a V-shaped metal tube. The tube's opening is smaller than the conductor's diameter. When the conductor is inserted, it forces the tube's terminating blades to spread open and exposes the bare copper to the blades. The tensile strength of the metal resists the outward force, acting like a spring and creating a vibration-resistant connection with the conductor. Also, the terminating blades penetrate any existing film or oxide on the bare copper to form a gas-tight, corrosion-resistant connection (Norton, 2002).

FIBER-OPTIC-PATCH PANELS. As fiber-optic networks are increasingly used in I&C systems, fiber-optic patch panels have become a necessary addition to the plant-wiring apparatus. Patch panels are simply a collection of fiber cable connector panels in a common housing. They are often used as part of a structured network wiring scheme in compliance with the EIA/TIA 568 standards (Rosenberg, 2000). The Electronics Industry Alliance (EIA; www.eia.org) and Telecommunications Industry Association (TIA; www.tiaonline.org) publish American National Standards Institute (ANSI) accredited standards for telecommunications wiring.

CONTROL-PANEL SPECIFICATIONS

CSI FORMAT. The Construction Specification Institute (CSI) format makes panel specifications easier. It provides a master list of numbers and titles for organizing information about construction requirements, products, and activities into a logical, standard sequence. (For more information on the CSI format, see Chapter 15.)

The latest revision of the CSI Master Format, the 2004 edition, has new chapters pertaining to instrumentation. For example, it includes a Process Equipment Subgroup in which some of the material is new and some has been relocated from Divisions 11, 13, and 14 in the 1995 edition.

To be complete, panel specifications should include the following supporting documents:

- An I/O list for PLC or DCS equipment,
- A panel list,
- Installation location drawings,
- A panel layout,
- Control strategies for programmable equipment,
- A networking block diagram if the panel is part of an I&C system network, and
- Graphics requirements for panels with operator interfaces.

STANDARDS, CODES, AND REGULATIONS. Panels used in water and wastewater treatment facilities must comply with several standards and codes, depending on the panel's purpose and location. At a minimum, panels should comply with the latest National Electric Code (NEC; NFPA 70) requirements for wiring and electrical equipment. The panels also may need to comply with NEMA and Underwriters Laboratory (UL) 508 requirements.

SUBMITTAL REQUIREMENTS. To ensure that a panel will meet the specifier's or user's requirements, the supplier should submit documentation substantiating the panel design. The project team should review this documentation for compliance before approving the panel's fabrication. If the panel is a vendor's standard offering, the project team should still review the standard documentation to ensure that the product will be satisfactory. The vendor's documentation should include the following:

- Panel-layout drawings showing all front- and interior-mounted components;
- The panel's NEMA rating (clearly stated on the layout drawing);
- A bill of material corresponding to all panel components shown on the drawings;
- Panel-wiring schematics;

- Interconnecting wiring diagrams showing all external wiring to and from the panel;
- Paint specifications for painted panels;
- Manufacturers' data sheets for all panel components;
- All panel operations and maintenance manuals, including manufacturer manuals for all panel-mounted devices; and
- A PLC program list for PLC-equipped panels.

GENERAL CONSTRUCTION REQUIREMENTS

To determine each panel's construction requirements, engineers first must consider its application. Panel location typically dictates general construction requirements. In water and wastewater treatment plants, for example, panels in process areas typically are exposed to corrosion, water, or other hazardous conditions.

Other panel construction details that engineers should consider include the following:

- Detailed mounting requirements for free-standing, wall-mounted, or support-mounted panels;
- Device and component placement so connections can be easily made and there is ample room to service each item;
- Location of interior access doors (front or rear);
- Component-mounting supports, which should restrain all devices and prevent movement;
- Wiring from all field instruments and equipment, which should terminate at clearly identified and numbered terminal blocks;
- Panel accessories, which should include 120 V AC duplex grounding receptacles, panel lighting with a corresponding switch, and circuit breakers or fusing for power feed;
- Unused panel space; for example, the bottom 305 mm (12 in.) of free-standing panels should be free of all devices, including terminal strips, to make installation and testing easier; and

Control Panels or Stations

- Component installation height; for example, no device should be mounted less than 914 mm (36 in.) above the operating floor unless otherwise specified. Components that need to be frequently accessed by an operator from the operating floor must be conveniently located without having to bend down. Similarly, equipment should not be installed too high from the operating floor, for example, to allow an operator access without the use of a ladder or other equipment.

IDENTIFICATION. All panels should include appropriate nameplates consistent with the plant's nomenclature and the controlled equipment's designation. Nameplates should be permanent and affixed to the panel as described in the specifications. Instrumentation, Systems, and Automation Society Standard RP60.6-1984 provides guidelines for identifying equipment. Panel specifications should provide the following nameplate details:

- Construction material and size [e.g., 2.3-mm-thick (0.09-in.-thick) laminated phenolic plate with white matte finish and engraved black lettering—13 mm high (0.5 in. high) for the panel identification and 6.4 mm high (0.25 in. high) for panel components];

- Mounting method (e.g., screws or adhesive);

- Location;

- Nameplate contents (e.g., the instrument or equipment tag number and descriptive title as shown or specified in the contract documents).

ENVIRONMENTAL REQUIREMENTS. *NEMA Rating.* In the United States, NEMA rates equipment based on the environment where it may be safely used. For example, a NEMA 4X rating indicates that the equipment may be used in corrosive areas, and a NEMA 7 rating indicates that the equipment is explosion-proof. A complete description of NEMA ratings can be found at the association's Web site (www.nema.org). The panel specifications should state the NEMA rating required. (For more information on enclosures and NEMA classifications, see Chapter 5.)

Typically, all panels in protected indoor areas (e.g., electrical rooms and control rooms) can be NEMA 1, but many utilities opt for NEMA 12 because it provides more protection. Panels in noncorrosive, wet process areas should be at least NEMA 4. Those in corrosive areas should be NEMA 4X.

In the United States, the NEC published by the National Fire Protection Association as NFPA 70 defines hazardous areas "depending on the properties of the flammable vapors, liquids or gases or combustible dusts or fibers that may be present, and the likelihood that a flammable concentration or quantity is present" (Art. 500.5, NFPA 70, 2002). The National Fire Protection Association's classification method defines areas by classes, divisions, and groups. In Class I areas, for example, flammable gases or vapors are or may be present in the air in quantities sufficient to produce explosive or ignitable mixtures. Division 1 areas are those where flammable hazards are present under normal operating conditions; Division 2 areas are those where the hazards are present under abnormal or accidental conditions (e.g., rupture of a process line). The Groups (A through D for Class I) define the types of gases found in the hazardous areas. A typical wastewater plant, where combustible gases may be present because of methane production in the digesters, may be classified Class I, Division 1 or 2, Group D. (For more information on NFPA classifications, see NFPA 70.) Panels in hazardous areas must comply with the NEC standard for such areas. For example, NEMA 7 explosion-proof panels must be used in Class I, Division 1, areas. Explosion-proof enclosures are bulky, difficult to work with, expensive in larger sizes, and so should be used sparingly, if possible. The multiple bolts used around the enclosure must be properly torqued and removed to access the equipment inside for maintenance. All wiring to and from panel must be via approved rigid metal conduits with seal fittings, and the enclosure penetrations must be done at the factory because field-drilling the enclosure is neither practical nor desirable.

Thermal Management. Panel design requires careful consideration of heat and humidity levels inside the panel, which will depend on the panel's location. Measures must be taken to protect the electrical and electronic equipment within. Outdoor panels must compensate for seasonal temperature variations. Those exposed to sun, for example, should include a sun shield. Indoor panels must be prepared to handle the ambient effects of normal and abnormal conditions (e.g., a ventilation system failure). Indoor control panels in process areas may also need humidity controls. A strip heater can maintain humidity at a desired value and temperature at least 6°C (10°F) above ambient.

To calculate the temperature rise in an enclosure, engineers must know the heat generated by the devices inside. This information can be obtained from the device manufacturers. Heat input values are usually given in watts but may also be in British thermal units (BTU)/hour. The physical size of the enclosure will be the primary factor in determining its ability to dissipate heat. The larger its surface area, the lower the temperature rise. Aluminum and stainless steel enclosures will have higher temperature rises because of their poor radiant heat-transfer effects. Nonmetallic enclosures have heat-transfer characteristics similar to those made of painted steel. Major panel manufacturers have charts or programs for calculating the temperature rise for their panels, as well as related cooling-equipment recommendations.

Heat can be dissipated from enclosures via static or dynamic ventilation. Static ventilation relies on radiation, convection, or specially positioned ventilation slots. Dynamic ventilation relies on forced air circulation in sealed or vented enclosures (e.g., heat exchangers and closed-loop air conditioners). Air-to-air or air-to-water heat exchangers, for example, can cool and recirculate clean air inside the enclosure while maintaining the protective seal against contaminants with little or no maintenance. So, panels can maintain their NEMA 4 or 12 rating and still cool their components. In panels where control-system failure is unacceptable or detrimental to the process, engineers should include a temperature sensor and an alarm that notifies personnel of any impending thermal problems.

ELECTRICAL SYSTEMS. The control panel design must address the electrical systems, including power source, power conditioning, internal power distribution, and wiring.

Power Source. Control panels should receive power from a low-voltage power source (e.g., 120 V AC, 208 V AC, or 24 V DC); this can be either a single- or three-phase power source (power distribution breaker panel) for AC.

Equipment suppliers typically furnish panels with their equipment—including motor-driven equipment, which requires a higher voltage (e.g., 480 V AC, three-phase)—and put the motor starters in the control panel, claiming that it is better because users only need one power feed for the packaged equipment. If at all possible, low-voltage operator controls should not be combined with motor-starter equipment because doing so unnecessarily exposes non-electrical personnel (operators) to higher voltages. Most water and wastewater plants include motor-control centers (MCCs), which centralize motor-control and other power-distribution equipment in a controlled environment (Crockett, 2002). If the utility does not use MCCs, then it should specify that the vendor supply a separate enclosure for motor-control equipment.

Safety Concerns. Employee safety is critical. The Occupational Safety and Health Act (OSHA) includes many electrical safety standards (by reference), including the

National Electrical Code (NFPA 70) and the National Electrical Safety Code (NFPA 70E), which addresses the electrical safety requirements for workplaces to safeguard employees. For example, Section 2-1.3.3.2 of NFPA 70E defines a flash protection boundary and provides the equation for calculating this boundary. For systems that operate at 600 V or less, the flash protection boundary should be 1.2 m (4 ft, 0 in.) based on the product of the clearing times of six cycles (0.1 second) and an available bolted fault current of 50 kA, or any combination not exceeding 300 kA cycles (5000 A-s). There is another equation for systems that operate above 600 V.

According to Section 2-1.3.3.3 of NFPA 70E, employees in flash hazard areas must wear flame-resistant clothing. Clothing made from synthetic materials (e.g., acetate, nylon, polyester, or rayon) either alone or blended with cotton, should not be worn.

Panels should include an internal power-distribution panel with enough circuit breakers to distribute power to the panel's components. Spare breakers also should be provided. Fused terminals can also be used. Each panel also should include a main power switch or breaker to remove power from the panel. Panels often receive wiring from various power sources, which may prevent the main disconnect from being able to remove all power from the panel. If so, the front of the panel should include a label clearly alerting staff to the situation.

Motor-Control Centers. A motor-control center is a large enclosure designed to house standard motor control equipment. They basically consist of two parts: section and units. A typical section (or structure) measures about 2290 mm (90 in.) high and houses one or more plug-in units (motor controls). A motor-control center may contain any combination of equipment, including

- Full-voltage reversing or nonreversing combination motor-control units;
- Solid-state industrial controllers (e.g., adjustable-speed drives, programmable controllers, and protective relays);
- Incoming line equipment (e.g., main lugs, fusible switch, isolation switch, or air circuit breaker);
- Control or lighting transformers; and
- Special equipment assemblies (e.g., pushbuttons, selector switches, indicating lights, and auxiliary contacts) that are an integral part of the above units (NEMA ICS 18-2001).

The advantages of MCCs include the following:

- Only one set of incoming power wires are needed because the MCC distributes power to all devices via its power bus network;

- They are prewired at the factory and only need wiring to the motor and external controls;

- Installation simpler than for individual motor-starter panels, which require several wall-mounted enclosures;

- They can be easily expanded to accommodate new units or sections;

- They provide a central location for all motor controls;

- A faulty unit can be unplugged without de-energizing or disturbing other units, allowing a spare unit to be installed; and

- They look better than the maze of conduits needed for separate starter enclosures.

Motor-control centers now include intelligent devices and networks that provide diagnostics and predictive failure information. Power-monitoring devices integrated with software show real-time data, trending, component history, wiring diagrams, user manuals, and spare parts. Motor-control centers can easily communicate with the process-control system via these networks or a separate process-control network.

Wiring. The panel specifications should describe the internal wiring, including type of wire, minimum wire gauge, use of wireways, types of termination connectors, and labeling. Some spare capacity should be required in the wiring scheme for future changes or field modifications (e.g., terminal strips and fuses); typically, 20% is adequate. National Electric Code Article 310 describes various wiring methods, types of wiring, and suitable applications.

Wiring sizes typically should conform to NEC requirements for limits on safe conductor current carrying limits, or ampacity. Type MTW stranded copper wire with thermoplastic insulation is commonly specified; it is rated for 600 V and a maximum operating temperature of 90 °C. Shielded cable should be used for signal wiring. Wiring for AC controls and power should be separated from wiring for DC signals. Wiring should follow the color code defined in the specifications.

Parallel groups of wiring in covered wireways should be bundled for a neat appearance, but wireways should not be overfilled and the bundles should not exceed 25 mm (1 in.). All wiring must be adequately supported and restrained to prevent sagging or other movement. Where wires pass through the panel walls, suitable bushings should be provided to prevent the insulation from being abraded or cut. Also, the wiring should be installed so if wires are removed from one device, power will not be disrupted to any other device. Provisions also must be made for panel grounding: both safety and signal.

All field wiring should terminate in clearly labeled terminal strips with terminal numbers that match the panel-wiring drawings. The strips should include a separate terminal for grounding all signal-cable shields. Two common options are compression terminals and barrier terminal strips, which involve screwed connections and crimp-on connectors. Some users prefer compression terminals because they are less labor-intensive. Soldered terminals should be avoided because they do not provide any flexibility. Also, all powered devices should be individually fused via fused terminals or be fed from a separate breaker.

Surge Protection. Surge-protection devices are a cost-effective method for protecting critical, microprocessor-based equipment from the damaging effects of electrical transients. Transient voltage spikes may be the result of lightning, utility grid switching, electrical accidents, plant power generators switching online, copiers, welding equipment, and large motors. Surge-suppression devices quickly convert the high energy transients to levels that are safe for AC equipment. This not only prevents catastrophic failure but also extends the life of any electronic equipment.

Surge-suppression devices must meet UL 1449 standards. They should be provided for all panels with electronic instruments. Ground wires for all surge-protection devices should be connected to a good earth ground in compliance with NFPA 70 Article 250. Additional grounding guidelines can also be found in American National Standards Institute (ANSI)/ISA 12.01-1999 and ANSI/ISA 82.02.01-1999. PC-based consoles typically receive power from either a central uninterruptible power supply (UPS) distribution panel or individual UPS units mounted in the console.

FACTORY TESTING. All panels should undergo a factory test before being shipped to the job site. Whether this test should be witnessed depends on how complex the panel is. The factory test should verify that system components are functioning properly and meet the user's functional and performance requirements, as defined in the panel specifications.

The test procedure is developed by the panel supplier. It should identify all testing to be performed at the factory. Panel testing typically includes

- Verifying that all components are properly mounted, wired, and arranged according to the approved panel drawings;
- Confirming that all nameplates and tags are correct;
- Ensuring that all wire sizes, insulation, and color coding are correct and that power and signal wiring are separated;
- Confirming that terminal block types, rating, spares, and identification numbers are present and correct;
- Ensuring that the panel is properly grounded;
- Verifying that the enclosure rating, finish, and color are correct;
- Checking the AC/DC power for correct voltage and polarity for DC; and
- Testing all system components for functionality and proper operation.

Panels equipped with programmable devices (e.g., PLCs) must include testing of the device's software via a PC loaded with appropriate control system software. The testing should simulate actual performance conditions as closely as possible.

Panels equipped with operator interfaces must test the interface—including graphics displays—for correctness and functionality. Interface testing also should confirm the correctness and functionality of the following:

- Log in procedures, password assignments, and security levels;
- Alarm management and displays;
- Log and report generation (typical reports of each type should be generated and printed);
- Screen navigation and control actions via touchscreen, mouse, or keypad; and
- Graphic display color scheme and symbols.

A new trend has been the use of simulation software to test PLC- or DCS-equipped panels. (For more information on simulation software, see Chapter 3.)

INSTALLATION

Successful panel installation depends on careful planning. Panels to be installed in newly constructed areas, for example, should be scheduled for delivery after major mechanical, electrical, and other building infrastructure work is completed, and provisions must be made to protect the panel until startup. Floor-mounted panels should be installed on a concrete equipment pad at least 102 mm high (4 in. high). The control room should have a raised floor to make console wiring easier. The following items should be considered for all panel installations:

- Method of field cable penetration (i.e., top or bottom entry),
- Area classification,
- Maintenance access,
- Minimum clearance space around panel,
- Panel equipment security (provisions to prevent unauthorized access), and
- Panel shipping requirements (see ISA RP 60.11, 1991).

PANEL LIST. A panel list should be included with the panel specifications. The list identifies every panel required for the particular job, along with the following data:

- Panel description,
- Panel installation location,
- List of drawings associated with panel,
- Panel NEMA rating, and
- Panel supplier.

SPECIAL REQUIREMENTS. *UL 508: The Underwriters Laboratory Standard for Industrial Control Equipment.* So long as the equipment will be used in ordinary locations, as defined by the NEC, UL 508 covers industrial control devices for starting, stopping, regulating, controlling, or protecting electric motors. It also covers industrial control devices or systems that store or process information and have an output motor control function(s), as well as industrial control panels that are assemblies of industrial control devices and other devices associated with the control of motor-operated and -related industrial equipment. The devices must be rated 1500 V or less and intended for use in an ambient temperature of 0 to 40 °C (unless specifically indicated for use in other conditions). In other words, UL 508 covers such devices as

- Manual, magnetic, and solid-state starters and controllers;
- Pushbutton stations, including selector switches and pilot lights;
- Control circuit switches and relays;
- Time-delay relays and switches;
- Solid-state time-delay relays;
- Programmable controllers;
- Definite purpose controllers;
- Solid-state logic controllers; and
- Industrial microprocessor and computer systems.

Equipment intended for use in hazardous locations, as defined by NFPA 70 of the NEC, are covered by the UL 698, *Standard for Industrial Control Equipment for Use in Hazardous (Classified) Locations*. Electrical instruments are covered by UL 1437, *Standard for Electrical Analog Instruments—Panel Board Types*.

Options for Panels in Hazardous or Corrosive Areas. *Intrinsically Safe Systems.* Intrinsic safety is a wiring method that allows non-explosion-proof equipment to be used in hazardous areas. It involves putting barriers in a safe area and then wiring them to the field equipment (associated apparatus). This limits the allowable energy available to the hazardous area (Art. 504, NFPA 70, 2002; ANSI/ISA RP12.06, 2002; Flanders, 2000).

Purged Enclosures. Purged enclosures are another acceptable method for complying with the hazardous area rating (NFPA 496, 2003). It involves using a source of clean air or inert gas (nitrogen) to pressurize the enclosure so no flammable vapors can enter. The standard defines the following types of purges:

- *Type X Pressurizing.* Reduces the classification in the protected enclosure from Class I, Division 1, or Class I, Zone 1, to unclassified.

- *Type Y Pressurizing.* Reduces the classification in the protected enclosure from Division 1 to Division 2 or Zone 1 to Zone 2.
- *Type Z Pressurizing.* Reduces the classification in the protected enclosure from Class I, Division 2, or Class I, Zone 2, to unclassified.

National Fire Protection Association Standard 496 also requires the use of an alarm, indicator, or cutoff switch on all protected enclosures. These devices must either alert the user (Type Y and Z) or automatically cut off power to all protected equipment (Type X) when the protected enclosure(s) lose(s) positive pressure. The code also requires an alarm to indicate failure of the protective gas supply. In addition, NFPA 496 requires all protected enclosures to be marked with a "permanent label" in a "prominent location" near all doors and access covers. Labels for Class I locations must state:

WARNING: PRESSURIZED ENCLOSURE
This enclosure shall not be opened unless the area is known to be free of flammable materials or unless all devices have been de-energized.

According to Section 3-4 of NFPA 496, "Additional Requirements for Type Y or Type Z Pressurization":

> The protected equipment shall not be energized until at least four enclosure volumes of the protective gas ... have passed through the enclosure while maintaining an internal pressure of at least 25 Pa (0.1 in. water).
> Exception: Equipment shall be permitted to be energized immediately if a pressure of at least 25 Pa (0.1 in. water) exists and the atmosphere within the enclosure is known to be nonflammable.

According to Sections 3-5.1 and 3-5.2 of NFPA 496, "Additional Requirements for Type X Pressurization":

> A timing device shall be used to prevent energizing of electrical equipment within the protected enclosure until at least four enclosure volumes of the protective gas have passed through the enclosure while maintaining an internal pressure of at least 25 Pa (0.1 in. water).
> If the enclosure can be readily opened without the use of a key or tools, an interlock shall be provided to immediately de-energize all circuits within the enclosure that are not approved for Class I, Division 1, or Class I, Zone 1, locations when the enclosure is opened.

Additionally, according to Section 2-2.3 of NFPA 496 "Standard for Purged and Pressurized Enclosures for Electrical, Enclosures":

> In Division 1 locations, where the conduit or raceway entry into a pressurized enclosure is not pressurized as part of the approved protection system, an explosion proof conduit seal shall be installed as close as practicable to, but not more than 18 in. (458 mm) from, the pressurized enclosure.
>
> In Division 2 locations, an explosion proof conduit seal shall not be required at the pressurized enclosure.

However, while explosion-proof enclosures require conduit seals on all cable entries, in accordance with Sections 501-5 (D) and (E) of NFPA 70, protected enclosure cable entries for pressurized enclosures can be sealed in accordance with Section 501-5 (e) (4), which permits the use of compression gland fittings or other sealed cable entry fittings, rather than approved conduit seals, if the wiring method is otherwise suitable for the hazardous location.

Purging can also be used for panels in corrosive areas, although the requirements for hazardous areas would not necessarily apply. The purpose of purging in those areas is to keep the corrosive fumes out of a panel's interior, thereby extending the life of its components.

NEW DEVELOPMENTS OR TRENDS

OPERATOR INTERFACES. One trend in operator interfaces design practice is to eliminate the hard drive in the operator interface terminal. The HMI software is installed in the operator interface terminal with Windows CE operating system using flash memory. This eliminates problems with hard-drive component failures (e.g., fans, moving parts) and maintenance.

One of the latest developments is the use of thin client workstations, which rely on a central server for the software. It is described in the literature as

> "object–server architecture. This architecture is similar to the client–server arrangements except that all control information will be built into an object server. A control object will be a control loop point and all of its functionality, including a graphic of the valve and instruments, the process variable, the set point, the alarms, and every other characteristic of the loop. The object will

be available to every other program, such as the HMI client, to be dropped in a display, a historian, or a Web page" (Liptak, 2002).

Among other advantages, it simplifies software maintenance because all updates are done at the server.

TOUCHSCREENS. Touchscreens are becoming more common in operator workstations. According to G. K. McMillan (1999), "This is probably the simplest of the graphic input devices to use because it requires no associated button for triggering events. This can be done by sensing a change in the presence of a finger or some other object." Currently, six touchscreen technologies are available: resistive, capacitive, infrared, acoustic wave, near field imaging (NFI), and dispersive signal technology (DST).

Resistive. A resistive touchscreen typically consists of a flexible top plastic sheet and a sheet of glass separated by insulating dots (Figure 8.1). A transparent metal oxide coats the inside surface of both sheets. When voltage is added, a current is produced across each layer. Pressing the top sheet creates an electric contact between resistive layers, essentially closing a switch in the circuit.

Resistive touchscreens have many advantages. They are pressure sensitive, work with any stylus, have high touch resolution, and respond rapidly. They also are inexpensive; widely available; and not affected by dirt, dust, water, or light. Their disadvantages are low clarity (75%) and susceptibility to damage by a sharp object. The technology is available in four, five, or eight wires; the five-wire version currently is the most reliable.

Capacitive. Capacitive touchscreens consist of a transparent metal oxide bonded to the surface of a sheet of glass (Figure 8.2). When a bare finger touches the glass, its inherent capacitance draws some current from each corner of the touchscreen, where voltage has been applied. The system measures the capacitance of the person touching the overlay and the frequency changes to determine the X and Y coordinates of each touch. A conductive stylus can be used for signatures, on-screen annotations, or applications requiring precise input. Their advantages include high durability; high touch resolution; high image clarity (88%); and lack of susceptibility to dirt, grease, and moisture. Their disadvantages include costs and the need for shielding from internal electronics noise and external electromagnetic interference.

Infrared. Infrared touchscreens are based on light-beam interruption technology (Figure 8.3). Basically, the display surface has a frame with light sources or LEDs on

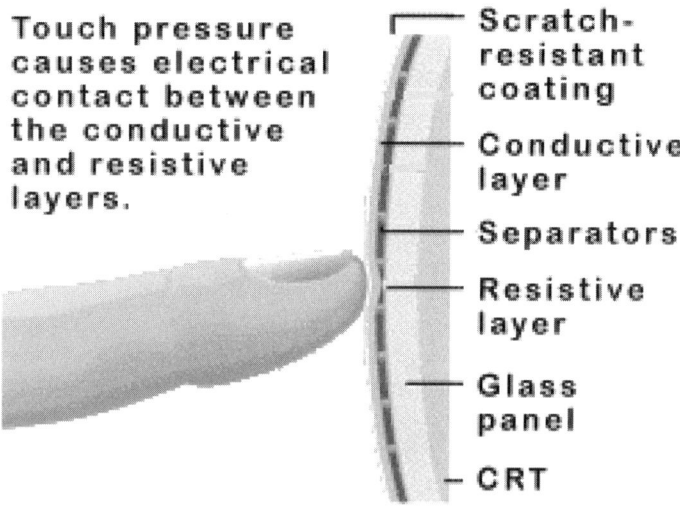

FIGURE 8.1 A schematic of a resistive touchscreen.

one side and photosensors on the opposite side, creating an optical grid across the screen. When an object touches the screen, it causes a drop in the signal received by the photosensors. This durable technology provides high image clarity at reasonable cost, responds rapidly to any touch, and can be used with a gloved hand. It can be used on large displays (e.g., plasma screens) and is less sensitive to damage than other touchscreen technologies (e.g., resistive and capacitive). However, it has lower resolution than other touchscreen technologies, so it is not recommended for Microsoft Windows applications Also, it cannot be washed down directly and is subject to activation by flying objects.

Acoustic Wave. There are two types of acoustic wave touchscreens: surface acoustic wave (SAW) and guide acoustic wave (GAW). Both use transducers mounted at the edge of a glass screen to emit ultrasound waves along two sides. In SAW technology, the waves travel across the glass surface, while in GAW technology, the waves travel

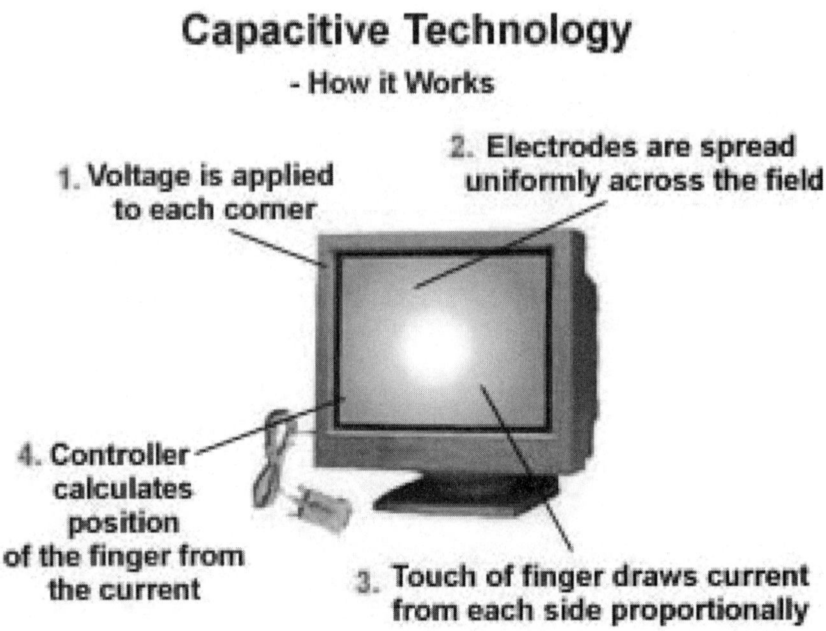

Capacitive Technology
- How it Works

1. Voltage is applied to each corner
2. Electrodes are spread uniformly across the field
3. Touch of finger draws current from each side proportionally
4. Controller calculates position of the finger from the current

FIGURE 8.2 A schematic of a capacitive touchscreen.

through the glass (Figure 8.4). The ultrasonic waves are received by sensors on the opposite sides. When a finger or soft-tipped stylus touches the screen, the sound energy is absorbed, causing the wave signal to weaken. Guide acoustic wave screens are relatively insensitive to surface effects because the acoustic energy travels in the glass, so they perform reliably in applications where acoustic touchscreens previously could not operate.

Because acoustic wave touchscreens have no layers, they are durable, clear, and therefore recommended for public information kiosks, training, and other high-traffic indoor environments. They also have a high touch resolution (90%). However, they must be touched by a finger, gloved hand, or soft-tip stylus; a hard object will not work. Also, they are not completely sealable and can be affected by large amounts of dirt, dust, or water.

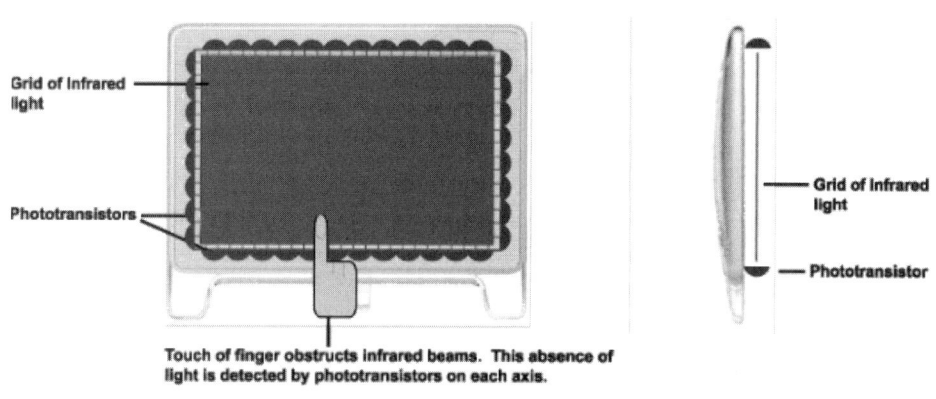

FIGURE 8.3 Schematic of an infrared touchscreen.

Near Field Imaging. A near field imaging touchscreen uses a sophisticated sensing circuit that can detect a conductive object (e.g., a finger or conductive stylus) through a layer of glass, as well as gloves and other potential barriers (e.g., moisture, gels, or paints). The touchscreen consists of two laminated glass sheets with a patterned coating of transparent metal oxide in between them (Figure 8.5). When an AC signal is applied to the coating, an electrostatic field is created on the screen's surface. When a finger or conductive stylus touches the screen, the electrostatic field is disturbed.

The system has high optical clarity (83% or more), and its operations are impervious to dirt, moisture, and other surface contaminants, as well as scratches and other surface injuries. It also performs well in harsh environmental conditions, including outdoors applications, withstanding considerable shock and vibration. without compromising performance or safety. The primary disadvantage of this technology is that it has less touch resolution than other types.

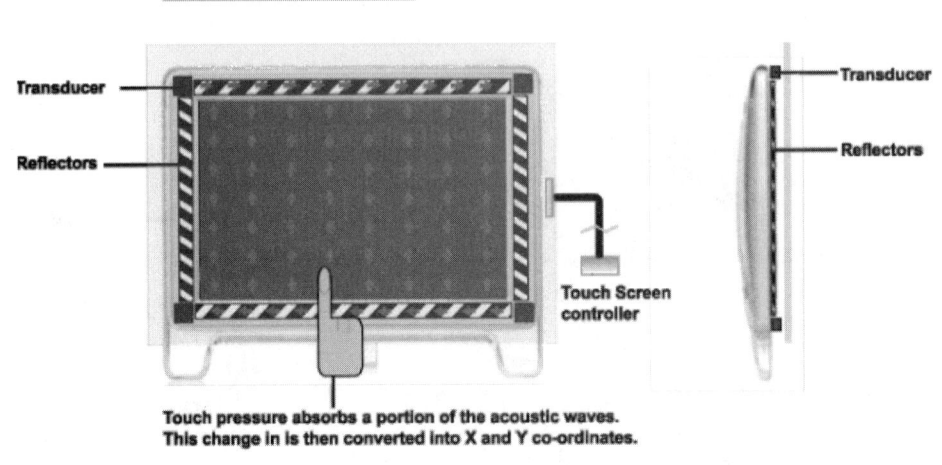

FIGURE 8.4 A schematic of an acoustic wave touchscreen.

Dispersive Signal Technology. A dispersive signal technology system basically measures the mechanical energy (vibrations) in a piece of glass that occurs when someone touches its surface. Typically, sensors on the back side of the glass convert the vibrations into electrical signals. Using advanced proprietary signal processing algorithms, the DST system determines the location of the touch, taking into account the profile of the vibration, glass dispersion effects, and other characteristics of the substrate.

Nearly any object can be used to activate the touchscreen, but the system will ignore hands or other objects resting on the touchscreen unless they tap the surface. Also, its performance is not significantly affected by contaminants (e.g., dirt, grease, and solids) or surface damage (e.g., scratches and gouges). It also scales readily, using similar electronics for all sizes of glass.

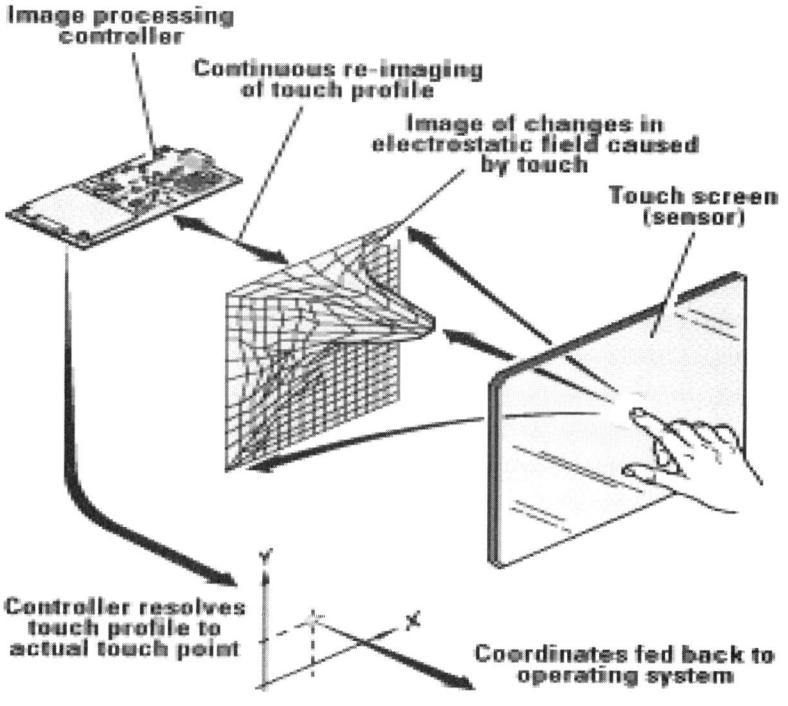

FIGURE 8.5 A schematic of a near field imaging touchscreen.

The substrate is pure glass, so it provides exceptional optical clarity and light transmission. Solar reflecting film can be added to block external light in high sunlight applications.

FLAT-PANEL DISPLAYS. Flat-panel displays are rapidly replacing cathode ray tube (CRT)-based technology in operator workstations, as they have in commercial-grade desktop PCs. The following discussion focuses on industrial-grade flat-panel displays, which are field-proven to withstand shock, vibration, dust, and washdown (NEMA 4/12). Some versions are suitable for hazardous areas.

There are basically two types of flat-panel displays: PC-based displays and monitors.

PC-Based Displays. Industrial-grade PC-based workstations are available in various sizes with an active matrix TFT LCD touchscreen, hard drive, and power supply. They include the latest Intel processors and Microsoft Windows operating system, a 10-Base-T Ethernet interface, and even expansion chassis that permit more peripheral component interconnect (PCI)/industry standard architecture (ISA) slots.

The PCI–ISA Passive Backplane Standard was developed by the PCI Industrial Computer Manufacturers Group's member companies to adapt desktop PCI standards for industrial applications. The PCI interface originally developed by Intel is a processor-independent interface (does not use the motherboard) operating on its own clock rather than that of the processor devices used for I/O communication in industrial PC-based control systems. In May 1994, the manufacturers' group was established to define an industrial PCI/ISA passive backplane and central processing unit (CPU) card interface specification. The group's technical committee generated a comprehensive specification for a passive backplane architecture to support both PCI and ISA I/O buses. The PCI–ISA passive backplane standard moves all of the components typically located on the motherboard to a single plug-in card (www.picmg.org).

The bright displays also can be read in outdoor or high ambient light. Utilities can purchase the workstations with the HMI software preloaded by the manufacturer. This option is cost-effective for standalone control systems, small applications, or vendor-packaged equipment.

Another trend is thin client PCs, which have embedded processors and operating systems (e.g., Windows CE and Linux), but no hard drive. Applications are processed by a remote server. The thin clients use less power, typically require no active cooling, and have no moving parts. They use protected flash memory for storage. Also, their wireless network bandwidth requirements are lower than those of standard PCs. In a thin client environment, only mouse movements, keystrokes and screen updates are transmitted to and from the end user. Because only the changes on the virtual display being generated on the server are sent to the client, bandwidth requirements are much lower and the display can be updated at the same speed as could a screen running on the server. A 10-Mbs network will easily handle the traffic between a thin client and the server. The "thin" in thin clients refers to a PC that is fanless and diskless, with no moving parts. Control system redundancy is less expensive because only one extra server is needed, rather than multiple workstations. Maintenance is also simplified, because software only needs to be updated and backed up at the server(s). The server(s) can be installed in a secure location, thereby reducing unauthorized changes or tampering. Thin clients are also easier to replace than traditional workstations.

Monitors. Monitors are PC-compatible devices that accept standard video graphics adapter (VGA) video signals and full-motion video. The VGA is a video standard introduced in 1987 for the IBM PS/2 series of PCs and can provide 640×480 pixels with 16 colors and 320×200 pixels with 256 colors. The VGAs are available in the same housing as the PC-based workstations.

Plasma. Plasma panels (Figure 8.6) are an array of cells called pixels. Each pixel consists of three subpixels containing phosphors that produce colored light (red, green, or blue) when they react with plasma. The phosphors are the same as those used in CRT devices (e.g., televisions and computer monitors). Each subpixel is individually controlled by advanced electronics to produce more than 16 million colors, so the overall display provides perfect images in rich, dynamic colors.

In addition to bright, crisp images, flat-panel plasma screens are less than 152 mm thick (6 in. thick) and absolutely flat, which reduces glare and broadens the viewing angle (more than 160 degrees), so no matter where people are in the room, they can see the display. The panels are backlit, so they work well in bright environments. Currently, large screen [1067 mm (42 in.) or more] plasma displays are finding their way into control rooms.

Digital Light Processing. Digital light processing (DLP) is a technology that accurately and precisely reproduces lifelike color images (Figure 8.7 and Figure 8.8). Digital light processing displays produce filmlike, all-digital images via a projection lamp, an electronic video signal from a source (e.g., a VCR or computer), and a thumbnail-size semiconductor light switch called a *digital micromirror device* (DMD).

Digital micromirror devices are micromechanical silicon chips that are less than 16 mm ($5/8$ in.) on each side but contain about 442,400 tiny adjustable aluminum mirrors and a wealth of electronic logic, memory, and control circuitry. Each mirror is mounted on a hinge so it can be individually tilted back and forth; it produces one picture element (pixel) and provides sharp quality and intricate detail. The mirrors are protected by a sheet of glass.

When the lamp and projection lens are positioned correctly, the DLP system processes the input video signal and tilts the mirrors accordingly to generate a digital

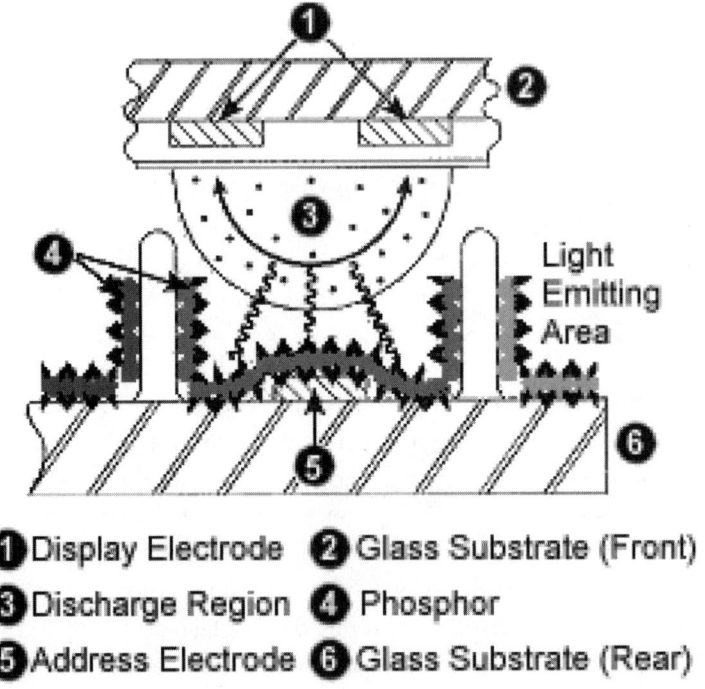

FIGURE 8.6 A schematic of a plasma display.

image (Yoder, 1997). The process takes 10 microseconds. These large projection displays can be used in treatment plant control rooms.

The digital light processing systems are more efficient than competing LCD systems because they are based on the DMD and do not require polarized light. Also, because the DMD is a reflective device, it does not absorb heat, even when projecting a fixed pattern over a long time, so "burn in" does not occur.

Digital light processing displays typically are 1270 mm (50 in.) or more (diagonal). Screens can be combined seamlessly for larger displays. The system uses color-correction circuitry for optimal blending between screens when in the stacked cube configuration typical of larger screens.

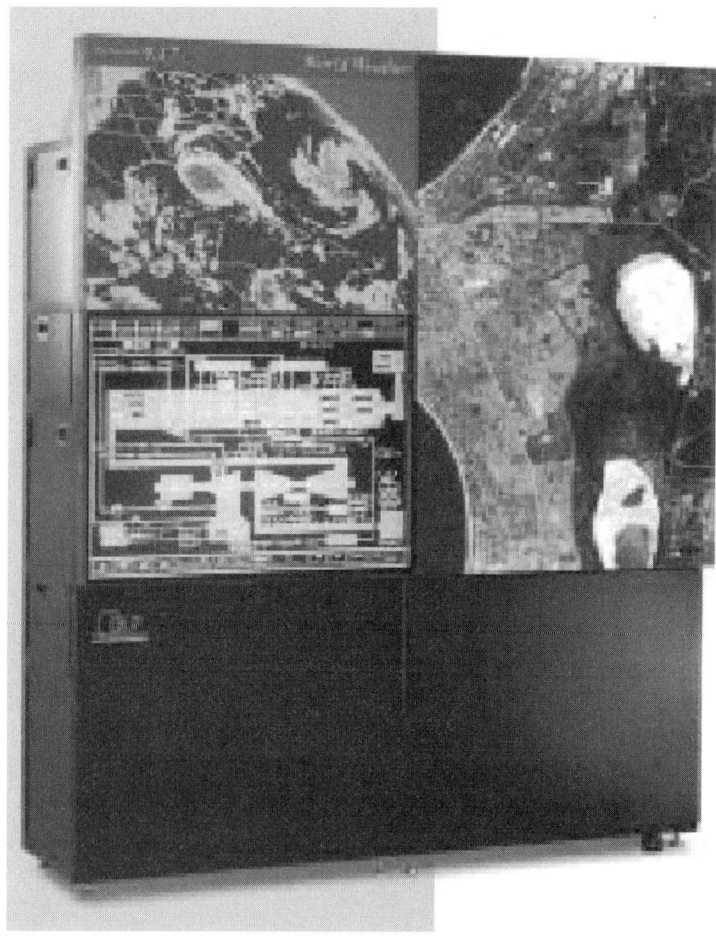

FIGURE 8.7 A large digital light processing display. (Courtesy of Mitsubishi.)

WIRELESS OPERATOR INTERFACE. Wireless operator interfaces enable operators to leave the control room while still overseeing the I&C system. A wireless operator interface can be a portable PC, handheld PC, pocket PC, or personal digital assistant (PDA) device with the appropriate wireless interface. The wireless interfaces take advantage of the new standards for wireless communications (IEEE 802.11b,

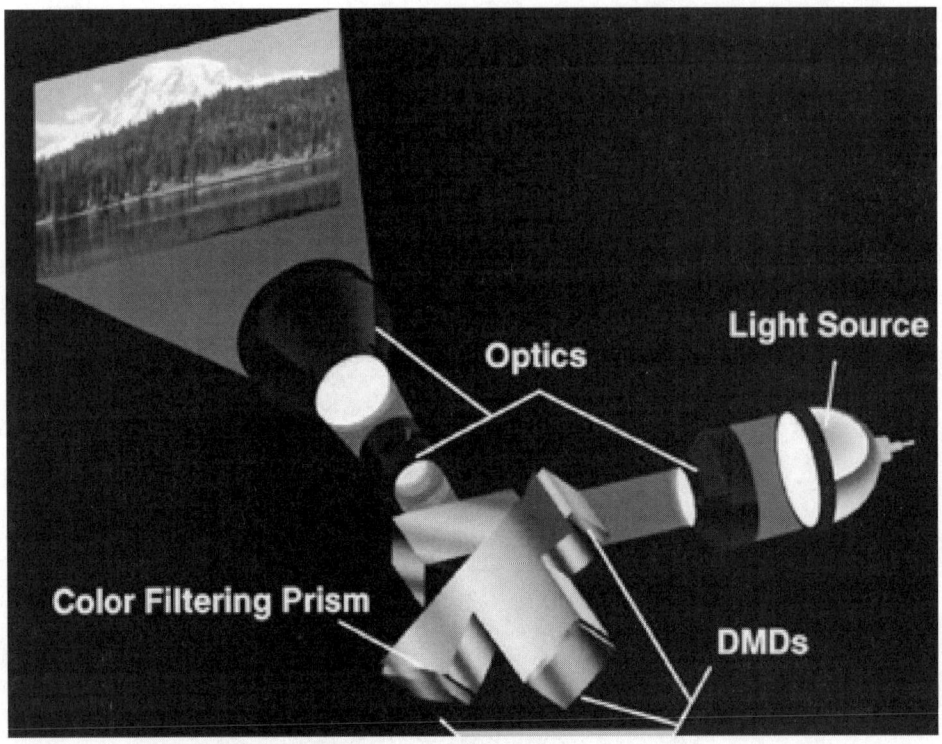

FIGURE 8.8 A digital light processing display. (Courtesy of Texas Instruments.)

802.11a, 802.11g) According to *Wireless Networks for Industrial Automation*, the new standard (IEEE-802.11g) has caused suppliers of wireless equipment to drop prices significantly, allowing the wireless interface to offer a very economically attractive alternative to the wired interface.

These interfaces are not replacing existing wired control-room interfaces; rather, they are complementing and enhancing them. For example, operators can now visually inspect a process and troubleshoot problems and keep the I&C system at their fingertips while they do so, even when they are away from the control room.

DIGITAL RECORDERS. Paperless chart recorders have become the preferred option in new and upgraded treatment plants. These recorders emulate the display of a traditional paper chart recorder but use a high-quality color LCD display, electronic recording media, and data logger storage methods. In addition to emulating traditional paper charts, the LCD can provide any number of trend-display variations, online reports, alarm annunciations, and intuitive menu-driven configurations. The recorders also enable utilities to store massive amounts of data via the following standard media storage :

- Solid-state "flash" storage cards to 2 GB,
- Hard drives (1 GB or higher),
- Secure digital cards (up to 1 GB), and
- Universal serial bus (USB) memory keys.

For example, one secure digital card can hold the equivalent to 60 rolls of chart paper, and it provides 60 times better resolution than a roll of paper.

DATA LOGGERS. Data loggers and digital recorders function in much the same way. Many manufacturers offer them as combined units. The technology used for data loggers and digital recorders is similar, but data loggers tend not to provide the visual displays and digital recorders are used more frequently to collect data for downloading to a PC for additional analysis.

SELECTION CONSIDERATIONS FOR RECORDERS OR DATA LOGGERS. Given the wide variety of instruments for recording or logging acquired data, the prospective user needs to review the key factors involved in making a suitable selection. To help, here are some key questions and basic points to consider:

1. Will the data loggers or recorders be portable or permanently mounted? Will the instrument be mounted bench-top or in a panel?
2. Is the mounting location in a process area, remote field location, or control room?
3. What type of instrument power is available?
4. How many inputs will be recorded or logged (channels, points, or records)?

5. What types of inputs are involved [voltages, thermocouple (T/C) or resistance temperature detector (RTD), 4-20 mA]?
6. Do different input types need to be handled by the same unit?
7. What type of recording or logging is required—continuous single-point or multipoint? And if the latter, what minimum scan cycle is required?
8. What degree of accuracy is required? Accuracy may be defined as the closeness to the actual signal that the measured value or trend position takes—stated in either a plus/minus percentage of full scale or percentage of reading?
9. Is a communication interface required to transmit measured data to a host computer? Is the instrument set up remotely? Does the instrument connect to an external printer?
10. Is log-type recording desired instead of or in addition to continuous-trend recording?
11. Is color differentiation desirable for multiple trend lines?
12. If the instrument is to perform alarm functions, how many alarm setpoints are needed per channel? What types of alarms are desired, for example, threshold, rate, or delta? Are relays contacts required to handle external alarm outputs? If so, how many?
13. What, if any, signal processing is required? These may include linear scaling (conversion to engineering units), thermocouple characterization, difference calculation, or square root calculation (as for a Venturi flowmeter).
14. Are any higher-level math functions required? These may include absolute value, logarithm, exponential, maximum, minimum, time average, group average, summation, standard deviation, and integration functions.
15. What secondary visual indications are desired: analog bar graph in percentage of full scale, analog scale indication in percentage of full scale, digital channel (point) number and measured value, alarm status, engineering units?
16. What operating modes will be required? Options include normal monitoring, which involves monitoring at a set scan interval, or logging at set intervals.

REMOTE ANNUNCIATION. Remote annunciation systems are currently used in many wastewater treatment plants (RACO, 1994). Staff can be notified of alarms via phone, fax, pager, or e-mail.

Alarm autodialing and remote monitoring systems can allow users to digitally record speech messages for later retrieval or for archiving to digital recording media. This can be of benefit to facilities in determining circumstances surrounding an alarm condition. Various hardware configurations are available (32 contact channels and up to 16 analog channels is typical). Remote autodialer alarm units interface with programmable controllers and gather alarm information from remote sensors, which then is passed on to the preprogrammed remote dialer units. The remote dialer unit then automatically starts calling a list of preprogrammed phone numbers, dialing them one by one until it gets an acknowledgment. The systems are remotely programmable via any touch-tone phone over a standard telephone line, so operators can turn equipment on or off and perform remote terminal unit functions from any phone.

Various manufacturers offer cellular-telephone-based units, which provide temporary or permanent cell phone service when conventional telephone lines are unavailable. The units can be powered by AC or DC, but DC-powered systems allow for solar-powered monitoring. These autodialing systems provide more security, as well as backup to a traditional phone network in case of a telephone line outage or failure.

REFERENCES

American National Standards Institute; Instrumentation, Systems, and Automation Society (2002) *Wiring Practices for Hazardous (Classified) Locations—Instrumentation, Part 1*, RP12.6.01 (R2002); Instrumentation, Systems, and Automation Society: Research Triangle Park, North Carolina.

Caro, D. (2005) *Wireless Networks for Industrial Automation*, 2nd Ed., Instrumentation, Systems, and Automation Society: Research Triangle Park, North Carolina.

Crockett, J. (2002) Motor Control Centers a Must. *Consulting-Specifying Engineer*, Dec 1. www.csemag.com (accessed March 2004).

Code of Federal Regulations (2005) *Occupational Safety and Health Standards*, 29 CFR 1910. www.osha.gov.

Flanders, P. (2000) Demystifying Intrinsically Safe Systems, *Proceedings of ISA Expo 2000*; New Orleans, Louisiana, Aug 21–24.

Instrumentation, Systems, and Automation Society (1991) *Crating, Shipping and Handling for Control Centers*, ISA RP 60.11; Instrumentation, Systems, and Automation Society: Research Triangle Park, North Carolina.

Instrumentation, Systems, and Automation Society (1992) *Annunciator Sequences and Specifications*, STD S18.1; Instrumentation, Systems, and Automation Society: Research Triangle Park, North Carolina.

Institute of Electronics and Electrical Engineers (1997) *Wireless Local Area Networks*, Standard 802.11; Institute of Electronics and Electrical Engineers: Piscataway, New Jersey.

Liptak, B. G. (1995) *Process Control*, 3rd Ed.; Chilton Book Co.; Radnor, Pennsylvania.

Liptak, B. G. (2002) *Process Software and Digital Networks*, 3rd Ed.; CRC Press; Boca Raton, Florida; Chapter 3.1.

McMillan, G. K. (1999) *Process/Industrial Instruments and Controls Handbook*, 5th Ed.; McGraw-Hill: New York.

National Electrical Manufacturers Association (2001) *Motor Control Centers*, NEMA Standards Publication ICS 18-2001; National Electrical Manufacturers Association: Rosslyn, Virginia.

National Fire Protection Association (1999) *Standard for Fire Protection in Wastewater Treatment and Collection Facilities*, NFPA 820; National Fire Protection Association: Quincy, Massachusetts.

National Fire Protection Association (2000) *Electrical Safety Requirements for Employee Workplaces*, NFPA 70E; National Fire Protection Association: Quincy, Massachusetts.

National Fire Protection Association (2002) *National Electrical Code*, NFPA 70; National Fire Protection Association: Quincy, Massachusetts.

National Fire Protection Association (2003) *Standard for Purged and Pressurized Enclosures for Electrical Equipment*, NFPA 496; National Fire Protection Association: Quincy, Massachusetts.

Norton, D. (2002) IDC – As reliable as screws and spring clamps. *Control Engineering* Europe, April. www.manufacturing.net.

PCI Industrial Computer Manufacturers Group Home Page.www.picmg.org (accessed June 2006).

RACO Manufacturing and Engineering Co. (1994) *WATER/Engineering & Management*, Bulletin 102; RACO Manufacturing and Engineering Co.: Emeryville, California.

Rosenberg, P. (2000) *Fundamentals of Telecom/Datacom*; EC&M Books, Prism Business Media: Overland Park, Kansas.

Shea, T. (2001) *North American Markets and User Needs for Operator Interfaces and Monitors in Discrete and Process Manufacturing: Are We Ready for the Internet Yet?* VDC report, 4th Ed.; VDC: Natick, Massachusetts. www. Vdc.corp.com.

SID (2004) 3M Touch Systems Unveils Break-through Touch Technology. www.3M.com (accessed June 2004).

Underwriters Laboratories (1999) *Standard for Transient Voltage Surge Suppressors*, UL 1449; Underwriters Laboratories: Northbrook, Illinois.

Underwriters Laboratories (2005) *Standard for Industrial Control Equipment*, UL 508; Underwriters Laboratories: Northbrook, Illinois.

Underwriters Laboratories (2002) *Standard for Industrial Control Equipment for Use in Hazardous (Classified) Locations*, UL 698; Underwriters Laboratories: Northbrook, Illinois.

Underwriters Laboratories (1998) *Standard for Electrical Analog Instruments—Panel Board Types*, UL 1437; Underwriters Laboratories: Northbrook, Illinois.

Yoder, L. (1997) *The Digital Display Technology of the Future*, Proceedings of INFOCOMM 97; Los Angeles, California, Jun 5–7.

SUGGESTED READINGS

Geib, J. L., Jr. (2000) *IDC is Changing Terminal Blocks*, connectorspecifier.com (accessed June 2004); Mar.

Construction Specification Institute Home Page, www.CSINet.org/MasterFormat (accessed June 2006).

Chapter 9

Connectivity Options for Process Control Systems

Basic Communications-System Terms	297	Communication System Networks	301
Client	297	Serial	301
Driver	298	*Modbus RTU and Modbus ASCII*	301
I/O Server	298	*DF1*	302
Master	298	*Other Proprietary Protocols*	302
Media	299	Proprietary	303
Peer-to-Peer	299	*Data Highway, Data Highway Plus, Data Highway 485*	304
Protocol	299	*Modbus Plus*	304
RS-232	299	*Genius*	304
RS-422/485	299	Fieldbus	305
Serial	299	*Speed of Operation*	306
Server	299	*Plant Layout*	306
Slave	299	*User Friendliness*	306
Transport	300	Ethernet	306
Basic Communications-Network Components	300	Ethernet in Depth	307
Physical Transport Media	300	*Copper Twisted-Pair Wires*	307
Network Transport Protocol	300		
Application Protocols	301		

Fiber Optic Cable	307	Wireless Solutions and Communications Software	317
Fiber Optic ≠ Ethernet	308		
Wireless Media	309	Communications and Connectivity Software	318
Bridging the Gap to Non-Ethernet Devices	309		
		Proprietary Drivers	318
Ethernet Encapsulation	309	*ActiveX Drivers*	319
Ethernet–Proprietary Network Bridges	312	*DDE Drivers*	321
		OPC Standards-Based Drivers	321
Common Ethernet Application Protocols	313	Communications Redundancy	323
Wireless Connectivity	313		
In-Plant Applications	314	*Instrument or Controller Network Interfaces*	323
Remote Sites	314		324
Wireless Security	315	*Communications Software*	325
Wireless Reliability	316	*HMI or SCADA Software*	326
		Web Connectivity	326

Communications are an integral part of the automation of water and wastewater treatment facilities. Modern communications technologies help make many of the benefits of automation systems possible. So, those designing and implementing automation systems should carefully evaluate the communications requirements early in the system design process.

 The limits and capabilities of communications networks and devices need to be considered to ensure that overall system requirements for functionality and performance are met. Users should not expect to purchase hardware and figure out the communications paths, connections, and software during system implementation. Failure to properly consider communications requirements when designing an automation system and to choose hardware and software that meet the design requirements will inevitably increase the cost and risk of failure. Those who do involve communications in their design process will reap the benefits that an effective and efficient communications system can deliver.

Instrumentation and control (I&C) system components typically communicate via any of the following paths (Figure 9.1):

- Instrumentation or field devices and a programmable logic controller (PLC) or a remote input/output (RIO) rack,
- PLC or RIO to a master PLC,
- PLC to an input/output (I/O) driver,
- I/O driver to a supervisory control and data acquisition (SCADA) system server,
- I/O driver to one or more historians,
- I/O driver to other applications,
- SCADA server to SCADA clients,
- SCADA server to historian(s), or
- SCADA server to other applications.

Following is a discussion of the most commonly used methods for establishing communication among I&C system components. The discussion is limited to PLC applications because they currently are the most common installations in water and wastewater treatment plants.

Note: To help readers become more familiar with communication technologies, the following discussion includes the brand names of some technologies. This use is in no way intended to advertise or endorse any particular vendor, but simply to help readers work more effectively with I&C system experts.

BASIC COMMUNICATIONS-SYSTEM TERMS

The following basic terms are useful to know when discussing communications systems and connectivity. The definitions provided here are intended to describe the fundamental concepts, not be 100% technically precise.

CLIENT. A *client* is a software application or device that wants to obtain information from or send information to another device. It typically initiates communications with a server device or software application.

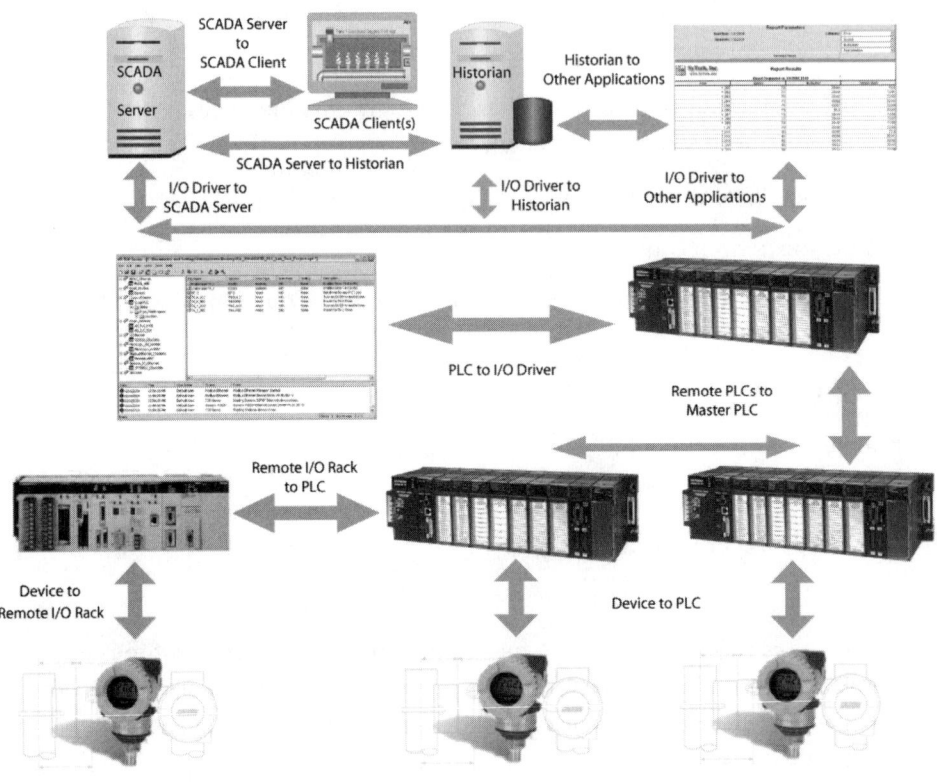

FIGURE 9.1 Communications is found throughout automation systems between a wide range of device and system types.

DRIVER. A *driver* typically is computer software used to communicate with a specific type of device, PLC, or control system.

I/O SERVER. An *I/O server* is a type of driver (although the two terms are often used interchangeably). The term comes from the concept of "serving up" the status of I&C system I/Os to a host system.

MASTER. In some communications networks, only one device is allowed to initiate conversations with all the others. That device is the *master*. The master device initiates communications with the other devices (slaves), and the slave devices respond when prompted.

MEDIA. *Media* refers to the cable (e.g., Ethernet or fiber optic) or other method for transporting information between two intelligent devices.

PEER-TO-PEER. In a peer-to-peer communications network, any device can initiate a conversation with any other device.

PROTOCOL. A *protocol* [e.g., RS-232 protocol and Modbus remote terminal unit (RTU) protocol] is a set of rules for enabling two intelligent devices to communicate with each other.

RS-232. *RS-232* is a standard that defines the pin connections and voltage levels needed to transmit data between two computer devices less than 15 m (50 ft) apart.

RS-422/485. *RS-422* and *RS-485* are standards that define the pin connections and voltage levels needed to transmit data among multiple computer devices that are 15 m (50 ft) apart or farther. Such a communication system is typically called a multidrop network because it enables one host computer to communicate with multiple devices over one cable.

SERIAL. *Serial* typically refers to communications involving RS-232 or RS-422/485 electrical protocols. In a serial connection, data are sent in a single bit stream—one bit after another—to the target device. The device typically responds by sending data in a single bit stream.

SERVER. A *server* is a software application that provides access to data and responds to requests from a client software application or device. Server software does not have to run on a computer hardware server or on a different computer than the client software.

SLAVE. In master–slave communications networks, a *slave* is a device that responds to requests from the master device. The slave never initiates communications.

TRANSPORT. *Transport* is the means of transferring data between communicating devices. It usually consists of a physical-cable or wireless-technology media and packaging protocol for packetizing and carrying data.

BASIC COMMUNICATIONS-NETWORK COMPONENTS

Communications networks consist of three basic layers: physical transport media, network transport protocol, and application protocols (Figure 9.2). The following definitions and railroad analogy should make these layers easier to distinguish.

PHYSICAL TRANSPORT MEDIA. *Physical transport media* include the cabling and the electrical specifications for signals sent via the cabling. It is defined by such standards as RS-232, RS-422, and 10-Base-T Ethernet. If the communications network were a railroad run by the postal service, physical transport media would be the train tracks.

NETWORK TRANSPORT PROTOCOL. *A network transport protocol* defines the basic "package" in which data is moved over the physical transport media. For example, transmission control protocol/Internet protocol (TCP/IP) basically is an Ethernet network transport protocol that enables data to be sent over 10-Base-T Ethernet cabling. If the communications network were a railroad run by the postal service, the network transport protocol would be the mail train.

Application Protocols – Ethernet/IP, Modbus TCP, GE SRTP = Contents of the Train Cars
Network Protocol Transport – TCP/IP Ethernet, FTP, HTTP = The train and cars
Physical Transport Media- RS-232, RS-422, 10-Base-T = The Train Tracks

FIGURE 9.2 The three layers that make up many types of communications networks.

APPLICATION PROTOCOLS. An *application protocol* defines how two devices can communicate with each other. It also defines the *metadata*—the data about the data (e.g., what the request is, how big it is, and which checksums can be used to find errors)—in the headers that precede the data itself. In other words, when Devices A and B use the same application protocol, Device A can send a request to Device B and expect a properly formatted reply. If the communications network were a railroad run by the postal service, the application protocols would be the language and format of the letters on the train and the data itself would be the letters' contents.

So if someone says, "I am using an RS-232 protocol," they are only telling part of the story. However, if someone says, "I am using a Modbus Serial RTU protocol on an RS-232 network", then they have given more complete information. Modbus Serial RTU is an open application protocol, and RS-232 is an open physical-transport-media standard.

Engineers should never accept "Oh, it's just RS-232" for an answer if they are responsible for creating an interface to the system involved. They should get the rest of the story.

COMMUNICATION SYSTEM NETWORKS

There are four major types of communication system networks: serial, proprietary, Fieldbus, and Ethernet.

SERIAL. Process-control systems have used serial communication networks since the 1970s. They enable controllers to talk to one another so long as only one request or response is sent at a time. These networks are based on RS-232 or RS-422 and RS-485 standards, depending on the number of devices involved (Figure 9.3). An RS-232 network uses the same communications ports as those found on the back of most personal computers (PCs). An RS-422 or RS-485 network can be connected to a PC via an interface board from such manufacturers as Digi, Sealevel Systems, Black Box, and B&B Electronics. If the PC does not have any available interface slots, a converter box can be used.

Some examples of serial application protocols include:

- *Modbus RTU and Modbus ASCII.* Originally created by Modicon for its PLCs, Modbus protocols are considered the de facto standard for serial communications. There are more than 6000 types of devices known to support Modbus, and this protocol has been open for more than 20 years.

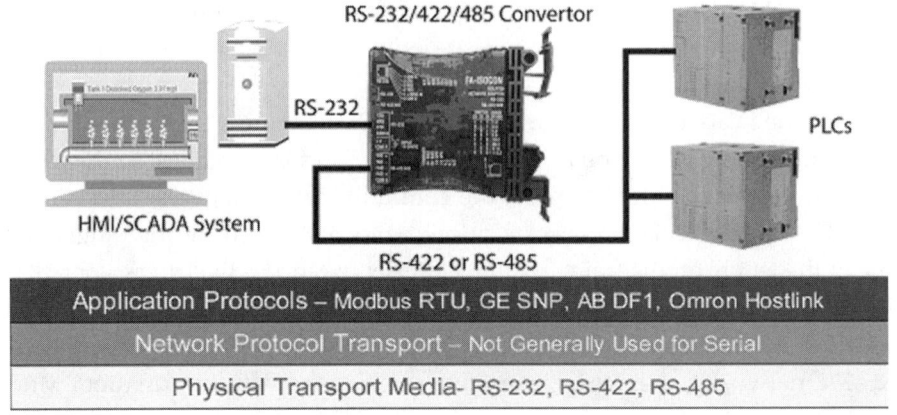

FIGURE 9.3 An example of a serial communications network, including RS-232 and RS-422 network.

- **DF1.** Documented by Allen-Bradley, which has the most installed PLCs in the United States, the DF1 protocol is often found in process-control systems. It is readily available but not nearly as pervasive as Modbus.
- **Other Proprietary Protocols.** Every PLC manufacturer has its own serial protocol (e.g., GE SNP, Omron Hostlink, Omron Fins, and Toshiba Computer Link).

Serial interfaces are fine for point-to-point, single-device applications involving moderate amounts of data at reasonable transfer rates. Most serial networks are limited to a data transfer rate of 38.4 kbaud or less, depending on the length of the connections. The amount of data that can be transferred typically is limited.

If multiple devices are involved, a multi-drop serial interface uses the master–slave approach, and the master device simply communicates with each slave device in sequence. This may slow the overall data-transfer rate. If so, the rate might be increased if the volume of data transferred can be reduced.

Serial networks typically are the slowest of the four communication options. Also, multidrop serial networks can be costly to install because of the number of conductors in the cables and the need to terminate the cable at each device. However, they may be a utility's only practical choice (e.g., if a device only has a serial interface or if radios are used to communicate with remote devices). Contractors can expect to see serial interfaces on various control devices for pumping stations and other treatment processes.

PROPRIETARY. To overcome the limitations of serial networks, many control-system manufacturers created their own networks for interdevice communications (Figure 9.4). Their goals typically were to provide higher data throughput and deterministic responses. In the process, the vendors typically designed their own physical transport media, network transport, and application protocols.

Proprietary networks often do a good job of delivering the results they were designed to provide. They often provide their best performance when all of the system's hardware and software is provided by the network creator. It is possible to integrate third-party systems into many proprietary networks, though, and obtain solid performance. Most proprietary network providers offer interface cards that can be inserted into expansion slots in PCs. Some businesses have been built on providing interface cards that connect Vendor A's proprietary networks to Vendor B's controllers.

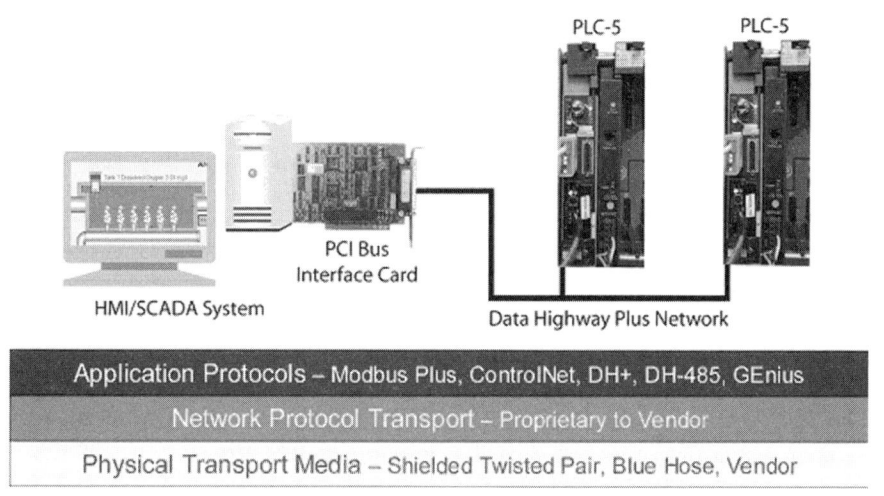

FIGURE 9.4 An example of a connection to a proprietary PLC network. Notice the network interface card required in the HMI/SCADA system computer.

The downside of proprietary networks is often cost. The cabling, for example, may be commercially available off-the-shelf (COTS) or it may only be available from the network provider, but it often is more expensive than other types of cabling (e.g., Ethernet). Likewise, the cost of software interfaces for the network can be higher because the vendor's volume of demand is less than that of an open software interface. The cost of interface cards for PCs can exceed $1000 per computer. Also, support for the network can be problematic if the utility has a mixed system and the network provider would prefer that it only used the provider's hardware and software.

Sometimes, the costs of proprietary networks are more than justified by the overall results, and they are the best solution. However, users should carefully evaluate all available options based on facts and a clear understanding of their project's requirements—and not be persuaded by marketing hype.

Many proprietary networks now are being replaced with Ethernet-based systems, as protocols have evolved from proprietary to nonproprietary "open" protocols. In fact, some protocols (e.g., Allen-Bradley's DeviceNet protocol) originally were proprietary and became "open" after multiple manufacturers adopted them. DeviceNet is a versatile, general-purpose FieldbBus based on controller area network (CAN) technology. Devices can be powered from the network, so wiring is minimized.

Some examples of proprietary protocols include:

- *Data Highway, Data Highway Plus, Data Highway 485.* Created by Allen-Bradley so its PLCs could communicate, this system is probably the most common proprietary network in the control systems industry. Personal computer interfaces are available, as are other interface modules for various third-party controllers and other hardware. The maximum data rate is 230 Kb.

- *Modbus Plus.* Created by Modicon as a means of obtaining faster throughput, this system is based on many of the same protocol concepts as Modbus Serial but is not totally open. It requires a PC interface card, and the maximum data rate is about 2 Mb/s.

- *Genius.* This system was created by GE Fanuc as a control network but also was used as a data-collection network before the general acceptance of Ethernet. It also requires a PC interface card. The maximum data rate is about 153 Kb.

If a water or wastewater utility manager wants to interface to an existing control system with a communications network installed before 1995, they probably will encounter one of these proprietary networks. The good news is that with PC interface cards and standard software solutions for interfacing to these networks, they need not remain closed.

FIELDBUS. Fieldbus networks are primarily used to provide digital communications between field instruments and controllers [e.g., PLCs, distributed control systems (DCSs), or computers]. However, they also provide power to instruments, gather process data, issue commands, and obtaining valuable diagnostic information about the instruments (e.g., level, temperature, pressure, and pH transmitters; valve controls; and solenoids).

Early Fieldbus systems were created simply to minimize field wiring by using one communications structure to talk to multiple field devices. Implementing these systems involved putting microprocessors in the field devices, thereby converting them from analog to digital. The embedded microprocessors and digital displays in the field devices also allowed engineers to distribute control to the devices and obtain intelligent data from them. The resulting control systems were more complex and enabled devices to provide deterministic responses. (Most water and wastewater treatment applications, however, do not require deterministic response.)

Standards for Fieldbus networks typically are open. Currently, more than 40 standards claim to be Fieldbuses or Fieldbus-related communications, including Profibus, Foundation Fieldbus, Hart, DeviceNet, AS-I, FIP, and Interbus. Many companies are building hardware, hardware subassemblies, and software for each Fieldbus standard. Specifications for a particular Fieldbus' physical network media, network transport, and application protocols are often available by paying a fee to join the industry organization that created the standard.

Because of the conflicting interests of controls manufacturers, engineers, and users, no single Fieldbus standard has yet dominated this market. Initially, the conflict was between the process control industry (which primarily used analog signals) and the manufacturing industry (which primarily used discrete signals). At first, Fieldbus development followed these two markets in different directions, but now several Fieldbus systems incorporate both analog and discrete signals and provide many of the same functions. Currently, the conflict seems to be among various international standards groups, industrial groups, and proprietary Fieldbus systems.

Nevertheless, Fieldbus systems have many advantages. For example, less wiring is needed; one wire links multiple instruments and controllers. Control data is more accurate and precise because sensors and analytical field instruments can provide better signal resolution via digital communications than via analog signals, which quickly degrade over long distances or in noisy environments. Smart instruments can execute programs traditionally requiring a controller, analyze their own "health", report problems via the Fieldbus, and be recalibrated via the Fieldbus. Embedded programs allow for standalone process control as part of a larger system. Some Fieldbus structures support peer-to-peer communications. Instruments designed for a specific Fieldbus can be interchanged with those from other manufacturers, provided they also are designed to meet the same Fieldbus' requirements. (Instruments designed for one specific Fieldbus typically are not interchangeable with those designed for another Fieldbus.)

The disadvantages of Fieldbus systems include less system support or more expensive system support because fewer people are familiar with various Fieldbus protocols. Also, Fieldbus structures are subject to changes due to advances in technology, so support personnel will require regular training updates to keep abreast of the latest revisions. In addition, the loss of the wiring path may halt system operations.

Not all Fieldbus systems were designed to meet wastewater treatment plant requirements, so engineers should consider the following when evaluating them:

- *Speed of Operation.* Wastewater treatment operations typically do not require continuous high-speed decisions. In fact, most plants work best when the control loops match the slower biological processes.
- *Plant Layout.* Unlike most industrial plants, wastewater treatment facilities can be spread out over a wide area. So, the Fieldbus system must provide reliable communications over long distances.
- *User Friendliness.* Fieldbus communication languages are system-specific. Engineers should select a Fieldbus system that can be easily supported by available personnel and local resources.

ETHERNET. The fastest-growing process-control system network is Ethernet. At the physical transport level, Ethernet uses the same COTS cabling and networking hardware used in most offices. The surge in Ethernet interest has been driven by dropping network hardware costs because of the Internet and the growth of Ethernet networks in offices. As the costs of 100-Mb Ethernet cards, cable, and switches have plummeted, more organizations have demanded access to this communications system for their process control systems.

For years, there has been much debate about the Ethernet's nondeterministic nature (i.e., the response time to a request is not predictable or consistent from one trial to another). However, when Ethernet switches are used instead of hubs, Ethernet becomes a full duplex network and can achieve near-deterministic response rates. When a 100-Mb Ethernet is run through a switch, it offers 100 Mb each direction in a full duplex connection, so the capacity available on an Ethernet network far exceeds that of all available proprietary networks.

Nowhere is Ethernet acceptance more obvious than in the use of Ethernet-based I/O systems. For several years now, a number of control-system manufacturers have been offering PLC-type I/O racks with a 10- or 100-Mb Ethernet interface to connect the I/O rack to the PLC. That control-system manufacturers trust the connection between the central processing unit (CPU) and remote I/O says much about the acceptance of Ethernet in process-control systems.

ETHERNET IN DEPTH

There are three types of physical transport media typically used in Ethernet installations: copper twisted-pair wires, fiber optic cable, and wireless media.

COPPER TWISTED-PAIR WIRES. The benefits of an Ethernet system based on copper twisted-pair wires include COTS components, standard conduit radius, and easy installation and maintenance. Category 5e copper cable is typically used for field installations because it offer speeds up to 100 Mbps and most PLCs and field devices operate at far lower baud rates. The cables can be installed in conduits and condulets with tight turns, and terminations can be easily done with the use of a specially designed compression tool. The most common termination is an eight-conductor RJ-45 connector. (Installers must be careful to use a standard color-code scheme when terminating wires.)

However, a switch or bridge must be installed every 100 m to re-amplify signals. Copper wires are also susceptible to inductive interferences that may result in transmission aberrations.

FIBER OPTIC CABLE. Fiber optic cable transmits light (optical) pulses over glass fibers. The benefits of fiber optic cable include high-speed transmissions, long-distance transmissions, and no inductive (noise) interference.

The optical cables and terminations require more skill to install and maintain than copper cable to avoid breaking the glass fibers. Installers must not exceed the cable manufacturer's maximum tension rating, especially when pulling the cable into conduit. Sharp bends should be avoided, and the cable must not be pinched or bent back. An experienced technician must perform the fiber optic terminations so there is minimal loss of light transmission at the connector.

Fiber optic cable lengths can range from 1 m to several kilometers, depending on the operating mode and type of fiber installed. One of the most commonly used types in process-control applications is 62.5/125-micron multimode fiber optic cable. The entire cable, including both end terminations, should not have a signal loss greater than 7dB. Experts suggest that multiple pairs of fiber cables be installed so if one fiber is damaged, another is already available. Also, because not all field deices can be directly connected to fiber optic cable, a special transducer may be required at one or both ends of the cable.

When trying to decide between copper and fiber optic cables, see Table 9.1.

FIBER OPTIC ≠ ETHERNET. Not all fiber optic cable is Ethernet-related. For example, Allen-Bradley's ControlNet protocol commonly uses fiber optic cable, but it does not include Ethernet. Fiber optic cable also can be used to synchronize PLC processors over proprietary protocols used in redundant PLC installations. Multiple interface cards are available that can send and receive data over fiber optic cable. So,

TABLE 9.1 An example of how to decide between copper and fiber optic cables.

Feature	Copper	Fiber optic
Most economical	Yes	No
Easy to install	Yes	No
Easy to maintain	Yes	Yes
Immune to inductive interference	No	Yes
Over 1000 Mbps transmission speed	No	Yes
Can be installed in short radius pipes	Yes	No
Can be installed in condulet fittings	Yes	No
Requires expensive termination tools	No	Yes
May require transducer for devices	No	Yes

if utilities choose fiber optic cable for their projects, they must be sure that the fiber optic hardware and software are compatible, whether the communications system is Ethernet-based or proprietary.

WIRELESS MEDIA. At the turn of the millennium, an explosion of wireless Ethernet technology became available. Most wireless Ethernet systems used the unlicensed radio frequencies between 900 MHz and 2.4 GHz. At the time, these frequencies offered connection speeds up to 2Mbps using Ethernet Protocol 802.11b. Since then, more unlicensed frequencies have become available to the public, and higher throughputs have been achieved (e.g., 802.11 g at 54 Mbps).

Wireless Ethernet can be installed in a multitude of system architectures. It is less expensive to install than wired systems—especially installations that involve cutting concrete, digging trenches, or crossing public or private property. The drawbacks, however, are security and reliability (see below).

BRIDGING THE GAP TO NON-ETHERNET DEVICES. Ideally, all control systems would have built-in Ethernet connections, use one or more of the three physical transport media, and use a common application protocol. In the real world, however, many devices do not have Ethernet connections or use a common application protocol. If the device uses a serial or proprietary communications protocol, there are some specific types of hardware that can connect it to a treatment plant's Ethernet system.

Ethernet Encapsulation. The term *Ethernet encapsulation* does not refer to a physical cable; it refers to a technique in which Ethernet transport protocols are used to encapsulate other protocols (primarily serial protocols) and transmit them over the physical transport media: copper wire, fiber optic cable, or wireless media (Figure 9.5). This technique provides a cost-effective Ethernet connection to a serial instrument.

When a control system "encapsulates" a serial application protocol (e.g., Modbus RTU serial), the serial data packets are placed unaltered in a software "envelope" written according to an Ethernet transport protocol (e.g., TCP/IP or UDP). On the instrument's end of the connection, a device (e.g., terminal server, Ethernet–Serial converter, or device server) does the encapsulation (Figure 9.6). The devices typically have both RS-232 or RS-422/485 and 10-Base-T Ethernet connections. They typically cost between US$100 and $500 or more, depending on the type of packaging, environmental ratings, number of serial ports, number of Ethernet connections, and type of Ethernet network media supported.

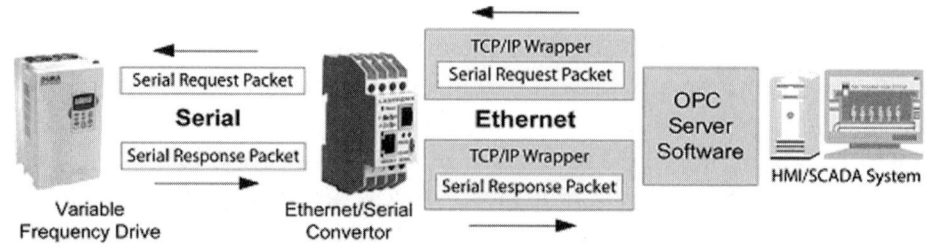

FIGURE 9.5 Ethernet encapsulation allows serial devices to communicate over Ethernet using cost-effective Ethernet–serial convertors when paired with OPC server software that supports Ethernet encapsulation.

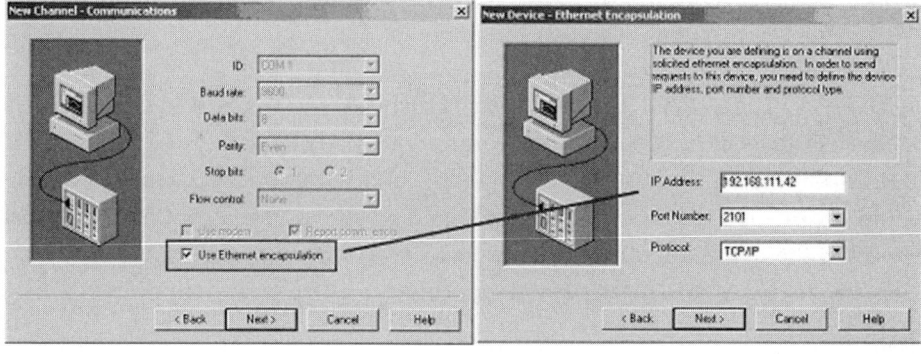

FIGURE 9.6 An example of enabling Ethernet encapsulation in an OPC server.

On the process-control system side of the Ethernet connection, the software must be able to support Ethernet encapsulation. Many Ethernet–serial convertors come with software that is intended to be loaded into the control system and make the Ethernet connection "appear" to be a communications port. This software may work well in commercial settings but cannot always be relied on in industrial applications. If this software is problematic, the control-system personnel have another failure point to deal with. Some Ethernet data-collection software already have the tools to "encapsulate" the serial protocol, and using this software eliminates a failure point, allows for more tightly coordinated timing, and yields a more robust solution.

Ethernet encapsulation hardware and software can provide astounding operations and maintenance benefits. Consider the following example. Let's suppose a treatment plant uses variable-speed motors for a blower. The motors are controlled by a variable-speed drive (VSD) with a traditional 4- to 20-mA control loop and the Modbus RTU serial protocol. The drive is connected to an Ethernet encapsulation device, which encapsulates the Modbus RTU protocol inside the TCP/IP packet (Figure 9.7). The encapsulated data is then transmitted over fiber optic cable to a PLC or PC. Because of the Ethernet encapsulation method, the utility can

- Eliminate the use of special analog I/O cards and wiring;
- Control the VSD from the central control room;

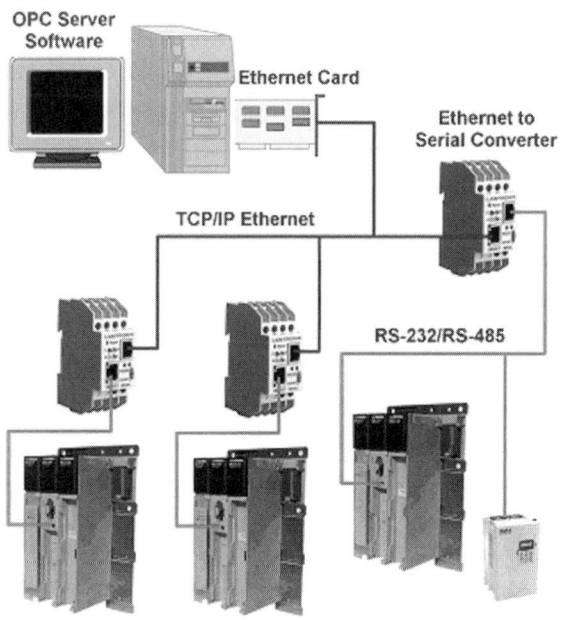

FIGURE 9.7 VSD and PLCs connected using Ethernet encapsulation.

- Obtain other VSD data, such as alarming and historical data, in the central control room;
- Calibrate and maintain the VSD from a central location; and
- Stop maintaining a notebook at the VSD for programming.

None of this would be possible if the 4- to 20-mA connection were used instead.

Ethernet–Proprietary Network Bridges. In addition to Ethernet encapsulation devices, there are hardware "bridges" that connect proprietary network instruments to the Ethernet (Figure 9.8). Bridge devices provide an Ethernet connection via one or more physical transport media and can connect to 1 to N devices on 1 to N proprietary networks, depending on the network and provider of the bridge device. However, rather than "wrapping" the proprietary data packet, a bridge device typically uses an internal processor to convert an Ethernet protocol into the proprietary protocol. For this to work, the Ethernet protocol must be able to provide a means for the requesting application to specify the address of the device in the proprietary network (Figure 9.9). The communications software driver must be able to support this additional information, which must be passed to the bridge device. So, when choosing Ethernet communications software for a control system that must connect to existing proprietary-system components, engineers must ensure that the software supports bridge devices.

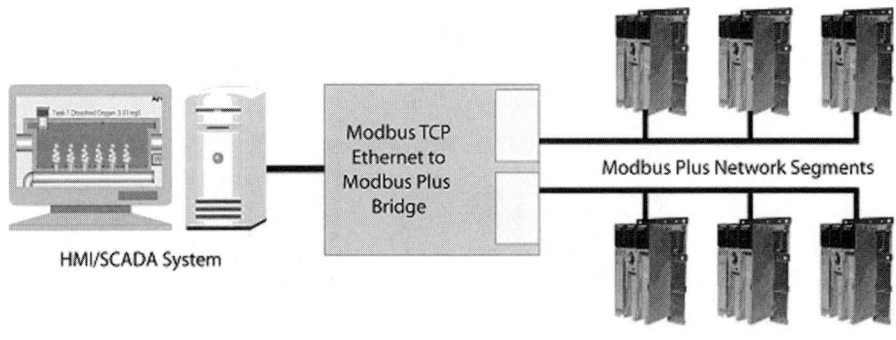

FIGURE 9.8 An example of architecture using an Ethernet-proprietary network bridge.

COMMON ETHERNET APPLICATION PROTOCOLS. As a reference when conversing with vendors of control system and networking hardware, below are some common protocol names:

- Modbus TCP,
- Ethernet/IP,
- AB Ethernet,
- GE Ethernet,
- Siemens Industrial Ethernet, and
- ProfiNET.

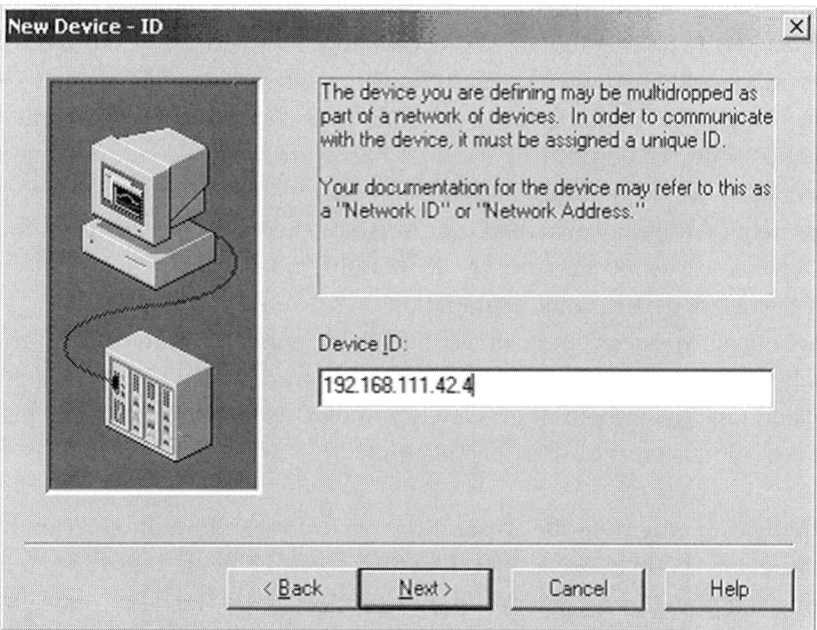

FIGURE 9.9 Configuring an OPC server to communicate through a Modbus TCP to Modbus Plus Bridge.

WIRELESS CONNECTIVITY. Wireless technologies are becoming more common in process control systems. They typically are used in two areas: in-plant and remote site communications.

In-Plant Applications. The goal in in-plant wireless applications typically is to avoid the costs and time delays involved in running network cable to reach a remote device. Sometimes, it may even be impractical to run more cable. With COTS wireless technologies used in commercial applications, plants can quickly create Ethernet connections from one part of a plant to another. Typical wireless technologies used are 802.11a, 802.11b, and 802.11g.

The range of an in-plant wireless system is site-specific. Wireless routers and access points may advertise that they can provide 108 Mbps speeds at up to 122 m (400 ft), but range can drop dramatically as signals have to pass through walls and other obstructions. The more obstructions between the device and the nearest wireless access point, the weaker the signal—so full speed may not be feasible. Some manufacturers provide repeaters or additional wireless access points to help with physical obstructions. Users planning to deploy wireless networks should first test signal strength by placing routers at proposed locations and then using a portable PC with a wireless access adapter to check the actual network bounds.

When implementing wireless networks, staff also should not assume infinite throughput. The rated speed of wireless hardware is its maximum speed under proper conditions; obstructions and interference can cause that speed to drop. Also, wireless networks are not switched networks, so when multiple devices talk concurrently, collisions can occur. In other words, the more traffic on the network, the greater the risk of performance degradation.

Wireless networks can perform well if implemented carefully and not treated as infinitely available. Wireless hardware manufacturers include Netgear, D-Link, Belkin, and Linksys. Industrial-grade wireless hardware vendors include Black Box, B&B Electronics, Digi, and Atop Technologies.

Remote Sites. For remote site applications, traditional information technology (IT) options are not a good fit because of their distance limitations. There are a number of emerging technologies for interbuilding or intercampus wireless connections that promise up to 48 km (30 mi) line-of-sight connectivity. Users needing remote site connectivity should stay abreast of emerging commercial technologies that could be adapted to industrial applications.

Presently, wireless technologies used for remote site applications are either private radio systems or public cellular-based systems. Only recently have public cellular-based technologies reached levels of performance and cost to make them practical alternatives to private radio networks.

Private radio networks are under the user's total control and can provide great security because of that control. However, they can be costly to set up because towers must be built, licenses may need to be obtained, and the entire network must be managed and maintained by the user. Water treatment facilities typically have used this method because they had many remote sites to manage and so it was cost-effective for them.

Meanwhile, the CDMA and global system for mobile communications (GSM) cellular technologies can offer up to about 105 kbps throughput on a public cellular network. At press time, several cellular providers are offering third-generation CDMA technology (1XEVDO). It offers throughput bursts up to 2 Mbps. The chief advantage of public cellular networks is that someone else bears the network's capital and maintenance costs. Utilities simply pay a monthly fee for each device with network access. Fees vary depending whether the data volume is metered or unlimited; unlimited-volume plans can be the most economical option as most cost less than $100 per month and discounts can be obtained for multiple-device contracts. Obviously, local pricing will vary and utilities should work with local carriers to negotiate a solution that meets their needs.

The downside of a public cellular network is that utilities do not control the network; they are sharing the network capacity with others, and there are greater security risks. The security risks can be mitigated with proper use of firewall software and hardware, data encryption, and VPN connections. Utilities should discuss capacity concerns with local carriers based on their requirements.

WIRELESS SECURITY. Given the nature of wireless technologies, utilities cannot block the interception of radio frequencies, so any transmission over radio frequencies is vulnerable. To make radio frequency transmissions secure, utilities must protect the contents of the transmissions. In other words, a potential troublemaker will be thwarted if he or she cannot decode, interpret, or alter the datastream sent via radio. (Of course, jamming the radio frequencies is another issue.)

Some radios achieve a secure connection by continually changing frequencies, thereby making the transmissions a moving target that is virtually impossible to intercept. This type of wireless Ethernet radio is called spread-spectrum frequency hopping (SSFH). To follow the signal transmission, both the sender and receiver must know the frequency-hopping sequence and transmission durations.

Another option is to use encryption and decryption methods at each end. There are encryption keys that are up to 128 bits or more and are known only to the sender and receiver. This encryption method is typically used to secure credit card transactions over the Web.

Virtual private networks (VPNs) can also be used to establish a secure, private connection between two nodes in a wireless network. This option is available from cellular telephone service providers and is purchased at a premium.

One of the biggest concerns of the IT staff is that a security hole will allow an unauthorized person access to the local area network (LAN) via the wireless Ethernet radio connection. To prevent this, the IT LAN could be totally isolated from the control system LAN, but this option may be problematic.

One of the most common security breaches occurs when the installer does not change the default security settings provided by the radio manufacturer. Hackers heavily rely on this mistake, which also applies to default firewall, router, and wireless router configurations. Commercial wireless hardware typically ships in a nonsecure state, with a default password. If a wireless router or access point is installed without changing the default password, an intruder can easily find the device and access the login screen. Nearly all wireless hardware "advertises" its make and model on its login screen. Then, the intruder can go to the Internet, download the manufacturer's manuals, obtain the default password, and take over the wireless network. By changing the default password, you prevent this.

Utilities should also change the default wireless network name in the router and tell the router not to broadcast its network name. This can help thwart potential intruders. They also should only buy wireless hardware that supports encryption and use the highest level of encryption available. When configuring the encryption setup in the hardware, never use the passwords in the hardware manuals or tutorials. Create passwords that are long and alphanumeric, just like any other secure network password. And never write passwords and passkey codes on the hardware as a means of convenience, on the assumption that intruders will never see it. Intruders can be insiders, not just random outsiders.

WIRELESS RELIABILITY. The reliability of a wireless Ethernet network depends on radio maintenance and the type of wireless installation. The cellular digital packet data (CDPD), CDMA (1xRTT), general packet radio service (GPRS) (GSM), and third-generation cell-phone technology (3G) (1XEVDO) services depend on the cell tower's signal strength (usually a factor of distance from the tower). If the signal strength is good, then the radio should be reliable. CDMA and 3G technologies associate with multiple cell towers, so if one tower fails, the radio will communicate with the remaining towers. Because cellular frequencies are regulated by the Federal Communications Commission (FCC), there is little interference by stray or unlicensed users, so the signal is reliable. Most cellular towers have uninterruptible power source (UPS) battery backup and emergency generators so the RF signal would not be lost during power outages. However, it is also up to the cellular service provider to maintain the equipment, but the companies typically are motivated to keep the systems operational so they will not lose market share.

On the other hand, if the utility installed wireless Ethernet radios that operate over unlicensed frequencies (e.g., 900 MHz or 2.4 Ghz), then the risk of interference is higher. A simple 2.4 Ghz cordless phone can knock out all the 2.4 Ghz radios within a certain radius. In many installations, radio repeaters are required because spread-spectrum radios generally require a line-of-sight installation. So, users may need to install UPS batteries and backup generators at each repeater site to increase network reliability. Otherwise, a power failure at a repeater station could cause loss of communication with multiple radios.

Most manufacturers of wireless Ethernet radios provide quality equipment with reasonable life expectancies. The largest threat to the life of the radio electronics is static electricity or lightning. All installations must be properly grounded and may even require an earth ground grid. Lightning arrestors must be installed between the antenna and the radio, and it must also be properly bonded to the grounding conductor. Finally, electrostatic dissipaters may be required to avoid an ionic charge buildup near the antenna that could attract lightning. Even with this protection, a direct lightning hit on an antenna rarely allows the radio to survive.

WIRELESS SOLUTIONS AND COMMUNICATIONS SOFTWARE

Regardless of the wireless technology chosen, it is important to ensure that the communications software you choose can work well with the wireless technologies you choose. You should integrate the selection of communication software with your wireless technology search because the software provider may know wireless hardware vendors that they have found work better than others. The communications software must also offer the flexibility to deal with the additional delays in transmission time that wireless technologies for remote locations can impose.

COMMUNICATIONS AND CONNECTIVITY SOFTWARE. Various types of software can provide connectivity to process-control networks, including proprietary drivers; ActiveX drivers; dynamic data exchange (DDE) drivers; and openness, productivity, and connectivity (OPC) standards-based drivers.

Proprietary Drivers. In any given software application [e.g., a human–machine interface (HMI) or SCADA application], the application provider may include its own drivers to connect to their own or to third parties' field devices (Figure 9.10). Because these drivers only work with the provider's software application, they are considered proprietary. However, a proprietary driver may use a nonproprietary protocol to communicate with a device.

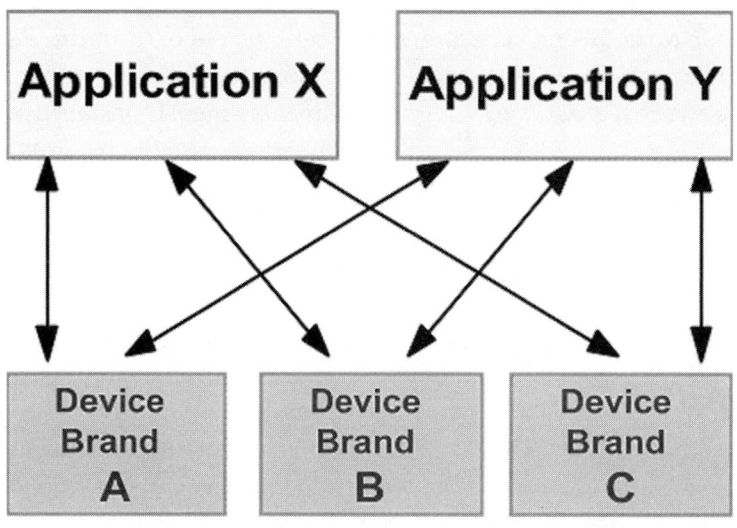

FIGURE 9.10 Device connectivity without OPC requires each application vendor to write its own drivers to talk to each device brand.

Sometimes, proprietary driver vendors will offer significant discounts on the drivers to sell their HMI or SCADA applications. However, if the utility later decides to change to another HMI or SCADA application, it will have to obtain new communications drivers and—more importantly—reconfigure them.

Using proprietary drivers also means the utility has only one point of contact for all technical support. This can be good or bad, depending on the quality of the vendor's technical support program. To support communications systems well, a vendor's technical support hotline must have staff with experience in communications and in troubleshooting system problems. Because a communications driver is in the middle of a system, there are many variables besides the driver that must be considered when trying to solve a communications problem. The support staff must know how to gather data and resolve the issue without seeming to "point fingers" or deflect the problem. (If nonproprietary drivers are used and that vendor depends on driver revenues, it is more likely that its technical support team will be staffed with dedicated personnel specializing in communications troubleshooting.)

Proprietary drivers may have limited functionality. Sometimes the connectivity software meets a basic set of requirements but does not support advanced functionality. So, before choosing a driver, control-system designers should determine whether it can access the data the user needs at the data rates and volume required.

For example, many proprietary drivers only support the minimum set of Modbus requirements and have only been tested with Modicon PLCs. More than 6000 types of devices are available that use the Modbus protocol but are not Modicon PLCs. Also, some areas of the Modbus protocol are open to interpretation, so a driver's configuration options must be able to account for Modbus implementation variances from one device to another. In addition, Modbus is heavily used in many remote devices (e.g., meters, variable frequency drives, and pump controllers) because it works well over radio and wireless networks, but the drivers also must be able to handle the settings necessary to work with modems and wireless networks. Few proprietary drivers for Modbus have the flexibility to deal with these networks.

ActiveX Drivers. For those who plan to use Visual Basic, Visual Basic.NET, Visual Studio, or other tools to develop their own control-system software, there are components (e.g., ActiveX control software or .NET components) that can handle communications to control-system devices. Basically, there are two types of control communications components: OPC connectors and device-specific.

Openness, productivity, and connectivity standards-based connectors are ActiveX software or .NET components that plug into a Visual Basic or Visual Studio.NET application and can connect to any OPC server (Figure 9.11). This option allows the custom-developed software to be run on any available OPC server, and it does not require software developers to know device-level protocols. It may have a higher per-unit deployment cost (e.g., runtime fees), but engineering costs can be lower.

To avoid all per-unit deployment costs, software developers should use device-specific ActiveX or .NET components. However, this option requires developers to know more about the device-specific protocols and to acquire an ActiveX or .NET component for each type of device in the control communications network. Before choosing

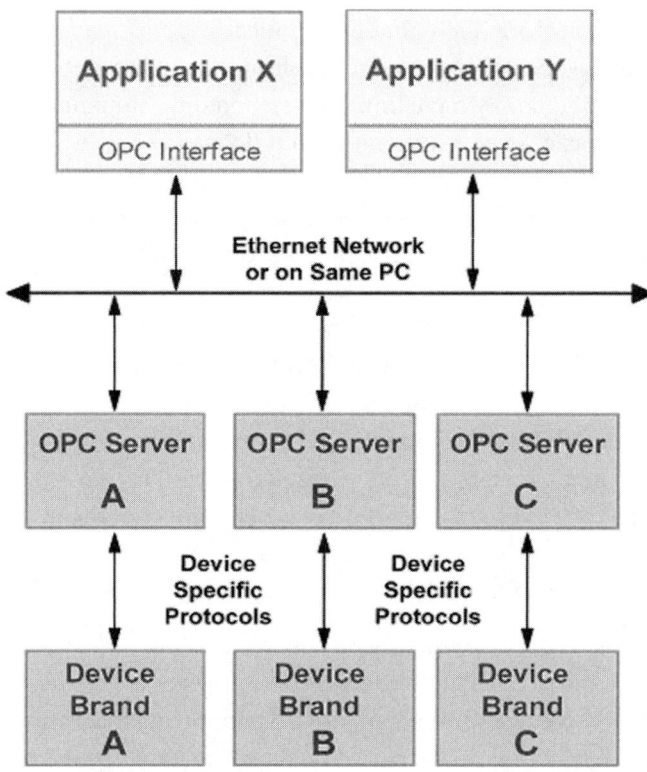

Figure 9.11 Use of OPC for device communications separates communications details from the application.

this option, users should assess their engineering team's skills and evaluate the total cost, including the maintenance and support of a custom software application.

DDE Drivers. Before OPC standards were developed for communications software, developers used DDE to write nonproprietary driver software. They would write a driver for a particular piece of hardware and provide a DDE software interface to connect to the HMI, SCADA system, or other software needing access to the hardware's data.

This option is based on technology that was state-of-the-art in 1990 and has serious performance limitations. To overcome these limitations and encourage third parties to provide communications software for their HMI systems, both Wonderware and Rockwell Software developed DDE extensions in the early 1990s called FastDDE and AdvancedDDE, respectively. Both companies provided tool kits to software developers to help them implement these versions of DDE. Both were successful, and numerous businesses were dedicated to developing communications software that would work with either HMI package.

However, FastDDE and AdvanceDDE could not be used with other vendors' products, and this shortcoming was a key factor in the development of OPC standards, which are now fully embraced by more than 300 vendors, including Wonderware and Rockwell Software. Because of the success of OPC, the use of DDE, FastDDE, and AdvanceDDE have rapidly declined since 2002.

OPC Standards-Based Drivers. The most widely used and fastest growing type of automation connectivity are communications software and drivers that adhere to the OPC standard. This open standard for software-to-software connectivity was created in 1996 by a group of process control industry manufacturers. They formed the OPC Foundation (www.opcfoundation.org), which manages the OPC standard. The foundation's primary mission is to foster open interoperability in automation software.

The first standard established was the OPC data access standard, which was designed to enable developers to create communications software (OPC servers; Figure 9.12) that could talk to various control networks and provide control data—via standard software interfaces—to applications that wanted to use them (OPC clients; Figure 9.13). The server handles the device-specific network topologies and hardware. The client only needs the appropriate tagnames and a standard method of reading and writing data to the OPC server. In effect, the OPC server acts as a translator and go-between for instruments, controllers, and the OPC clients in the HMI or SCADA system.

322   Automation of Wastewater Treatment Facilities

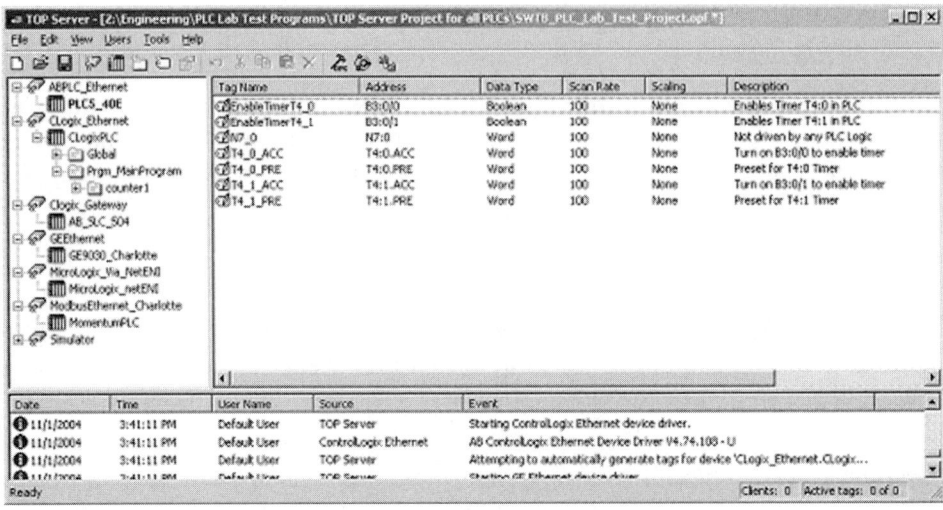

FIGURE 9.12 An example of an OPC server software user interface.

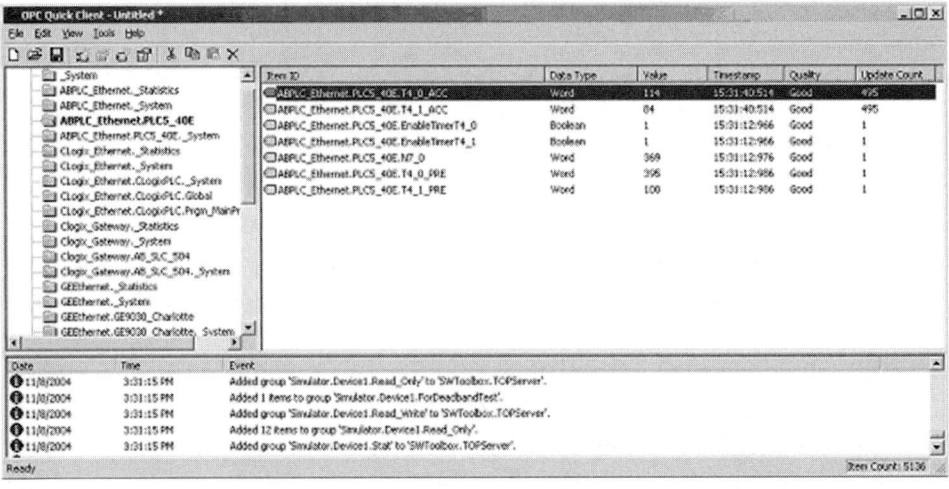

FIGURE 9.13 An example of an OPC client software user interface.

The OPC data access standard is not a replacement for the application protocols used in various process-control networks. It only addresses software-to-software communications. It simply enabled users to obtain control communications software from any OPC server and client vendors that could meet their needs. It also allowed OPC client developers to leave the business of connectivity to OPC server developers. (When discussing OPC-compliant applications, server and client are not hardware terms. A server application sends and receives data from instruments or controllers and makes these data available to client software, which need the data to perform their functions.)

Today, the OPC standards address a wide range of specifications for exchanging various information between instruments and process control software. This includes the exchange of complex, historical, batch process, and alarms and events data. The OPC-XML specifications, for example, provide a means for exchanging OPC data over networks using standard Web services and XML.

Using OPC standards-based products can reduce I&C system integration costs while incorporating areas that may once have been isolated.

COMMUNICATIONS REDUNDANCY. Sometimes, a temporary loss of communications between the instruments and the HMI or SCADA system can be tolerated. However, if the data are critical to operations, security, or safety, redundancy is needed to minimize communications disruptions. The level of redundancy required depends on regulations, plant policies, the type of operations, and the insurance underwriter.

Redundancy is a complex topic. This introduction addresses various types of redundancy and the related communications driver requirements.

The following types of redundancy can improve utility communications systems: the instrument or controller, the network cables or media, the network interfaces, the communications software, and the HMI or SCADA software.

Instrument or Controller. When redundant instruments or controllers are used, appropriate equipment is also needed to switch control from the primary to the backup unit. From a communications perspective, the HMI or SCADA software must know which controller is the primary controller so it can communicate effectively.

One method for switching control does not affect the communications software. In some PLC systems, the communications software connects to (in effect) a common Ethernet interface shared by the two PLCs. The PLC hardware automatically ensures that requests to and from the Ethernet interface are directed to the current primary unit, so the communications software thinks it is only talking to one PLC.

Sometimes, however, the communications software must be able to detect which CPU is "in control" and communicate with it. One method is for the communications software to switch to CPU B if communications with CPU A fail for X period of time. Another is for the communications software to monitor a "heartbeat" tag in the PLC and switch to the backup unit if that heartbeat stops changing for X period of time. Another is for the communications software to talk to both CPUs continuously and use the data from the secondary unit if communications with the primary fail.

When creating a redundant I&C system, utilities should carefully coordinate the hardware and software and get the software suppliers involved before the control hardware is purchased. This will enable the project team to investigate more options before selecting system components. The goal is to avoid installing redundant hardware that lacks a reliable means of communication with the HMI or SCADA system.

Network Interfaces. Sometimes controllers not only have redundant CPUs but also have redundant network cards (Figure 9.14). The control system's PC typically will also have redundant network media or cables and redundant network cards to protect against a cable break or failure by providing an isolated, alternate communications path from the PC to the instrument or PLC.

When installing redundant Ethernet cables, contractors should put them in separate conduits and, if practical, use different conduit routes to the destination. If using wireless technologies, different frequencies should be used, if possible. The goal is to ensure that if one network media fails, the other is not compromised simultaneously. Putting redundant Ethernet cables in the same conduit or even parallel conduits, for example, would not protect communications if a backhoe damaged the conduit(s).

Meanwhile, the communications software must be able to automatically switch from the primary to the backup network cards and media based on predetermined failure criteria. This switch should not require any HMI or SCADA system programming, which can be complex to set up and maintain, as well as unreliable. Also, moving the switch decision point further from the actual point of communications introduces other variables. For example, if the decision point is "no communications for 5 seconds" and communications are lost between the HMI and the communications software, the HMI could switch to the backup controller unnecessarily.

Timing can also be a problem. The HMI or SCADA system has many things to do besides worry about communications network redundancy and may not immedi-

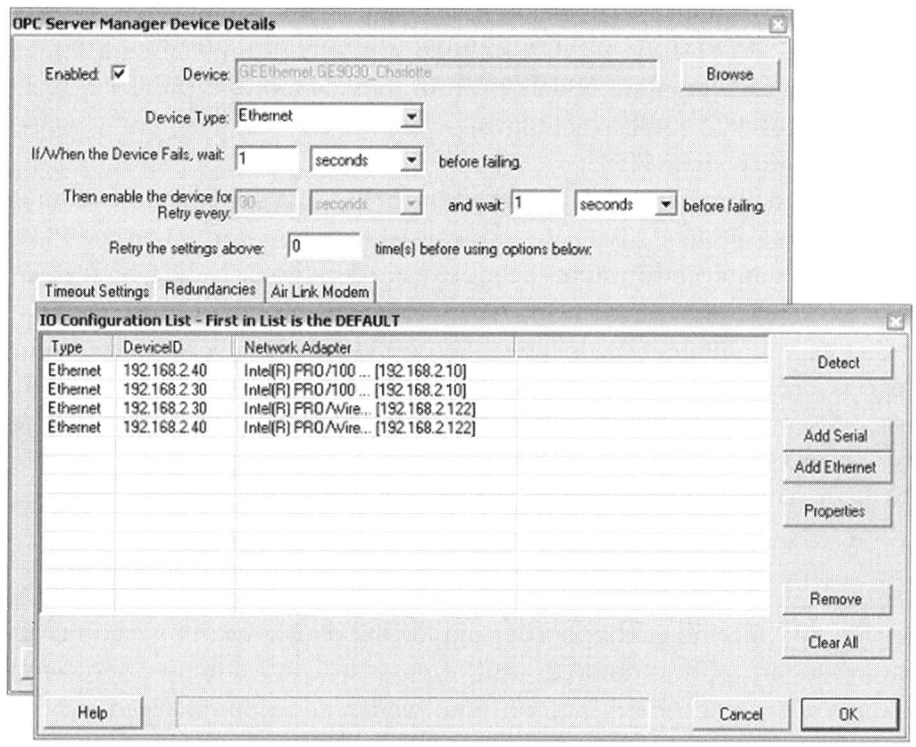

FIGURE 9.14   An example of configuring redundancy for network cards and media in an OPC server.

ately respond to lost communications. The delayed action could lead to constant switching, delayed switching when a rapid change is needed, etc.

Having the communications software decide when to switch to redundant connections reduces the risk of unnecessary changes. Communications packages vary in how they handle this. Some make this function an inherent part of the software, while others offer it as a dedicated plug-in option.

Communications Software. If all reasonable risk of communications downtime must be eliminated, utilities may want to have redundant copies of the communications software running on separate PCs, so if one fails, the other can take over. In these systems, the HMI or SCADA software typically is running on a third PC. Putting redundant communications software on two PCs helps protect against PC failure but is costly to implement (it involves duplicate PCs and duplicate licenses of all software used on both PCs).

Another challenge is interfacing the HMI or SCADA system to the redundant communications systems. Typically, both systems should not talk to the PLCs or instruments simultaneously because the extra network traffic could degrade performance too much. Some HMI or SCADA systems have built-in mechanisms for deciding which communications system to use and switching between them. Otherwise, HMI or SCADA system software is available that makes the two communications systems "appear as one". Users can decide whether to let the HMI or SCADA software think it is talking to one communications driver or to configure the software so it automatically switches between communications drivers, when appropriate, and is aware of the changes.

HMI or SCADA Software. Some utilities implement redundant HMI or SCADA systems. The switching mechanism depends on the vendor, but the communications software should duplicate effort as little as possible to avoid unnecessary network traffic and degrading performance. Software vendors should be involved early in the process to ensure that the I&C system is efficient and operates as intended.

WEB CONNECTIVITY. Control-system connectivity via Web browser-based technologies has taken many leaps forward in the last few years. Until recently, displaying process data in a Web browser was problematic. The most common method was to have the Web server read the control system data and deliver it to the Web browser as a static value. To update the data, users had to refresh the Web page.

Another option involved using software on the client PC or ActiveX controls or Java Applets in the Web browser to refresh data in the Web browser without refreshing the page. However, organizations' security measures often prevented software from loading in the browser, thereby preventing the systems from working.

Now, XML and Web Service have changed the way data can be delivered to a Web browser. Web Service is an intelligent application that runs on a Web server. The Web browser can load a Web page that will automatically pull new data from Web Service without refreshing the entire page. It will work on any PC with Microsoft Window and Internet Explorer 5.5 or higher.

There is also software available that will run on a Web server, connect to OPC server software, and publish process data on the Web server so users can see the live, continually updated process data in a remote Web browser without having to refresh the page or load software on the client PC (Figure 9.15). For example, the OPC Web Client is loaded on a Web server and talks to local or remote OPC servers. It allows users to read and write control-system data from their Web browsers (Figure 9.16).

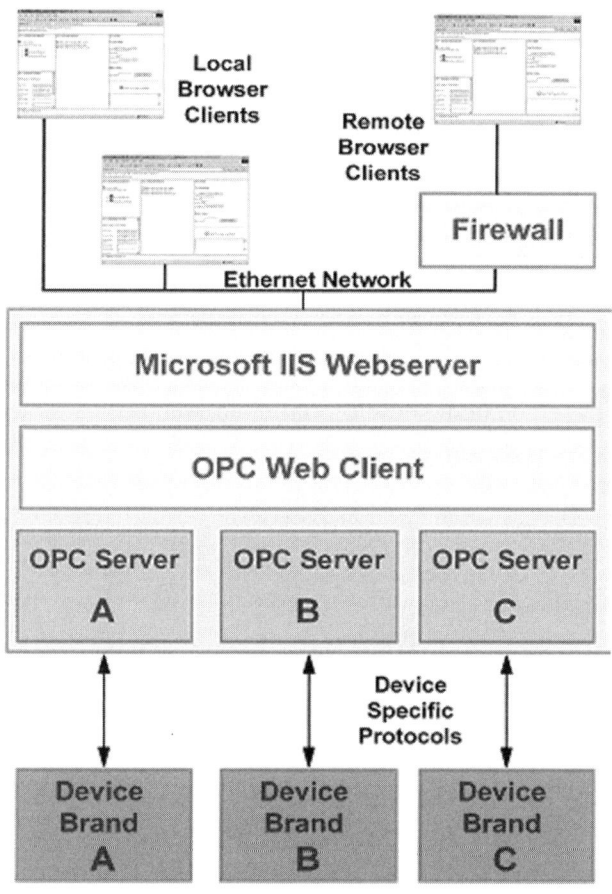

FIGURE 9.15 OPC Web client system architecture overview.

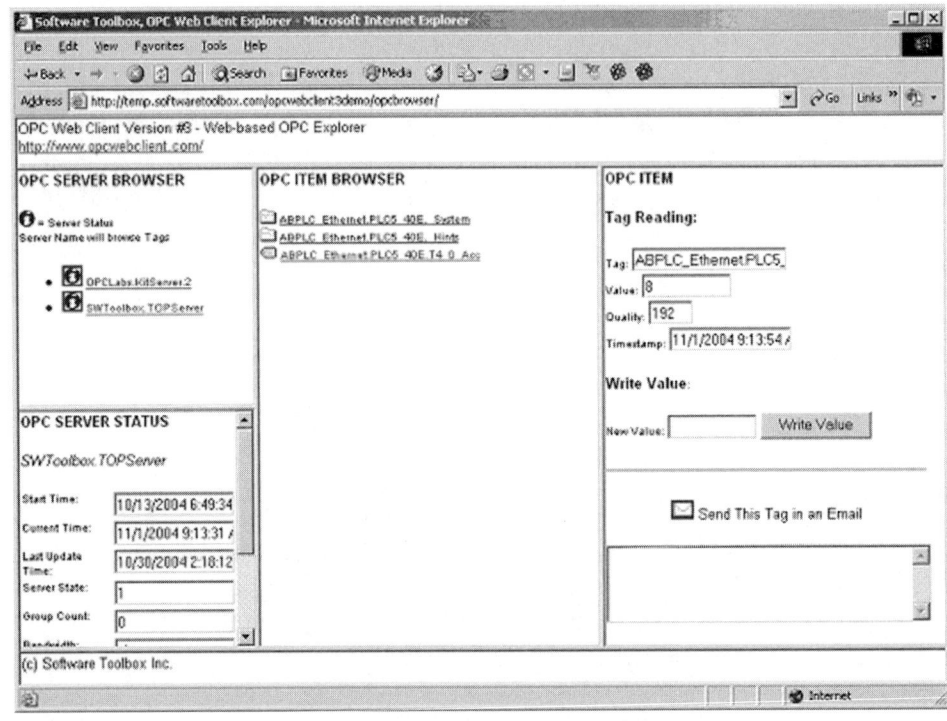

FIGURE 9.16   OPC Web client showing data in a Web browser.

Chapter 10

Automatic Process Control

Process Control Objectives	330	Combined Feedback–Feedforward	343
Benefits of Automatic Process Control	332	Advanced Options	343
Control Theory	332	Role of Microprocessors and Computers	345
Process Identification and Modeling	333	Process Control Systems	346
Control Strategies	336	Single-Loop Controllers	346
Sequential Process Control	336	Multiple-Loop Controllers	347
Regulatory Process Control	336	Supervisory Setpoint Control Systems	347
Controllability Problem	337	Distributed Networked Control Systems	347
Control Options	337	Control System Configuration	347
Feedback	338	References	348
PID	338		
PID Tuning	340		
Feedforward	342		

Process control is the regulation or manipulation of a process' conditions to bring about a desired change in its outputs (McGraw-Hill, 1984). In traditional process control at wastewater treatment plants, instruments continuously collect data about the state of a process, transmit these data to a database, and notify operators of unwelcome changes (Figure 10.1). Based on these data and notifications (alarms), operators modify the process to keep it running as efficiently as possible. An automatic process control system (PCS) is designed to do this more frequently and rapidly.

PROCESS CONTROL OBJECTIVES

Process control is needed because various disturbances disrupt a process' efficiency. Such disturbances include both known and unknown inputs and conditions (e.g., changing flows, chemical and biological compositions, temperature, and density). Process control devices (e.g., valves, flow meters, and pumps) also can introduce disturbances.

Once a process is disturbed, compensation is required. This compensation can occur as part of a manual or an automatic control system. Manual process control occurs when operators notice that a controlled variable has deviated from the desired setpoint and manipulate a process input (e.g., valve opening or pump speed) to try to return the variable to the setpoint. Under manual control, operators determine the timing, duration, magnitude, and direction of the required adjustment, based on their experiences with and knowledge of the process.

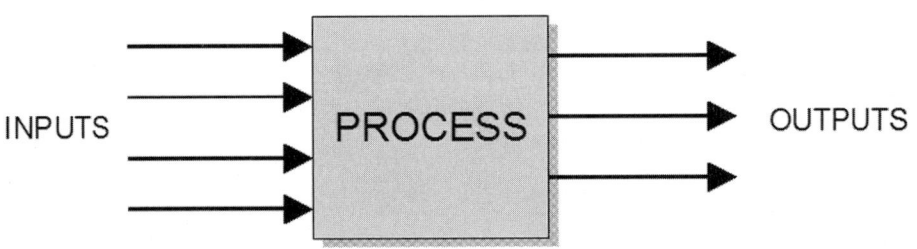

FIGURE 10.1 Information flow in a process.

Automatic process control involves the use of various pneumatic, electromechanical, and electronic equipment to emulate manual controls (Figure 10.2). The operator establishes the objective, but the automatic process-control equipment executes the required adjustments.

The goal (*control objective* or *objective function*) of any control scheme must be defined in terms of measurable output variables. It may be as simple as "minimize effluent biochemical oxygen demand (BOD)" or as complex as "minimize sludge disposal and energy costs while keeping BOD below effluent permit requirements". If some objectives (e.g., cost) conflict with others (e.g., effluent quality), engineers typically can assign weights to each objective or use some as constraints (e.g., permit requirements).

Often, the objective will be stated in terms of keeping one output variable as close to a desired value (*setpoint*) as possible. One example of a setpoint is "maintain dissolved oxygen level close to 2 mg/L". In other cases, the objective will be to minimize one parameter (e.g., energy use).

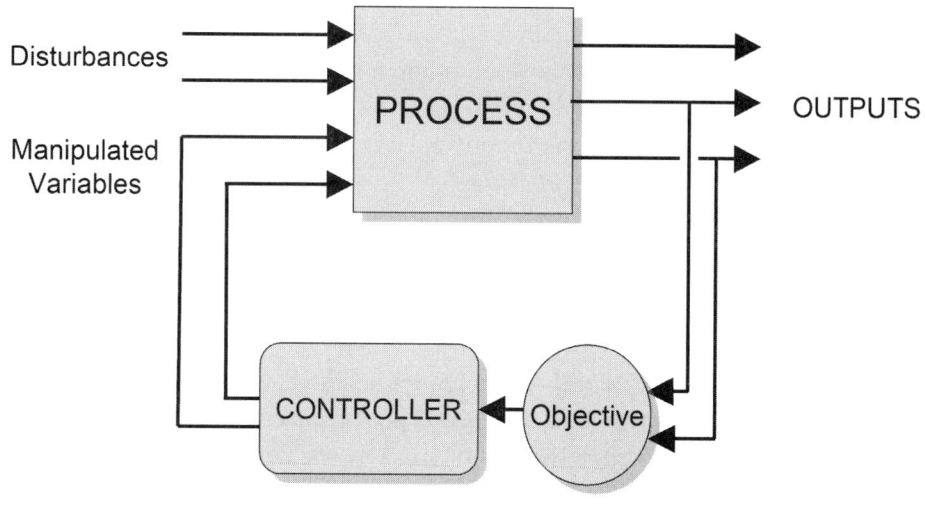

FIGURE 10.2 An example of information flow in feedback process control.

BENEFITS OF AUTOMATIC PROCESS CONTROL

There are numerous benefits of automatic process control. Two of the chief benefits are chemical and energy savings. Depending on the automatic control equipment selected and the treatment processes involved, chemical and energy savings of 10 to 20% and payback periods of less than 2 years are not uncommon.

Consistent operations is another benefit. When multiple operators manage the same process over time, the differences in their skills, insight, and experience will result in widely different process adjustments. Automatic control ensures that adjustments are based on the same logic and are the same magnitude, no matter which operator is in charge of the process.

When part of a plantwide, computer-based instrumentation and control system, automatic process control offers more benefits. One is the ability to maintain a permanent log of automatically time- and date-tagged alarms and events, noting process occurrences and control actions taken. These logs may legally prove when events occurred, as well as their duration and magnitude. Another benefit is the ability to sample operating data at high rates and then store, manipulate, and analyze them to better understand and optimize the process.

Automatic control also reduces the time that operators must spend specifically monitoring and adjusting the process to maintain the setpoint(s). Instead, an operator can take on different duties or be responsible for several processes simultaneously.

It is a common misconception that automatic process control will substantially reduce the number of operators needed. Rather, it allows existing operators to focus on process analysis, troubleshooting, and other tactical duties. Also, new skills may be needed to maintain the automatic control equipment in good working order, so any reductions in the number of process operators may be offset by the need for new maintenance personnel with different skills.

CONTROL THEORY

The better a process' behavior and inputs are understood and measured, the more its results can be controlled. However, a process' performance can be improved even if its behavior is only approximately known (i.e., a mathematical model of it has been created). Process models typically are used to develop a PCS, and the task of determining which model and parameters to use is called *process identification*.

PROCESS IDENTIFICATION AND MODELING. Although complex, detailed models are available for many processes, PCS designers typically only need a simplified model: steady-state, first-order lag, or second-order lag. [And, perhaps, a dead-time (transport) lag.]

In a steady-state process, the rate of change of all measurable quantities is zero. Any process is considered steady-state if its inputs change more slowly than its dynamics. An acid–base reaction in a flash mixer, for example, is a steady-state process. Its output (pH) depends only on its inputs (amounts of acid and alkalinity), not on the time elapsed since changes were made.

The model for a steady-state process is expressed algebraically, typically as a simple proportional relationship. The proportionality constant is called the *process gain* (K_p). If a process' input is X and its output is Y, then

$$Y = K_p X \tag{10.1}$$

For example, the sludge production in an activated sludge process may be proportional to the BOD load for a certain range of operating conditions. The process gain is the observed yield [kilograms (pounds) of sludge produced per kilogram (pound) of BOD removed]:

$$\text{Sludge production} = K_p X \text{ (BOD removal rate)} \tag{10.2}$$

If a process has appreciable dynamics, however, a steady-state model is inadequate. When input conditions change in such processes, the outputs adjust gradually to steady-state, not virtually instantaneously. The mathematical models of dynamic systems may include differential equations.

The simplest type of dynamic process is a first-order process, which is characterized by capacity and resistance. One example of a first-order process is a completely mixed, constant-volume (V) tank in which some nonreactive substance (C), such as table salt, has been added to the water (Q; Figure 10.3). When a first-order process' input changes, the process approaches a new steady-state—rapidly at first, and then more slowly. This type of response is called *exponential decay*.

The dynamics of a first-order process can be described by two parameters: process gain and process time constant (T_p). The process gain of a first-order process (once it reaches steady-state) is defined as the ratio of output changes to input changes. In this case, the gain is 1 (i.e., 1 unit change in input concentration results in 1 unit change in output concentration). Chemical reactions could change this gain.

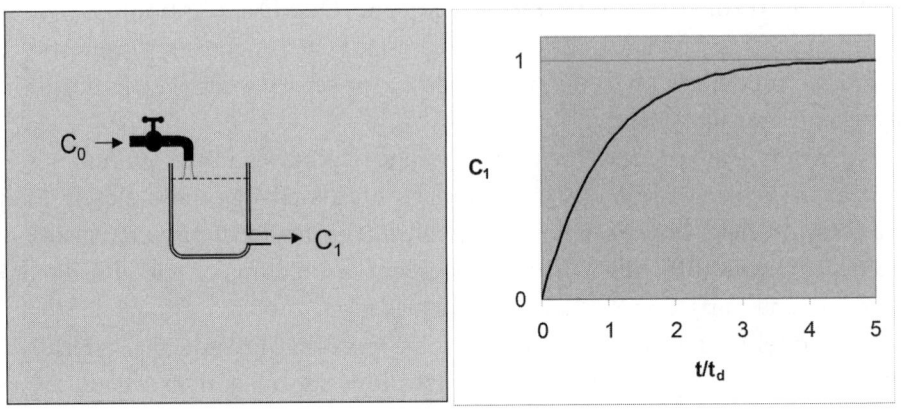

FIGURE 10.3 A first-order system and the response curve for a change in input concentration from 0.0 to 1.0.

The process time constant in this example is equal to the process' detention time (V/Q). The higher the flow rate, the faster the output responds to changes in influent concentration. In the presence of a chemical reaction that produces the salt, the dynamics would slow and T_p would increase.

Numerically, T_p equals the time required to achieve 62.3% of the difference between the initial and final conditions. After three T_p intervals, a first-order process will have completed about 95% of the time needed to achieve steady-state. (In an activated sludge process, the T_p for changes in biomass equals the sludge age.)

Many processes have more complex behavior than first-order processes. Processes with two capacities and resistances, such as two first-order processes connected in series, are called *second-order processes* (Figure 10.4). The response of such a process to an incremental change in input would resemble a sigmoidal (S-shaped) curve. The process would be described by process gain and two time constants.

Another type of second-order process is characterized by capacity, momentum, and, typically, resistance. One common example of this is a shock absorber.

Any second-order process can oscillate around the steady-state value. For example, if someone pushes down on a car fender, the spring will return the car to the original position. If the shock absorber is not providing sufficient resistance (i.e., the process is *underdamped*), momentum will cause the car to overshoot its original position. The car will then vibrate up and down with decreasing amplitude until it settles to the steady-state position. If the shock absorber is providing enough resistance (i.e., the process is *critically damped*), the car will not overshoot its original position (Figure 10.4).

Another common factor in a process' dynamic response is dead time (transportation lag). Let's suppose a first-order process is connected to a 305-m-long (1000-ft-long), 0.15-m-diam (6-in.-diam) pipe with a volume of almost 5700 L (1500 gal). If the flow through the pipe is 6 L/s (100 gpm), its discharge would be representative of what entered the pipe 15 minutes earlier. The response measured at the end of the pipe is described as first-order plus dead time. Dead time is significant because it is difficult to account for in typical process models.

Higher-order processes exist, but they often can be approximated by first- or second-order systems with dead time. Any process consisting of several processes in series would have one or more time constants for each subunit.

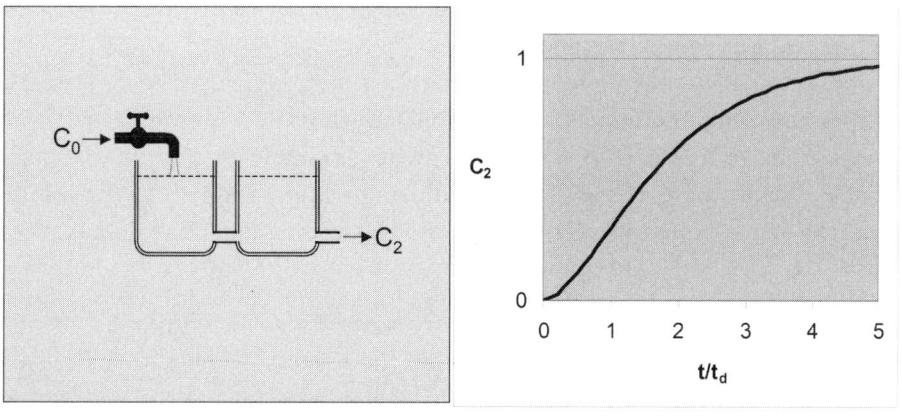

FIGURE 10.4 An example of a second-order system and a response curve for a change in input concentration from 0.0 to 1.0.

CONTROL STRATEGIES. Once engineers have determined which model and parameters best characterize a process, they can devise methods to control it. The variables that describe a process are called input variables, output variables, and process parameters. Input variables include *disturbance variables*, which cannot be controlled, and *manipulative variables*, which can be controlled. In an activated sludge process, for example, flow and BOD are typically disturbance variables, while return flowrate, air flowrate, and waste sludge flowrate are manipulative variables. (Although the waste flowrate is an output in the system's mass balance, it is an input in the control scheme.) Any change to an input, whether imposed or not, is called a *forcing*.

Output variables, which respond to changes in inputs, include the compositions of flows within and leaving the process. In an activated sludge process, for example, effluent suspended solids and BOD are outputs, as is the mixed-liquor suspended solids in the aeration tank. [The suspended solids concentration of waste activated sludge (WAS) is an output, but its flowrate is an input.]

Process parameters describe the process' behavior. These parameters may be empirical values (e.g., gain and time) or fundamental values (e.g., the maximum growth rate of microbes).

Using the process model and variables, engineers can design a controller using standard mathematical techniques, such as those found in Stephanopolous (1984).

Sequential Process Control. A sequential control system organizes a process into a series of events or states based on time and the process' previous states. This system is also called *logical control* because each control action or status change is based on the solution of logic sequences and matrices.

An example of sequential control is the start–stop logic used to operate a constant-speed centrifugal pump. A combination of pushbuttons, electromechanical control relays, timers, and pressure and limit switches enable an operator to start and stop a large centrifugal pump—without causing water hammer—by pressing one button. The same sequential control logic may be implemented via microprocessor software.

Regulatory Process Control. A regulatory control system continuously manipulates or adjusts specific variables to maintain the process' setpoints. This system is also known as *closed-loop control* because the control loop for a particular parameter does not require an operator to manually adjust the manipulated variables.

Regulatory control involves real-time algorithms that attempt to mathematically model a process and thereby predict its responses. Its process adjustments are continuous, not periodic, so deviations from setpoints are quickly detected, compensating adjustments are made, and process parameters are kept within tight operating limits.

Controllability Problem. For the control method to be effective, engineers must solve the *controllability problem* (i.e., find a process model that relates inputs to outputs). This can be challenging when working with biological treatment processes. There is no activated sludge model to predict effluent suspended solids concentrations based on the food-to-microorganism ratio, WAS, or other inputs. So, operators typically choose an indirect control method. Instead of effluent suspended solids, for example, the chosen output parameter is sludge age. The choice assumes a functional relationship between the controlled output and the desired objective. The actual relationship between effluent suspended solids and sludge age is unknown, but operators can determine an appropriate setpoint based on the trial-and-error method.

Another challenge when controlling biological processes is a possible change in process parameters as a result of both measured and unmeasured input changes. The inputs typically measured are gross parameters [e.g. BOD or chemical oxygen demand (COD)], but biomass may respond differently to wastewaters with the same BOD but different chemical compositions. In addition, a natural succession of microorganisms could change the process' response even if the inputs remain constant. Operators understand this and respond to unexpected changes in plant performance by continuing their trial-and-error search for optimum conditions.

CONTROL OPTIONS

Based on sensor measurements, control loops change manipulated variables via some type of actuator. They convert a mathematical control strategy into a mechanical or electronic system. For example, a home thermostat senses a room's temperature via the bending of a bimetallic strip and responds by using a switch to open or close the heater's circuit.

Control options may be classified as feedback, feedforward, or combined feedforward–feedback. Feedback control loops are based on process outputs, while feedforward control loops are based on process inputs. The following discussion is limited to single-input–single-output control loops, but multiple-input–multiple-output control loops (e.g., cascade controllers, override control, and split-range control) also exist (Stephanopoulous, 1984).

FEEDBACK. Feedback control involves adjusting a process' input variable based on measurements of one of its output variables. The input variable is the manipulated variable. Examples of feedback control include

- Adjusting pump speed to control the level in a wet well,
- Adjusting lime dose to control pH,
- Controlling dissolved oxygen concentration via airflow,
- Controlling sludge age via waste flowrate, and
- Controlling temperature by switching a heater on and off.

Processes without any systematic form of feedback are called *open-loop systems*. Those with a feedback control loop are called *closed-loop systems*. (In effect, an operator is a feedback controller when he or she makes manual adjustments based on periodic observations of process performance.)

The simplest form of feedback control is an on–off switch ("bang-bang" control). One example of this is an ordinary thermostat. When the temperature in a house drops below the setpoint, the thermostat responds by turning the furnace on. When the temperature has risen sufficiently, it turns the furnace off. The thermostat is designed not to allow the on–off cycles to occur too rapidly, but its chief drawback is that the temperature constantly oscillates around the setpoint.

PID. Aside from the on-off switch, the most common control mechanism in the wastewater treatment industry is the proportional-integral-derivative (PID) controller. To understand how it works, consider its use in controlling the speed of an automobile via a cruise control system.

If a car maintaining a speed of 88 km/h (55 mph) on a flat road slowed down because of a change in wind or road slope, the proportional (*P*) control loop would press the accelerator in proportion to the error (*E*):

$$E = PV - SP \tag{10.3}$$

Where

PV = actual speed, and
SP = desired speed.

The accelerator position is the manipulated variable (M). This type of control can be expressed as

$$M = M' + K_C \cdot E \tag{10.4}$$

Where

M' = the setting of the manipulated variable when E = 0, and
K_C = the controller's proportional gain (e.g., how far the accelerator is pressed or released per increment of error).

The response of the proportional control loop depends on the value of the gain. Let's suppose that K_C is 3.9 mm per km/h (0.25 in. per mph) above or below the setpoint speed. If this is too large, a drop in speed would cause the cruise control system to press the accelerator too far, increasing the speed too much, and then the system would release the accelerator too far, slowing the car too much. If it exceeds a critical value, a disturbance could cause the system to oscillate. So, K_C affects the system's stability, which is a crucial consideration in control-system design.

If, on the other hand, K_C is too small, a drop in speed would not sufficiently be compensated for, and the car's speed would settle at a new value with a nonzero error. This steady-state error is called the *controller offset* (droop), which cannot be completely eliminated while maintaining a stable system—except by an integral control loop.

An integral control loop continuously adjusts the manipulated variable at a rate proportional to the amount of error until the offset is eliminated. So, when the car starts moving up a hill, the proportional control loop presses the accelerator a preset distance, which partially compensates for the incline. The integral control loop continues pressing the accelerator until the setpoint speed is reached.

The combination of proportional and integral control (*PI control*) is described as follows:

$$M = M' + K_C \cdot \left[E + \frac{1}{t_I} \int E \cdot dt \right] \tag{10.5}$$

Where

t_I = the integral time constant (*reset time*), and
$1/t_I$ = minutes per reset (i.e., how fast the controller increases its action in proportion to the amount of error).

If $1/t_I$ were large, the controller would press the accelerator rapidly whenever the speed dropped. So, the car would still be accelerating when the setpoint was reached and would overshoot the desired speed. During each overshoot, the controller would correct itself, causing the system to oscillate around the setpoint. Depending on K_C and t_I, the oscillations would either decrease until the system settles at the setpoint, or they would increase, indicating that the system is unstable.

The third type of control in a PID controller is a derivative control loop, which adjusts the manipulated variable in proportion to the rate of change of the process variable. So, when the car's speed changes, the accelerator is moved in proportion to how fast the speed is changing. This helps dampen changes in response to large disturbances. (Derivative control is rarely needed in wastewater treatment applications.)

The equation for combined PID control is

$$M = M' + K_C \cdot \left[E + \frac{1}{t_I} \int E \cdot dt + t_D \cdot \frac{dE}{dt} \right] \quad (10.6)$$

Where t_D = the derivative rate parameter (how much the controller responds based on the error's rate of change).

So, in a PID controller, the proportional control loop responds to the control variable's current value, the integral control loop responds to the control variable's history, and the derivative control loop anticipates the control variable's future values.

PID Tuning. Selecting appropriate proportional-gain, reset-time, and rate-of-change parameters for a PID controller can be difficult. One option is to calculate the parameters based on a simple analytical model of the process using the basic control theory found in textbooks. Another is to tune the control system experimentally by introducing a disturbance to the treatment process, observing the dynamic response, calibrating a simplified model based on that response, and then calculating the parameters based on the model, again using basic control theory. To compare the effectiveness of various controller settings, engineers typically use standardized controller-performance measures, such as minimum offset, one-quarter decay ratio, or minimum integral square error (Stephanopolous, 1984).

Some of the most common methods for tuning a PID controller are: experience-based principles, on-line trial and error, Cohen–Coon, Ziegler–Nichols, analytical methods, and computer simulation. Experience is the tuning method of choice for common control loops (e.g., flow, level, pressure, or temperature). In flow control, for example, engineers typically set the proportional gain low to reduce the effects of noise, which are inherent in many flow meters (Luyben, 1973). They also set the integral reset time low to respond quickly to changes in setpoint error.

The on-line trial-and-error, Cohen–Coon, and Ziegler–Nichols methods are experimental. In on-line trial and error, engineers repeatedly double K_C until the process starts to oscillate (Luyben, 1973). The value of K_C at this point is called the *ultimate gain*. They then set K_C at half the ultimate gain. Then, they repeatedly double the integral control loop by repeatedly halving t_I until the system begins oscillating again. They then set t_I to twice that value. Finally, they increase t_D until signal noise begins to affect the system and then set t_D to half of that value. Engineers repeat this procedure using smaller changes in controller settings until the desired controller performance is achieved.

In the Cohen–Coon method, engineers first allow the process to achieve steady-state without the controller (Stephanopolous, 1984). They then change the manipulated variable and plot the process variable's response over time as the process returns to steady-state. This plot is the *process reaction curve*, from which two measurements are made. This curve is used to estimate the process gain, the process time constant, and the dead time (a first-order relationship with dead time is assumed). Engineers then use these values to calculate the controller tuning parameters K_C, t_I, and t_D.

The Ziegler–Nichols method involves using formulas to calculate the tuning parameters based on the K_C measured in the trial-and-error method.

Both the Cohen–Coon and Ziegler–Nichols methods have some practical shortcomings. They are not always accurate and must be followed by trial-and-error refinement. Also, the experiments required may be infeasible. For example, it may be impossible to achieve an initial steady-state for the Cohen–Coon method. If so, then the analytical or computer-simulation methods can be used.

Both the analytical and computer-simulation methods require a fairly reliable mathematical model of the process. The analytical method involves sophisticated mathematical procedures, called *Laplace-domain synthesis* and *frequency-domain synthesis*, to calculate the values of the tuning parameters. If the model is too complex, the analytical method may be impractical. If so, then a computer can use the model to simulate the process and the PID controller. Engineers then can find the tuning parameters via the trial-and-error method, using the simulation rather than the actual process.

Feedback control loops and PID controllers can control dissolved oxygen levels in activated sludge reactors (Corder and Lee, 1986). They also can control sludge age in an activated sludge process by manipulating the waste flowrate (Vaccari et al., 1988).

FEEDFORWARD. The basic difference between feedback and feedforward controls is that feedback controls compensate for changes once they occur, while feedforward controls anticipate them.

Let's suppose a utility wants to automate its chlorine residual controls. An effective chlorine dose depends on both the water's chlorine demand and its flowrate. A feedback control system would use a sensor to measure the final chlorine concentration at the point where the water typically would be about 15 minutes after the chlorine dose, so any remaining chlorine demand could be determined. It then would adjust the chlorine dose accordingly. Every time a change is needed, however, there would be a volume of water (that between the chlorine meter and the sensor) that would be inadequately treated. In other words, the delay between dose, measurement, and response could result in an instable process.

If the chlorine demand were fairly constant, a feedforward control system would simply adjust the chlorine dosage in proportion to the water flowrate. This method, called *ratio control*, maintains a constant input-to-measured-variable ratio by adjusting the manipulated variable, as needed. It ignores process dynamics, so it is best suited for situations in which the reactions occur so quickly that near equilibrium is achieved.

In chlorination, however, changes in flow also change the detention time in the chlorine contact chamber. So, even if chlorine demand and dose were constant, the concentration being discharged would change with the flow. When dynamics cannot be ignored, a dynamic feedforward control system could be used. It would measure the process' input variables and current state, and then use a model of the process' dynamic response to calculate the value of the manipulated variable that would keep the process at its setpoint. The mathematics involved are model-specific.

In theory, feedforward controls should maintain the setpoint despite input disturbances. In practice, however, several complications can hinder their effectiveness. First, disturbances can be difficult to sense adequately. It is difficult to precisely measure BOD, for example, in a biological treatment process.

Second, there may not be a suitable model of—or set of model parameters for—a process, especially a biological process with dynamic parameters.

Third, many processes respond more slowly to changes in the manipulated variable than to changes in inputs. In an activated sludge process, for example, sludge age may change rapidly in response to influent BOD changes, but slowly to changes in the waste sludge flowrate. Exact feedforward control is impossible in these situations, although partial control may be feasible.

Another example of feedforward control is a proposed design to control dissolved oxygen in an activated sludge plant (Corder and Lee, 1986). In this case, the control system would be based on the BOD loading rate.

COMBINED FEEDBACK–FEEDFORWARD. Sometimes, combining the feedforward and feedback controls results in better overall process control. The feedforward controls make the feedback controls more stable by reducing influent variations. The feedback controls reduce the perturbations that escape the feedforward controls.

This method works well in chlorination systems. The feedforward loop is designed to maintain a predetermined chlorine:water ratio, and the feedback loop uses a residual chlorine analyzer to refine the chlorine dose. The feedback loop compensates for changes in chlorine demand and the residual dynamic effects of flow. However, because of the dead time in the loop, its response is slower than the feedforward loop, which prevents flow-induced disturbances from affecting the chlorine residual.

This method also is proposed for controlling the methanol dose in a denitrification process (U.S. Environmental Protection Agency, 1985). As in the chlorination system, the feedforward loop is designed to maintain a predetermined chemical:water ratio, and the feedback loop is based on concentration. However, the concentration being tracked in the feedback loop is effluent nitrate, not methanol. So, the feedback loop would compensate for actual process performance.

ADVANCED OPTIONS. With digital computer support, more sophisticated types of control are feasible. Such advanced control methods include direct digital control and adaptive control.

Direct digital control can be implemented like analog control, except the measurements are made at discrete time intervals, not continuously. If the intervals are smaller than the system's and controllers' time constants, digital control can simply mimic analog. However, the computational capability of digital control allows for better control.

344 Automation of Wastewater Treatment Facilities

In adaptive control, the tuning constants can change over time in response to process changes (Astrom and Hogglund, 1988). Essentially, the controller (e.g., a self-tuning regulator) monitors the system's response to disturbances and forcings and modifies its tuning parameters, as needed. This type of control is useful when the process is not well known or may change significantly over time (e.g., biological wastewater treatment processes).

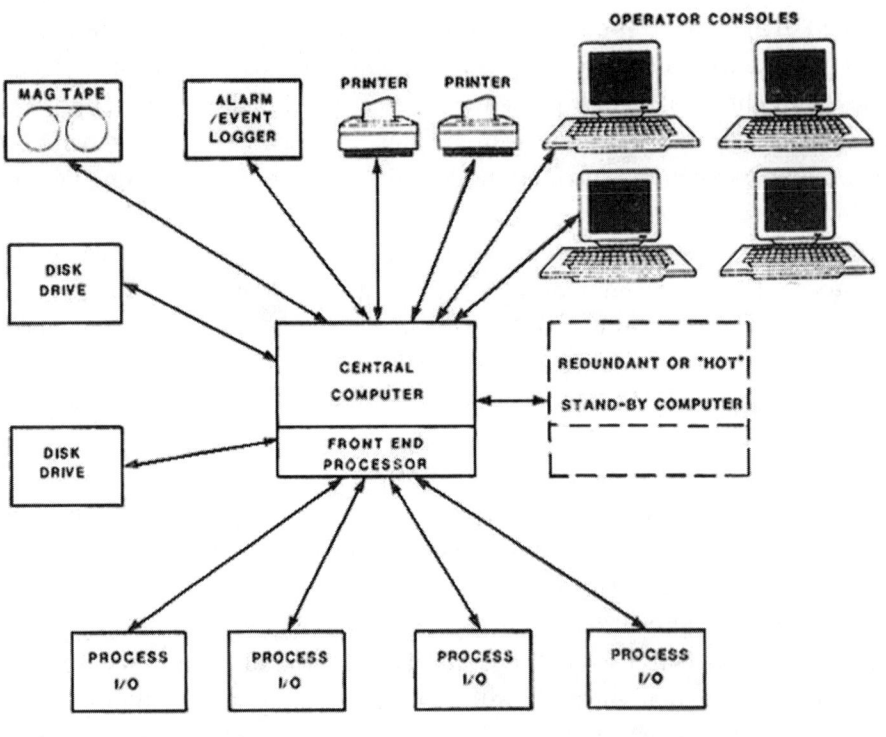

FIGURE 10.5 Direct digital control system.
(Water Pollution Control Federation, 1986)

ROLE OF MICROPROCESSORS AND COMPUTERS

Microprocessors now enable nearly all PCS components—from field sensors and control-panel devices to personal computers (PCs)—to function as specialized or general-purpose computers. They are routinely used to implement all levels of process monitoring, control, data collection, archiving, and analysis.

Before microprocessors became widely available, computers were only used at the highest level of a PCS, providing direct digital control of treatment processes from a central location (Figure 10.5; Water Pollution Control Federation, 1986). These systems were costly and typically only used at larger treatment facilities, where the investment could be justified.

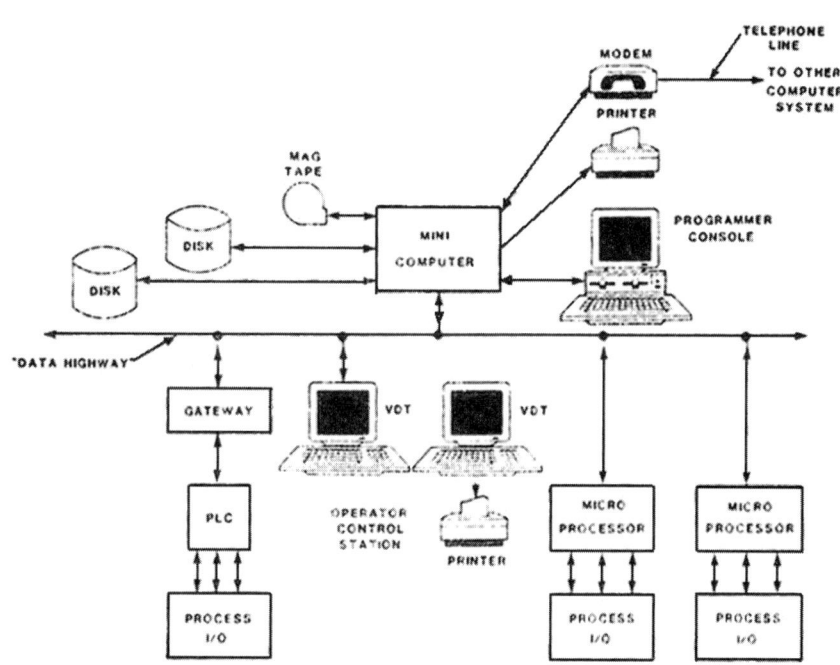

FIGURE 10.6 Distributed control system architecture. (Water Pollution Control Federation, 1986)

Now intelligent digital processing can occur nearly anywhere in the field, and distributed control systems (DCSs) are gaining popularity (Figure 10.6). A distributed system is simpler, more reliable, and less expensive than a centralized system. It also can be configured in a building-block approach, making future repairs, upgrades, and expansions relatively easy.

PROCESS CONTROL SYSTEMS

Functional standards for PCSs are available from the Instrumentation, Systems, and Automation Society (ISA). These standards address microprocessor-based devices that function as

- Single-loop controllers (SLCs);
- Multiple-loop controllers;
- Supervisory setpoint control systems; and
- Distributed networked control systems.

SINGLE-LOOP CONTROLLERS. A single-loop controller is dedicated to controlling one loop (e.g., a chemical flowrate), regardless of the status or condition of other controls. Such controllers can be implemented using microprocessor-based sequencers, timers, automatic regulatory control stations, and programmable logic controllers (PLCs). When combined with traditional control-panel devices (e.g., pushbuttons, selector switches, lights, meters, and annunciators), multiple SLCs can enable operators to control an entire process from one location.

Single-loop controllers may be programmed via standard regulatory process-control algorithms or more advanced algorithms in customized or proprietary software. If they include microprocessors, SLCs can self-tune or adapt to changing conditions. If equipped with communication-network interface hardware, they can connect to other controllers, computers, or communications links and send and receive data (e.g., setpoints).

Control systems based on SLCs are very reliable, because an SLC failure only affects the loop it controls. However, not every loop needs a dedicated controller. Using too many SLCs can make a PCS excessively redundant and unnecessarily expensive to configure. A shared- or multiple-loop controller may offer sufficient reliability less expensively.

MULTIPLE-LOOP CONTROLLERS. A multiple-loop controller basically consists of several SLCs that use one power supply and share other processing resources (e.g., communication links and memory). A programmable logic controller or PC can be configured to function as a multiple-loop controller. If a PC, or a PLC with a dedicated PC, serves as a multiple-loop controller, a video-based human–machine interface can be added for graphical data monitoring and process control.

SUPERVISORY SETPOINT CONTROL SYSTEMS. Supervisory setpoint control systems combine several single- or multiple-loop controllers with data logging, data archiving, analysis, and other capabilities. The controllers are distributed throughout the process and receive setpoints and sequential process-control commands from a centralized (supervisory) computer. Process data from sensors is input to the single- or multiple-loop controllers and passed on to the supervisory computer, which consolidates and stores the data for reporting and analysis.

DISTRIBUTED NETWORKED CONTROL SYSTEMS. A distributed networked control system transmits data and process-control information to and from user-configured, video-based control stations via a wired or wireless communications network. Basically, it is a combination of SLCs, PLCs, PCs, operator stations, and minicomputers. Depending on how the system is configured, the SLCs may be used as standalone devices or as data conduits.

Distributed systems range from those handling 10 to 20 control loops up to those handling 10 000 inputs/outputs (I/O) and several hundred control loops. It does not require a central minicomputer, as is needed with a direct digital control system or a supervisory setpoint control system. Instead, its control functions are strategically located throughout the plant. Overall system reliability is high because of this distribution.

When creating a distributed control system, designers should be careful not to overtax the communications network's ability to transport data, which would reduce the system's response time. Each component should be chosen to reduce network traffic, ensuring that the network will be able to handle both normal and abnormally heavy data transfers (e.g., during power failures).

CONTROL SYSTEM CONFIGURATION

Today, control system choices are almost unlimited. When configuring a plant's control system, design engineers should consider the following (Manning and Dobs, 1980):

- The facility's geographical size, the size of each process area, and the distribution of operations and maintenance staff;

- Expansion plans that will affect the system's functionality and equipment distribution;

- The utility's operations and control philosophy (e.g., whether control equipment should be close to the process equipment being controlled, in an area central to a particular process, or in an area central to the overall treatment process);

- The level and location of backup control equipment if primary controls fail (e.g., a dedicated SLC for automatic level control in an area subject to flooding);

- Control system security (for protection against unauthorized access or changes); and

- Information management processing and reporting requirements.

REFERENCES

Astrom, K. J.; Hogglund, T. (1988) *Automatic Tuning of PID Controllers*; Instrumentation, Systems, and Automation Society: Research Triangle Park, North Carolina.

Corder, G. D.; Lee, P. L. (1986) Feedforward Control of a Wastewater Plant. *Water Res.*, 3, 301.

Luyben, W. L. (1973) *Process Modeling, Simulation, and Control for Chemical Engineers*; McGraw-Hill: New York.

Manning, A. W.; Dobs, D. M. (1980) *Design Handbook for Automation of Activated Sludge Wastewater Treatment Plants*; U.S. Environmental Protection Agency: Washington, D.C.

McGraw-Hill (1984) *McGraw-Hill Dictionary of Engineering*, S. P. Parker, ed.; McGraw-Hill: New York.

Stephanopoulos, G. (1984) *Chemical Process Control: An Introduction to Theory and Practice*; Prentice-Hall: Englewood Cliffs, New Jersey.

U.S. Environmental Protection Agency (1985) *Process Design Manual for Nitrogen Removal*; U.S. Environmental Protection Agency: Washington, D.C.

Vaccari, D. A.; Cooper, A.; Christodoulatos, C. (1988) Feedback Control of Activated Sludge Waste Rate. *J. Water. Pollut. Control Fed.*, 60, 1979.

Water Pollution Control Federation (1986) *A Primer for Computerized Wastewater Applications*; Manual of Practice No. SM-5; Water Pollution Control Federation: Washington, D.C.

Chapter 11

Human–Machine Interfaces

Real-Time Display and Control Devices	354	*Annunciators*	357
Types of Control Panels	354	*Status-Indicating Lights*	357
Wall-Mount Vertical Panels	354	*Switches and Pushbuttons*	359
Free-Standing Vertical Panels	354	*Strobe Lights and Horns*	359
Console Panels	354	*Bar Graphs and Loop-Controller Displays*	360
Breakfront Panels	354	Control-Panel Distribution	361
Device Arrangement	354	*Local Control Panels*	361
Put Devices within the Operator's Line of Sight	356	*Vendor-Supplied Control Panels*	361
Group Displays and Associated Controls by Equipment Function	357	*Area, Master, and Main Control Panels*	362
		Graphic Panels	364
Put Frequently Used Devices in the Center	357	*Sandwich Displays*	364
		Mosaic Tile Displays	364
Put Displays above Related Controls	357	Computerized Displays	365
		Text Displays	365
Make Labels Easy to Read	357	*Proprietary Graphic Displays*	365
Avoid Abbreviations	357	*PC-Based Displays*	367
Provide Test Functions	357	*Video Projectors*	368
Visual and Aural Stimuli	357		

Large, Multisegment Flat-Panel Displays	368	Printers	380
Browser-Based HMIs	368	Audible Alarms	380
Wireless HMIs	369	*Autodialers*	380
HMI Software	370	*Control-Room Alarms*	380
Configuration Standards	370	*Pagers*	381
Object-Oriented Programming	371	Real-Time and Historical Trends	381
Screen Navigation	371	Recorders	381
Screen Layout	375	Displays	381
System Navigation Bar	376	*Local HMI Stations*	382
Graphics Area	376	*Historian*	384
Alarm Summary	376	Update Times and Sampling Intervals	384
Colors and Shapes	377	*Deadband*	384
Graphics and Bitmaps	377	*Smoothing*	385
Screen Refresh Rate	378	Control Rooms and Human Engineering	385
Limit the Number of Active Tags on a Screen	378	Ergonomics	390
Avoid Bitmap Graphics	378	Equipment Layout	390
Stagger Process-Loop Updates	378	Physical Security	390
Calculations	378	Fire Protection	390
Alarm and Event Communications	378	HVAC	391
Software Alarms	379	Lighting and Electricity	391
Alarm Triggered	379	Consoles and Furniture	391
Alarm Acknowledged, but Alarm Condition Still Present	379	*Console Subsystem*	392
		Console Surface and Appurtenances	392
		Chairs	393
Alarm Acknowledged, and Condition Has Returned to Normal	379	Security	393
		Security Vulnerabilities	393
		Network/PCS Security	395

Separate the SCADA System from Other Networks	395	Implement Backup and Recovery Procedures	398
Separate SCADA and Security Functions	397	HMI Access	398
Control Access to SCADA Equipment	397	Level I: View Only	399
		Level II: Shift Supervisor/ Control Room Operator	399
Secure Remote Connections	397	Level III: Process Engineer	399
Implement Firewall Protection	397	Level IV: Software Engineer	399
		Level V: Administrator	399
Leverage All Logging Features	398	Level IIA: Secondary Systems Operator	399
Install Virus-Protection Software	398	Level IIB: Solids Handling Building Operator	399
Implement the Highest Operating-System Security	398	Security Policy	400
		References	400

To be successful, a process-control system (PCS) must be able to interact with operators, engineers, managers, and others effectively. A *human–machine interface* (HMI) is, collectively, all the displays and controls that allow humans to interface with the facility's processes and equipment. The term is typically used to describe personal computer (PC)-based systems and software, but applies equally to distributed control systems (DCSs), which use PCs, and to the simple lights, switches, and pushbuttons on a field panel.

Knowledge about human capabilities and behaviors should be applied when designing an HMI. An optimal HMI should effectively communicate routine information (e.g., analytical measurements, flow values, process parameters, and equipment status). It should alert plant staff to abnormal conditions (e.g., process deviations, control system failures, and regulatory violations). It also should provide tools to help streamline process operations. The following aspects of HMI are discussed in this chapter:

- Real-time display and control devices;
- Alarm and event communications;
- Real-time and historical data trending;
- Control rooms and human engineering; and
- Personnel safety and security.

REAL-TIME DISPLAY AND CONTROL DEVICES

Real-time display and control devices allow humans to interface with the PCS and, ultimately, control plant operations. They should be designed to be as intuitive and effective as possible.

TYPES OF CONTROL PANELS. Control panels (control centers) are a fundamental part of an HMI. They should be designed for easy use by operations and maintenance (O&M) staff and easy access for maintenance, expansions, and modifications. Several types of panels are available (Figure 11.1):

- *Wall-Mount Vertical Panels.* Wall-mount vertical panels are available in many standard sizes. They typically cost less than other types of panels.
- *Free-Standing Vertical Panels.* Free-standing vertical panels should be used when larger panels are needed. They are typically mounted on a concrete base to provide stable mounting, permit access to electrical conduits, and to facilitate cleaning floors.
- *Console Panels.* Console panels should be used when O&M staff need to see the process on the other side while using the controls (e.g., in front of a tertiary filter during backwash operations). The console surface typically has a 45-degree slant.
- *Breakfront Panels.* Breakfront panels should be considered if so many displays are required that displays and annunciators would have to be put over the operator's head. These panels are typically custom-built and more costly.

DEVICE ARRANGEMENT. The devices on a control panel should be arranged so operators can comfortably find and use what they need (Figure 11.2). Also, the

Human–Machine Interfaces 355

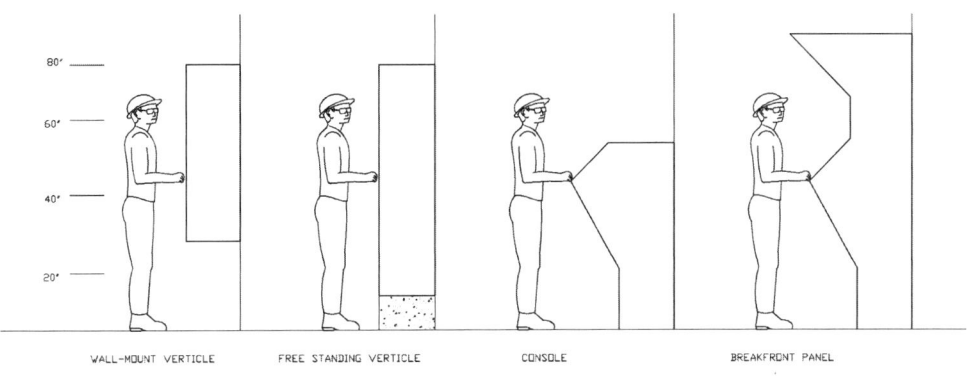

Figure 11.1 Panels designed for human interaction.

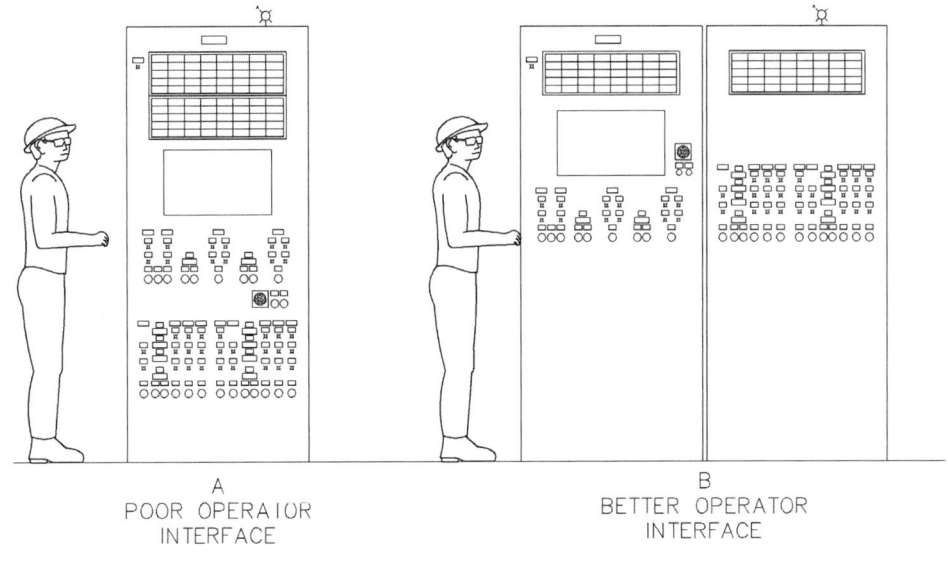

Figure 11.2 Control device layout.

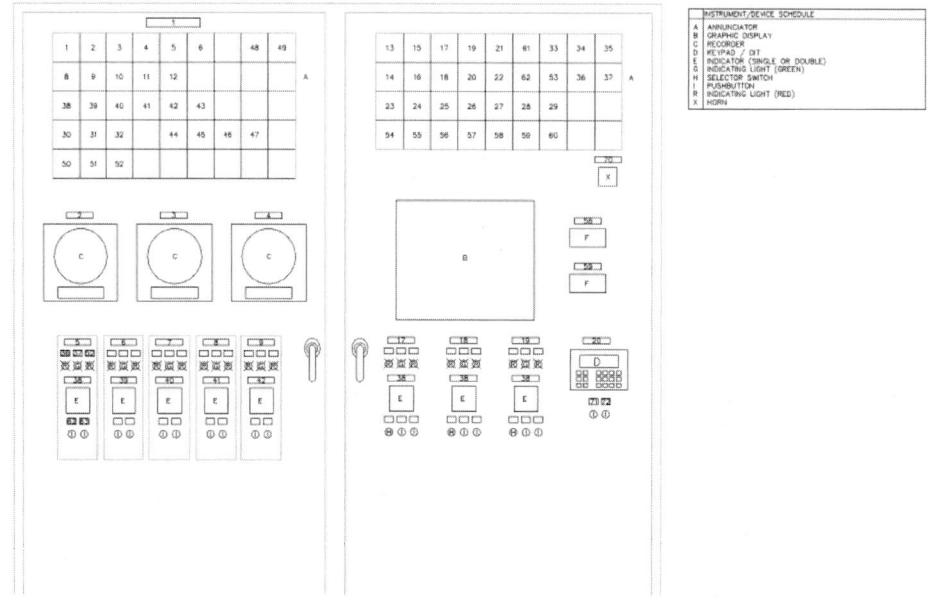

FIGURE 11.3 An example of the device layout on a control panel.

panel should be large enough to accommodate all the displays required, so operators do not have to strain to reach the controls.

When laying out a control panel, design engineers should observe the following guidelines [Figure 11.3; Instrumentation, Systems, and Automation Society (ISA), 1985]:

- *Put Devices within the Operator's Line of Sight.* Studies have shown that, when standing, human eye level typically is between 140 and 165 cm (55 and 65 in.). Studies also show that the optimal angle for human interaction with equipment is within 45 degrees above and below the line of sight. So, all status-indicating lights and switches should occupy the panel space between 1.2 and 1.8 m (4 and 6 ft) off the ground. Devices placed too low or too high may actually prevent personnel from noticing problems or responding to alarms.

- *Group Displays and Associated Controls by Equipment Function.* A consistent arrangement of indicators and pushbuttons will help operators respond quickly if a problem occurs. If possible, spaces should be left between groups of devices. A bewildering, haphazard array of lights and pushbuttons could lead to confusion and costly mistakes.

- *Put Frequently Used Devices in the Center.* For example, a device that must be frequently read should be placed directly at the operator's line of sight. Analog recorders and displays typically are put in the center of the panel. Annunciator systems that display multiple alarms are put at the top of the panel so they can be seen from anywhere inside the area.

- *Put Displays above Related Controls.* When operating a control, an operator's hand or arm should not obstruct the related display (for safety reasons).

- *Make Labels Easy to Read.* Operators should be able to easily read all display and control nameplates under normal operating conditions. Recommendations for sizing nameplate lettering can be found in ISA-RP60.6.

- *Avoid Abbreviations.* Abbreviations should be used sparingly and, if used, should be consistent throughout the PCS.

- *Provide Test Functions.* Test buttons should enable operators to verify that all status-indicating lights and audible signals work.

VISUAL AND AURAL STIMULI. Visual stimuli enable operators to distinguish among and react to the information displayed. People typically respond to changes in color (e.g., lights), position (e.g., gauges, recorders, and other analog readouts), pattern (e.g., bar graph, trends), and values (e.g., digital readouts).

Annunciators. Annunciators—typically multiple lighted squares with engraved text—are a simple way to signal many alarms. Design engineers should ensure that the lettering can be read from a reasonable distance. A typical annunciator sequence is defined in ISA Standard 18.1 (Figure 11.4). Because annunciators require frequent maintenance to ensure that all indicating lights are working, designers should include an appropriate test function for them.

Status-Indicating Lights. Status-indicating lights (pilot lights) are ubiquitous and a primary way that humans interface with plant processes and equipment. The designer should make sure that the colors used throughout the HMI are consistent

358 Automation of Wastewater Treatment Facilities

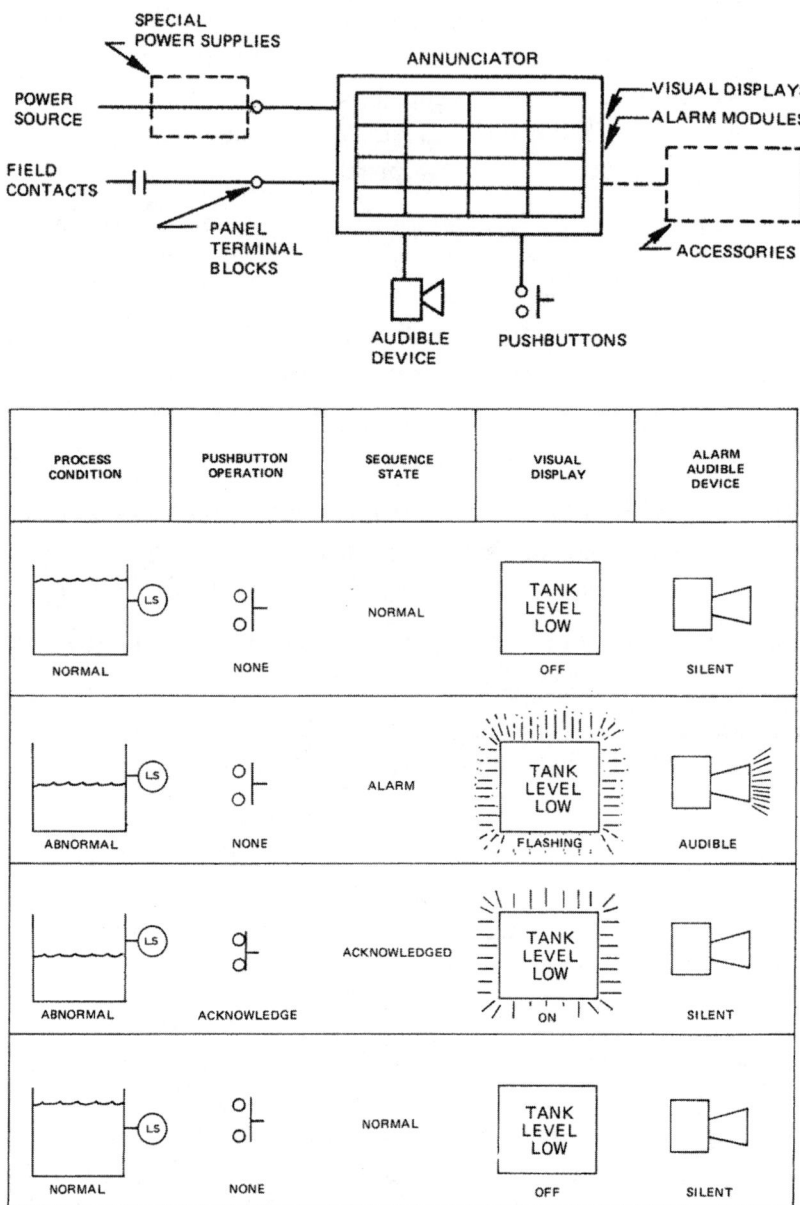

FIGURE 11.4 Annunciators and alarm annunication sequence.
(Reprinted with permission from ISA-18.1-1979 (R2004), *Annunciator Sequences and Specifications*, copyright 1979 ISA.)

and conform to facility standards. The following are some common color assignments:

- Green is typically used to indicate that equipment is off (in a safe, unpowered state), although some plants use green to indicate that equipment is on (powered).
- Red is typically used to indicate that equipment is on, although some plants use red to indicate that equipment is off. The design engineer should verify which color standard is used by the plant early in the design process.
- Blue is typically used to indicate that the control panel has power.
- Yellow is typically used to indicate that a valve or device is in an intermediate position.
- Amber is typically used to indicate that an alarm condition is present, requiring operator intervention.

Switches and Pushbuttons. Control devices should be placed so operators can reach and use them without undue strain. When arranging switches and pushbuttons on a control panel, designers should

- Include sufficient clearance for easy access and operation;
- Use guarded or protected pushbuttons with lockout features for critical equipment, emergency shutdown, and trip functions;
- For spring-return switches, which may need to be held for a long time, consider designs that reduce strain on the operator's hand;
- In areas where gloves are required, adjust the layout to compensate; and
- Use identical control panel layouts for duplicate equipment.

Strobe Lights and Horns. Some critical events require extra attention-getting measures to avoid injury or death to someone near the event. So, both strobe lights and horns or sirens are used. Horn output will typically be adjustable, with a range of about 80 to 100 dB at 10 ft/3 m for both indoor and outdoor applications. Siren output can be as much as 125 dB at 100 ft/30 m. Output level should be selected considering the level of ambient noise and the degree of hearing protection required in the area. Designers also must include a method for acknowledging the alarm and thereby silencing the horn or siren.

360 Automation of Wastewater Treatment Facilities

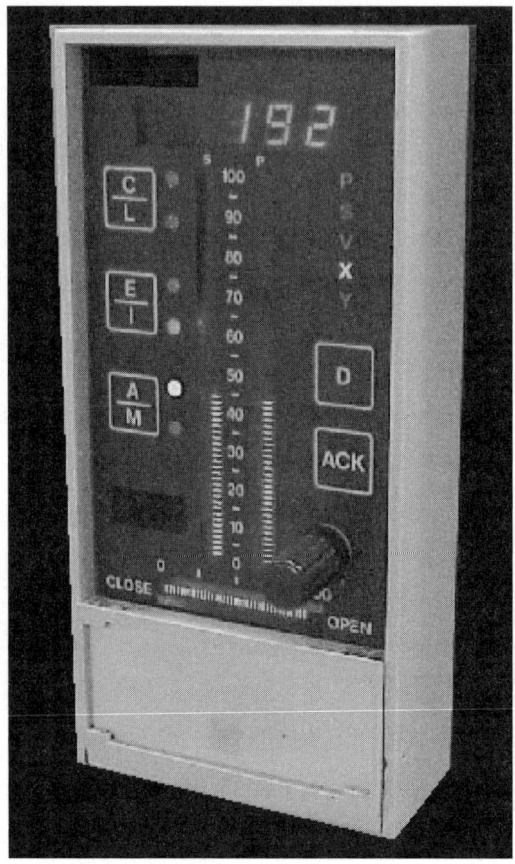

FIGURE 11.5 Display for a single-loop controller.

This combination typically would be used at chemical fill stations and in fire-protection systems, gas monitoring systems, and critical shutdown or emergency stop sequences.

Bar Graphs and Loop-Controller Displays. Bar graphs and loop-controller displays effectively illustrate the relationships among key process parameters. Displays for single or multiloop controllers show the process variable(s) measured, the controller output, and the loop setpoint(s) (Figure 11.5). Bar graphs convert numerical measurements of two or more parameters into appropriately sized bars and display them side by side (because the human brain can distinguish a pattern more rapidly than it can compare two or more numbers).

CONTROL PANEL DISTRIBUTION. The proper distribution of control panels and associated annunciation devices will ensure that operating personnel have the right information and control capability needed to respond to alarms and events.

Local Control Panels. Local control panels are key components of the HMI system. They typically have equipment Hands-Off-Auto (HOA) switches to allow for automatic (Auto) or manual (Hand) control of equipment, are on or next to their associated piece of equipment, and are frequently supplied by that equipment's manufacturer. Panels typically includes motor starters, indicator lights, and protective and safety interlock logic, which will shut down equipment independent of remote control commands to reduce the possibility of injury to personnel or damage to equipment. In computer-based PCSs, the panel also includes a local–remote or local–computer switch. This switch allows the equipment served by the local panel to be controlled remotely via an area control panel, main control panel, or computerized PCS.

Vendor-Supplied Control Panels. Equipment manufacturers typically supply control panels for major mechanical equipment (e.g., belt filter presses, centrifuges, and blowers). These panels often include a programmable logic controller (PLC) with control strategies and logic created and thoroughly tested by the manufacturer to ensure that the equipment functions effectively and has a long life. The equipment's warranty may require that its controls be left unchanged from the date of purchase. (During installation, the PCS should allow for enough local control so vendors can prove that their mechanical equipment meets the contract specifications independent of the PCS.)

It is preferable for the equipment manufacturer to supply this control panel and set up all protective interlocks. The question for PCS designers is whether a vendor's standard control panel would be best for the project, or whether a custom panel is necessary. For smaller equipment, a standard panel with protective interlocks is typically appropriate. As the equipment becomes more complex or operation requires monitoring of more process parameters than just those related to the piece of equipment (e.g., temperature and vibration), large custom-designed panels with computerized displays may be better. Designers should work with the vendor and plant staff to make this decision.

For plant managers intent on standardizing equipment, all vendor-supplied panels should be fully integrated into the PCS. Standard Fieldbuses, OLE for Process Control (OPC), and Ethernet communications now enable common PLCs to be easily integrated into any PCS. However, using a PLC of the same make and manufacture as in the rest of the PCS may be preferable to reduce equipment, training, and spare parts needs. Expect to pay a premium for the code to be programmed into your favorite PLC, but the long-term savings in reduced training and spare parts inventory may justify this cost. In any case, the amount and type of control these panels furnish should be consistent with the utility's overall control philosophy.

Area, Master, and Main Control Panels. Area, master, and main control panels (MCPs; control centers) typically include real-time displays, control devices, communications with multiple local control panels, and supervisory control logic for sequencing equipment operations. They can be used to control a particular treatment process or the entire treatment plant. For example, a disinfection control center may include indicators and controls for the sodium hypochlorite feed system, sodium bisulfite feed system, dilution water pumps, sample pumps, chemical tanks, chemical fill stations, and eyewash stations. To be effective, an MCP should have enough space for operators to conveniently view all indicators and access all controls.

These control centers typically are installed in a control or electrical room. Whether they are put near each process being controlled or in a central control room depends on how the overall plant is intended to be monitored and controlled (ISA, 1995). When determining control panel locations, designers should consider the following (Figure 11.6):

- The utility's control philosophy,
- Locations of existing controls,
- Types of instrumentation involved,
- Operator experience,
- Frequency of operator presence,
- Accessibility and visibility,
- Proximity to equipment being monitored and controlled,

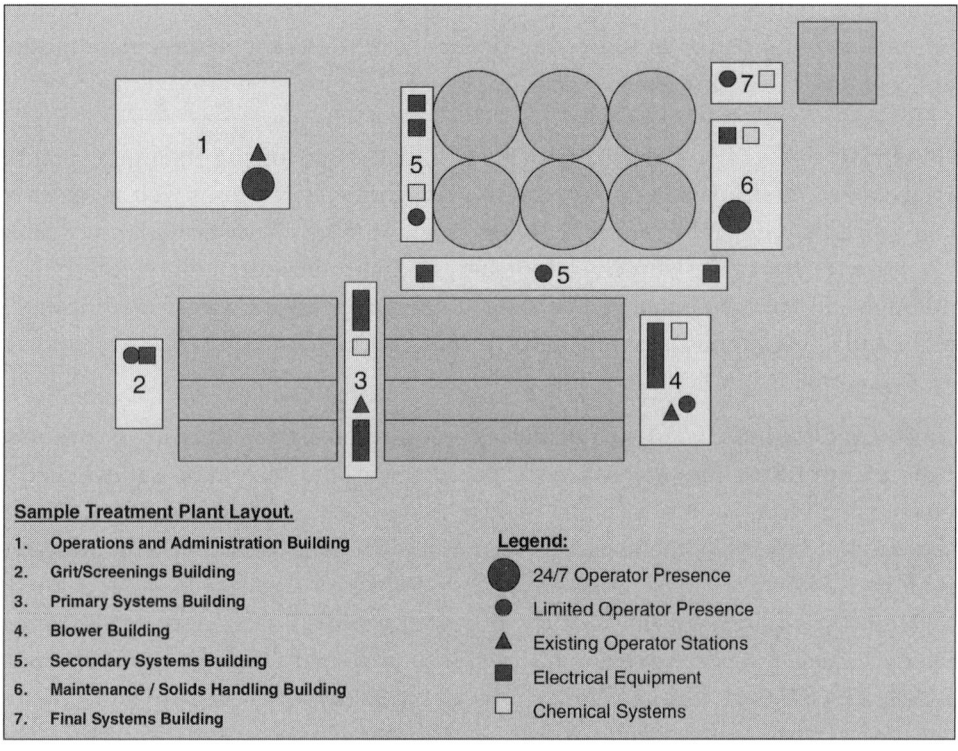

FIGURE 11.6 Issues to consider when determining control panel locations.

- Proximity to electrical and chemical systems, and
- Location of possible hazards.

Now that rugged, industrial PCs with membrane keypads or touchscreens are available, more treatment plants are using computerized displays in their MCPs. Plants with computerized PCSs typically interface signals with the PCS at their MCPs. However, when planning control-system redundancy, designers should avoid duplicating control logic at an MCP and the plant's main control system. A better approach to redundancy would be to use either fault-tolerant processors or redundant communication links to the PCS.

GRAPHIC PANELS. Graphic panels are illustrations of the treatment plant's process flow and display equipment status and alarm conditions, as well as selective analog signals. They range from simple strip graphics to sophisticated, modifiable mosaics with lights, instruments, switches, and annunciators embedded within. Although still available, such displays are falling out of favor because modifications are difficult and a significant portion of processor inputs/outputs (I/O) are required simply to provide information to the display.

Sandwich Displays. Sandwich displays typically consist of a drawing or printout of the plant process mounted on translucent Mylar and inserted ("sandwiched") between two Plexiglas sheets (nonglare Plexiglas on the front and clear Plexiglas on the rear). Alternatively, the process diagram may be made of vinyl with adhesive backing.

Main flow lines typically are more prominent than secondary feeder lines. Analog values typically are shown via multidigit, seven-segment, light-emitting diode (LED) displays. Digital signals typically are shown via a 1-amp lamp with a translucent plastic lens and color-coded vinyl border. Liquid crystal displays (LCDs) are gaining favor because they use less electricity and last longer, but they are difficult to read in bright light.

If mounted at the top of a high panel, this display should be tilted 15 degrees from the vertical.

Mosaic Tile Displays. Mosaic tile displays consist of multiple 2.5 cm × 2.5 cm (1 in. × 1 in.) tiles arranged to form a process flow diagram of the plant. The tiles are typically a molded compound (e.g., Valox) with an integral self-clamp. Some tiles are rear-lit and translucent or contain a translucent lens. One benefit of this display is that the tiles can be removed, replaced, or reoriented to reflect process changes.

The symbols used typically are site-specific. The graphics are airbrushed, engraved or silk-screened onto the tiles, and the lettering is typically a die-cut vinyl with a waterproof adhesive. Indicators may be incandescent lights, although backlit LEDs are more durable. Analog values are typically displayed at an LED readout assembly, with characters ranging from 1.6 to more than 2.5 cm (0.625 to more than 1 in.) tall.

The demand for such mosaics has declined in the wake of color graphic video displays. However, if the treatment process will not change often, these large displays can provide useful information. Designers should ask operators whether they prefer such mosaics and carefully compare the costs to those of computerized displays.

COMPUTERIZED DISPLAYS. The most common method for providing real-time data trends and process control in the field is a computerized display. Unlike traditional control panels, video display terminals do not have to be physically modified or rebuilt to accommodate process changes or additions. They also are less expensive and help staff respond to process problems more quickly. Video display terminals can be incorporated into existing control panels or directly connected to a plantwide PCS.

Design engineers should carefully consider whether to use computerized displays in the field. Rugged, industrial PCs are now available that can be implemented in virtually any environment. Many treatment plants are replacing large, traditional control panels with a small panel or console that includes a computerized display.

The decision should depend in part on the utility's operational philosophy. Some utilities want operators to regularly walk around the plant and observe process equipment. If so, then computerized field displays connected via the PCS network may be useful. Other utilities have fewer staff and prefer centralized monitoring of key locations.

Text Displays. Text displays basically are small devices with two- to four-line displays and a keypad (Figure 11.7). They typically are used to enter setpoints, provide alarm messages, or note event information in control panels where few changes are anticipated (e.g., vendor-supplied or local-equipment control panels). For example, if operators only occasionally need to enter several timing or loop setpoints at a particular field panel, then a text display may be more cost-effective than an industrial PC.

Proprietary Graphical Displays. Proprietary graphical displays were created for PCS designers who wanted to provide limited graphics and an operator interface at a site with a harsh environment. The displays are available with touchscreen or membrane keypad interfaces and 10- to 43-cm (4- to 17-in.) flat screens (Figure 11.8). Although they may look like mini-PCs, these displays are developed exclusively for control systems. They do not run Windows or UNIX operating systems; their proprietary operating systems are optimized for use in a PCS. They are very reliable because they are connected directly to a process controller, not the PCS network.

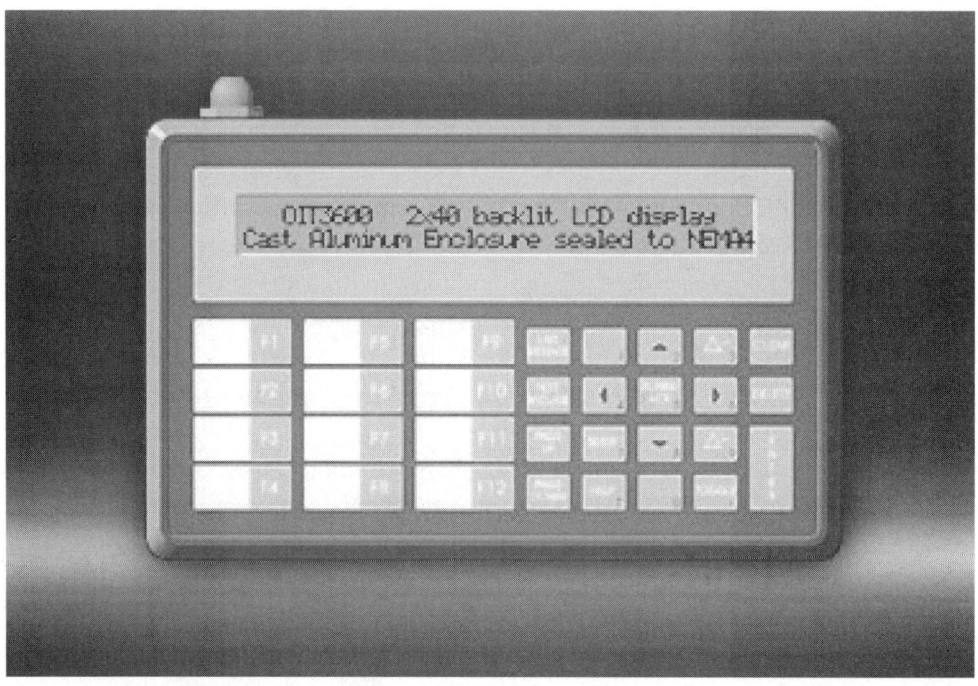

FIGURE 11.7 Text display.
(Courtesy of Maple Systems, Inc.)

FIGURE 11.8 Proprietary graphical displays.
(Courtesy of Rockwell Automation.)

FIGURE 11.9   Industrial PC panel-mount display.
(Courtesy of Xycom Automation.)

However, these devices' software is different than that used for the plantwide PCS, so any changes to the process may require software changes in multiple places. Also, operators will have to be trained to configure and operate another piece of hardware and software. In fact, unless a small display is desired, a better option may be an industrial PC rated for harsh environments that can run the plant's HMI software.

PC-Based Displays. Two types of PC-based displays are available: cathode-ray tube monitors and flat-screen LCD monitors. Cathode-ray tube monitors are inexpensive and available in many control rooms, either on desktops or inside consoles.

Flat-screen LCD monitors are more expensive but provide clearer pictures, occupy less space, and are rapidly gaining popularity.

As a result of improvements in flat-screen technology and the availability of industrial, panel-mounted PCs (Figure 11.9), PC-based displays are replacing proprietary graphic displays in field control panels. The displays help simplify the PCS because they can run the HMI software and can be connected over an Ethernet network. No additional programming software is required.

Video Projectors. A video projector can provide a comprehensive view of the process without the need for a hardwired graphics panel. The projector is typically ceiling-mounted at least 2.1 m (7 ft) above the control-room floor and can project an image on a large surface 3 to 6 m (10 to 20 ft) away. It basically works by using an internal LCD to form a color image and shining a light through the LCD to project the image onto a viewing surface. The image may be the result of one or several active LCD matrices whose images are superimposed over each other.

Video projectors are fairly inexpensive. Projecting a large PCS image on a wall, for example, enables operators and engineers to discuss system problems in the control room without having to crowd around a small computer screen. Projectors also are useful training tools.

However, projector screens can only display a limited amount of information—whatever can be graphically depicted on a computer screen. Also, if the projector is constantly on, its lightbulbs will have to be replaced frequently.

Large, Multisegment Flat-Panel Displays. As a result of improvements in digital video technology, more control rooms are using large, multiscreen flat-panel displays based on plasma or digital-light processing (DLP) technology. These displays have wider viewing angles than conventional flat-panel LCDs and work well in bright environments, producing less glare. Some manufacturers provide stackable multisegment displays, in which the screens can be digitally combined to function as a comprehensive display. The graphics drivers for these systems allow live video feeds and other media to be combined with other data and displayed anywhere on the display. The designer should carefully consider which type of display is warranted based on the type of information to be displayed and the layout of the control room where the display is to be installed.

Browser-Based HMIs. Web browser-based HMIs allow users to access PCS information from any computer via a utility's intranet or extranet, or even across the Internet (using appropriate security). This application is platform-independent; users

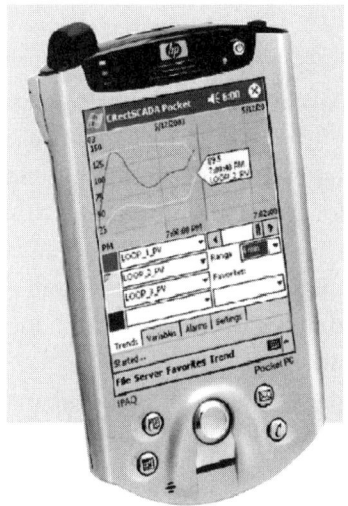 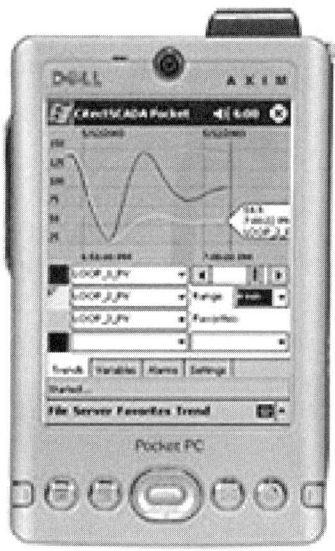

FIGURE 11.10 Wireless pocket PC HMI display.
(Photo of Citect Pocket HIM courtesy of Hewlett-Packard and Dell Computers.)

only need browser software, not special PCS-related software. Most HMI packages include Web-server capabilities and can convert an HMI graphic screen into a Web-based applet so the screen can be animated and viewed from anywhere via a standard browser.

This is a cost-effective and useful method for transferring data elsewhere (e.g., town or city offices) or monitoring several treatment plants from a central location. Potential drawbacks include its very accessibility, as well as the usual security concerns related to sending information over the Internet.

Wireless HMIs. Most process-control HMI packages now have wireless networking capabilities, so real-time information can be distributed to field staff via a Pocket PC

(Figure 11.10) or personal digital assistant (PDA). Likewise, O&M staff can access the PCS network via a laptop, PDA, or pocket PC and obtain real-time data while walking throughout the plant.

There are several wireless network protocols available. Bluetooth and Wi-Fi, for example, both provide connectivity via radio waves as long as the user is near an access point. Bluetooth was essentially designed to replace cables. It provides a wireless, point-to-point "personal area network" for laptops, printers, and other devices within 10 m of each other. Its access speed is limited (720 kbps), and it can only be used by Bluetooth-enabled devices. Several major PCS manufacturers now make Bluetooth-enabled systems for configuring and downloading data from instrumentation, drive systems, and programmable controllers.

Wi-Fi is designed to provide wireless, high-speed access to a local area network (LAN). It can provide access to an Ethernet network at up to 11 Mb/s as long as a wireless access point (node) is within 100 m.

Wireless networks can be useful for monitoring dispersed facilities (e.g., lift stations). For example, if a collection system's PCS is Bluetooth-enabled, staff could use a standard Bluetooth-enabled laptop or PDA to monitor any lift-station equipment, access stored data, and make any necessary adjustments.

However, security is a concern when using wireless HMIs. Staff must take care to protect laptops and handheld devices from theft. Also, a utility should implement strict security guidelines and properly authenticate anyone using the wireless hub or access point before granting him or her access to the PCS network.

HMI SOFTWARE. Although the PCS can provide a lot of data to the HMI, these data will only be useful if they are easy to obtain, relevant, and properly displayed. Otherwise, staff will be confused and dissatisfied with the HMI.

An effective PCS converts the plant's standard operating procedures into computer control logic so well that the transition from manual to automated operations is smooth. This is the most critical step in any plant automation project, and success depends on close communication between the control-system engineer and plant personnel.

Configuration Standards. First, the system designer and plant personnel should jointly develop a set of configuration standards to help programmers produce useful graphic displays and controls for the utility. These standards will define the appearance of all display screens and associated controls, and should be followed by all PCS programmers. (Enforcing these standards is especially important if the PCS is a multiphase project involving multiple contracts over many years.)

The configuration standards should define the following items:

- A preliminary list of required displays;
- The hierarchy of displays and methods for navigating the system;
- Display layout template(s);
- Symbols to be used in graphic displays;
- Colors assigned to indicate "on", "off", "alarm", "available", etc.;
- Alarm priorities and groups (usually by process area);
- Levels of access and associated environments (see below);
- Security requirements; and
- Requirements for collecting and storing historical data and conventions for displaying and trending data.

Object-Oriented Programming. Object-oriented programming should be used to create the graphic displays and associated controls because this programming method only requires each unit process or piece of equipment to be defined (e.g., symbol, colors, alarming scheme, setpoints, and control functionality) once. Duplicate equipment uses the same definition, so programming and troubleshooting are less time-consuming.

Screen Navigation. To simplify PCS navigation, design engineers should establish a logical hierarchy of displays (screens) and, if possible, limit the "depth" or "tiers" of screens so operators can reach any information within three or four clicks of a mouse button (Table 11.1). A convoluted navigation scheme only discourages HMI use.

A typical HMI hierarchy for a wastewater treatment plant begins at the plant overview level (first tier), which only contains basic information about the entire plant (Figure 11.11). This is the default screen that appears whenever a workstation is booted up or a previous user has logged off. The overview screen may consist of:

- A three-dimensional image of the plant noting key process information,
- An image of the plant, or
- A process-flow schematic of the plant.

TABLE 11.1 Sample HMI screen list for a wastewater treatment plant.

1. Plant Overview	
Initial Screen	1 Screen
Wastewater Process Overview	1 Screen
Solids Process Overview	1 Screen
2. Analytical Lab / Calculation Sheet	1 Screen
3. Generator Building	3 Screens
4. Preliminary Treatment	
Screen Building	2 Screens
Grit Collectors\Pumps	2 Screens
Grit Building	1 Screen
Primary Sludge Pumping System	1 Screen
Chemical Feed Systems	3 Screens
5. Main Pumping & Influent Equalization	
Main Pump Building	2 Screens
Main Pumps	One Per Pump
Control Screen	1 Screen
Equalization System	2 Screens
6. Primary Tanks & Pumping Stations	
Overview	4 Screens
Control Screen Overlay	4 Screens
7. Blower Building	
Blower Building Overview	1 Screen
Blowers	One per Blower
Control Screen	One per Blower
8. Aeration Tanks	
Aeration Tank Batteries	4 Screens
Control Screen	1 Screen
Foam Collection System	1 Screen
Surface Waste System	1 Screen
Trend Screen	1 Screen
Chemical Feed Systems	4 Screens
9. Final Settling Tanks and Secondary Pumping Stations	
Overview	1 Screen
Secondary Clarifiers	One per clarifier
RAS Pump Control Overlay	4 Screens
WAS Pump Control Overlay	4 Screens

TABLE 11.1 Sample HMI screen list for a wastewater treatment plant. *(continued)*

10. Tertiary Filtration System		
	Filter Facility Overview	1 Screen
	Filters	One per Filter
11. Disinfection Systems and Post Aeration		
	Chlorination	2 Screens
	Dechlorination	2 Screens
	Post Aeration System	1 Screen
12. Thickening Centrifuge Building		
	Thickening Centrifuges	3 Screens
	Polymer Feed	1 Screen
13. Gravity Thickeners		
	Process Overview	1 Screen
	Secondary Process Screen	1 Screen
	Polymer Feed	1 Screen
	Gravity Thickeners	4 Screens
14. Digesters		
	Process Overview	1 Screen
	Heating	1 Screen
	Recirculation/Mixing	1 Screen
	Digester Gas	1 Screen
	Digester Transfer	1 Screen
	Digester Feed	1 Screen
15. Sludge Storage		
	Overview	1 Screen
	Pump Control Overlay	1 Screen
16. Odor Control Systems		
	Overview	1 Screen
	Odor Control Systems	2 Screens
17. Electrical Systems		
	Plant One-Line	1 Screen
	DC/MCC One-Lines	6 Screens
18. Diagnostics/Maintenance		
	Equipment Runtimes	5 Screens

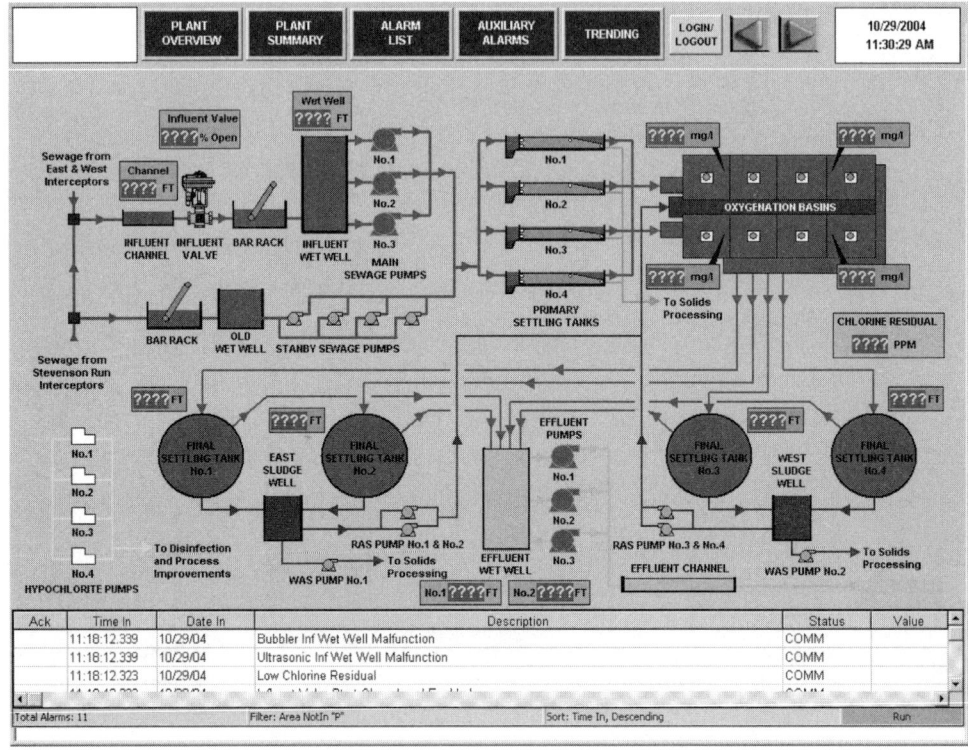

FIGURE 11.11 An example of an HMI plant overview schematic.

Design engineers should consult plant O&M staff about which format to use and how much information to display. The overview level typically should provide minimal information. The level of detail should increase as operators move further into the hierarchy (Figure 11.12).

Operators should be able to access all process screens (second tier) from the overview screen. Each process screen should contain information about a specific treatment process. Operators then could access equipment screens (third tier) detailing related equipment data. Authorized operators could then access control

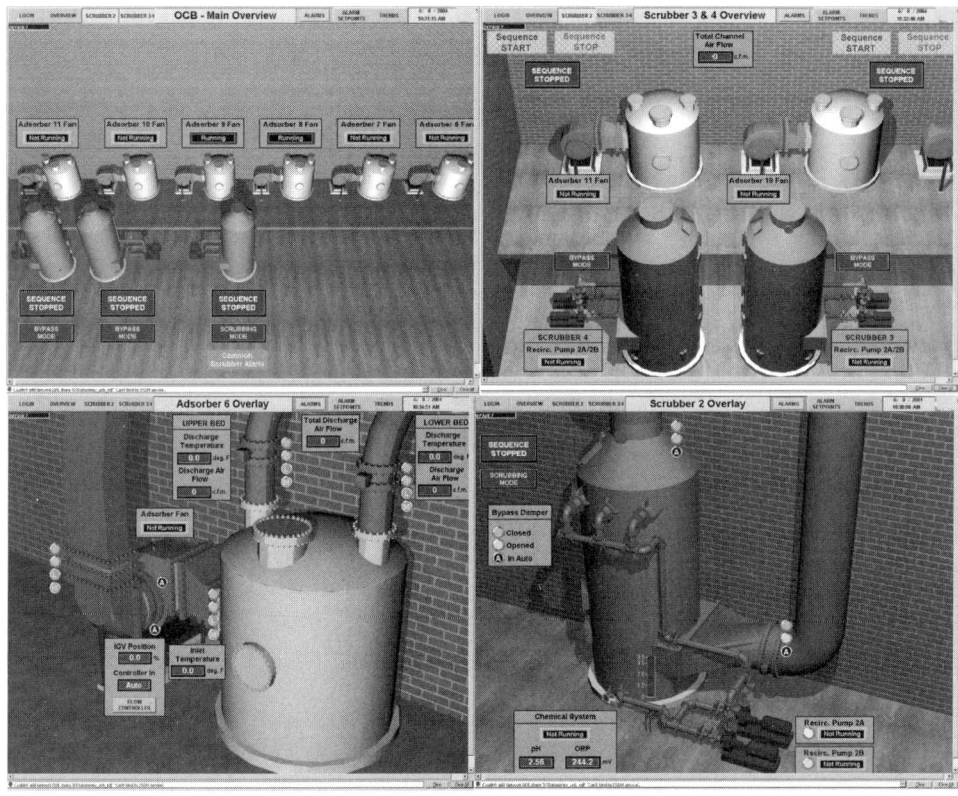

FIGURE 11.12 HMI screen hierarchy. Clockwise from top left: odor control system overview; subsystem detail; adsorber equipment detail screen; scrubber equipment detail screen.

screens (fourth/last tier), which include faceplates and popup screens for adjusting setpoints and performing other control activities.

Screen Layout. Design engineers should develop appropriate templates before creating any HMI screens (Figure 11.13). Each screen should have the following three main zones:

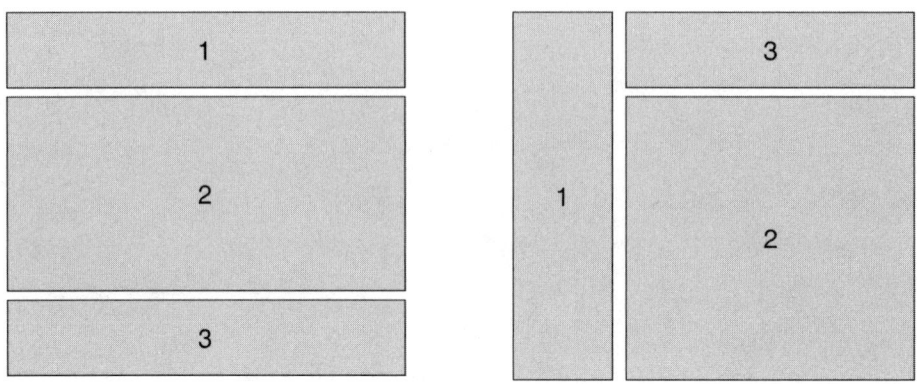

FIGURE 11.13   HMI screen template. (1) System/Navigation Bar; (2) Graphics Area; (3) Alarm Summary.

System Navigation Bar. The navigation bar basically consists of a column or row of pushbuttons that provide links to related screens in the hierarchy. Because consistent navigation tools are critical, a standard set of buttons should be used. At the appropriate access levels, the bar also should provide access to controls and other programs. At all levels, it may also include the date, the time, a message field, and other relevant information.

Graphics Area. This area contains all related graphic and status displays. It typically is the largest area of the screen.

Alarm Summary. This area (usually at the top or bottom of the screen) shows the most recent alarm conditions, summarizing the abnormal conditions requiring immediate responses. This summary should complement other alarm-annunciation methods.

Some treatment plants also include a banner on every screen that provides such data as login information, modes of operation, and key process information.

These screens typically do not include control information that is rarely accessed or requires a security clearance. Instead, the standard practice is to display such information via "pop-up" screens, when appropriate. Typical pop-up-screen contents include setpoint adjustments, tuning screens, equipment-control faceplates, and related reference information (e.g., process calculations or photographs).

If effective operations depend on real-time trends (e.g., flows or chemical usage for the last half hour), most HMI software packages allow them to be included in the display screen. This avoids the need to call up historical data, which, depending on the system architecture, can be time-consuming.

Colors and Shapes. Color is an effective way to communicate information rapidly. Designers should keep in mind, however, that some operators may have trouble distinguishing certain colors (e.g., red and green).

Typically, HMI graphic displays use the same color for a process that the plant uses for the process' piping and American National Standards Institute (ANSI)-designated colors for any processes not color-coded by the plant. To ensure that each color is consistent throughout the PCS, designers should follow a color standard, such as FS595B, the U.S. government standard for color matching (U.S. General Services Administration, 1994). The standard nomenclature indicates

- Luster (gloss, satin, or matte),
- Color (e.g., brown, green, red, blue, orange, gray, or yellow), and
- Reflectance (the lighter the color, the more reflectance).

The federal standard's list notes colors and attributes as they would appear on a bright day at noon. The standard also includes red, green, and blue (RGB) equivalents (in hexadecimal or decimal) for each color. Because monitors mix red, green, and blue to present any color, HMI programmers can use the equivalents to ensure that colors displayed anywhere in the PCS will be identical.

Likewise, designers should use one shape to represent similar equipment (e.g., pumps). Most HMI packages provide standard symbols for use in developing screen graphics.

Graphics and Bitmaps. Although graphics can help users intuitively understand a process, HMI programmers should avoid excessive use of them. For example, photographs can be useful reference material, but on an HMI display they will slow response time and could slow the screen refresh rate.

Screen Refresh Rate. *Screen refresh rate*—the time needed to update the tags in the HMI screen—is an important aspect of an effective PCS. They should be less than 1 second. Longer refresh rates do not provide the instantaneous feedback typically required for effective process operations and can cause operators to lose confidence in the PCS.

To improve screen refresh rates, HMI programmers should do the following:

- *Limit the number of active tags on a screen.* If a screen takes a long time to load and refresh, it is better to divide the screen into several screens with less information.
- *Avoid bitmap graphics.* Programmers should use the standard shapes provided or create shapes with the toolset provided by the HMI software manufacturer. If pictures or bitmaps are required for operator reference, programmers should put them in reference ("pop-up") displays instead.
- *Stagger process-loop updates.* Slow process loops should be updated every several minutes, not seconds.

Calculations. Programmers should include important process calculations (e.g., sludge volume index and sludge retention time) in the HMI. Any formula used should be indicated in the display, along with a definition of any parameter that is not intuitively obvious.

The HMI also should perform calculations related to equipment performance and "wear and tear" (e.g., theoretical versus actual pump flow, power consumption, and efficiency calculations). The software's calculating ability is chiefly limited by the data available from the field instrumentation.

ALARMS AND EVENT COMMUNICATIONS

Alarms are an extremely important part of an HMI. They should be consistent and concise to avoid unnecessary delays in reaction to critical events that could disrupt the overall wastewater treatment process.

Alarms include both process and system alarms. Process alarms associated with the process operation and control are usually indicated on the associated process graphic. System alarms include hardware and network failures. Process and system alarms should be subdivided into different alarm groups so not all alarms are displayed at all workstation areas.

TABLE 11.2 Examples of PCS alarms.

Type of alarm	Priority
Affects life or safety	1
Permit or regulatory	1
Equipment failure or malfunction	2
Control system or network	2
Equipment failed and standby started	3

SOFTWARE ALARMS. Typically, a PCS can distinguish among multiple process-alarm levels, with Priority 1 being the most critical and Priority 5 or higher being the least critical (Table 11.2). The software's alarm-and-operator-acknowledgement sequence should mimic that of traditional hardwired annunciator panels. It typically occurs as follows:

Alarm Triggered. An alarm situation occurs, and the alarm summary display flashes the appropriate alarm message, indicating the time the problem occurred and a description of the problem. The computerized display button linked to the alarm screen typically blinks, alerting operators to check that screen for information. On the treatment process graphic, meanwhile, the symbol of the affected equipment flashes in the alarm color to alert operators to the alarm location.

Alarm Acknowledged, but Alarm Condition Still Present. Once operators have acknowledged the alarm, the alarm display button and related equipment symbol will stop flashing but remain the alarm color. An "alarm acknowledged" message is sent to the alarm or event historical file and printer (if used).

Alarm Acknowledged, and Condition Has Returned to Normal. Once the situation returns to normal, the alarm display button shuts off, the related alarm message is deleted from the alarm summary display, the related equipment symbol reverts to its normal color, and a "return to normal" message is sent to the alarm or event historical file and printer (if used).

System designers should consult operators when determining which colors to use for "unacknowledged alarm", "acknowledged but active alarm", and "return to normal" messages.

PRINTERS. A traditional method for displaying alarm and event information is a dot-matrix printer with continuous-feed paper. It typically is configured to print one line of text whenever an alarm occurs. However, if the printer jams, all alarms and events may be printed on the same line and become illegible. Designers can avoid this problem by configuring a backup electronic file to process all alarms and events. In fact, most plant control rooms now use electronic alarm logs instead of printers and back up the logs regularly.

AUDIBLE ALARMS. Designers can integrate audible alarms into the PCS alarm system by adding a voice-annunciation or telephone-controller unit to area control panels or the control room. This controller enables the PCS to announce alarms or important events remotely via a standard telephone line or public address system.

Autodialers. System designers should incorporate autodialers in the alarm system to ensure that important alarms are communicated to the right people. If such alarms occur, the device will dial the designated phone number(s) and play a pre-programmed, digitized voice message when the phone is answered. Some units also include such advanced features as call prioritization, password protection, touch-tone alarm acknowledgement, and remote programming.

Autodialers are easy to install and maintain. They can reduce routine trips to unattended sites and enable the utility to concentrate personnel in the plant's more problematic areas. They can be especially useful at remote collection-system sites (e.g., pumping stations). Such units should be able to handle between two and 16 alarm contacts, and provide a voice message via cellular telephone to on-call staff.

Control-Room Alarms. Operators are not constantly looking at the PCS display screens while in the control room; they may be processing laboratory reports, taking samples, and performing multiple other duties. So, alarms should be both visible and audible. Many control panels use horns, but in a control room, a loud horn can be disruptive. A low beeping or chirping sound typically will suffice.

At any given time, however, control-room staff may be dealing with multiple acknowledged and unacknowledged alarms with differing priorities. Under such circumstances, simple tones are not especially helpful. Instead, designers should consider adding a speaker in the control room that would be used to announce critical alarms requiring immediate action. If a critical alarm is not acknowledged within a certain amount of time, the system would repeat the broadcast a predetermined number of times. If it is still not acknowledged, then the alarm could be announced via another method (e.g., autodialer or portable radio).

Pagers. There are many forms of paging available. Many large treatment plants use plantwide broadcast systems to announce alarms and simultaneously communicate with staff throughout the facility. Pagers can also be used. Several commercial software packages can be integrated into a PCS to provide vocal or alphanumeric messages to operators' pagers or cellular phones.

REAL-TIME AND HISTORICAL TRENDS

Displays of real-time (live) and historical trends allow users to observe responses to setpoint changes and to understand how processes behave over time.

RECORDERS. Many wastewater treatment plants use paper-circle or -strip chart recorders to track data. These recorders can use multiple colored pens on one chart, so several process variables can be tracked simultaneously. The charts help operators make decisions by showing recent process data and a certain amount of historical trending. A circular chart can provide up to 7 days of trend data, while a strip chart typically lasts much longer.

Computer-based recorders are now replacing the paper-based ones. The computer recorders are simple to configure, and the data can typically be exported to other programs for manipulation. Several recorder manufacturers offer a computerized circular chart recorder that functions much like the pen-and-paper model (Figure 11.14). Designers should consider paperless strip and circular chart recorders as components of the HMI.

DISPLAYS. Computer technology has made data trending simpler and, therefore, more useful. Ready access to real-time and historical data trends (Figure 11.15) can help staff better optimize treatment processes, manage resources, and troubleshoot problems. The two types of trends basically differ in purpose, duration, and data-storage location (e.g., local HMI stations and the historian).

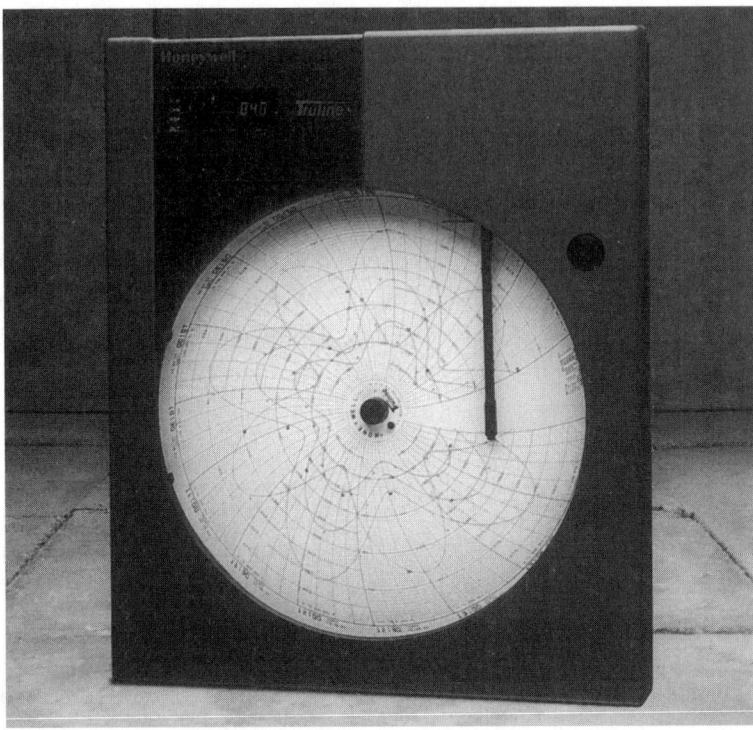

Figure 11.14 Electronic circular chart recorder. (Courtesy of Honeywell International Inc.)

Local HMI Stations. A real-time trend shows current process changes (e.g., the last several minutes or hours) using data from the local HMI station. Each station typically stores raw (recent, uncompressed) data for a limited time before transmitting them to the historical database. This information is recent, of short duration, and typically contains uncompressed, raw data. Such temporary storage prevents all data from being lost if the historian fails.

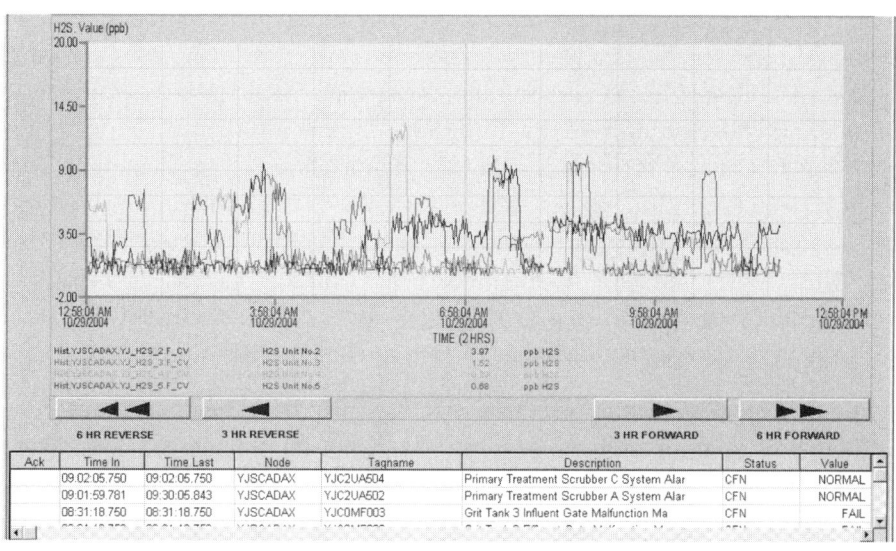

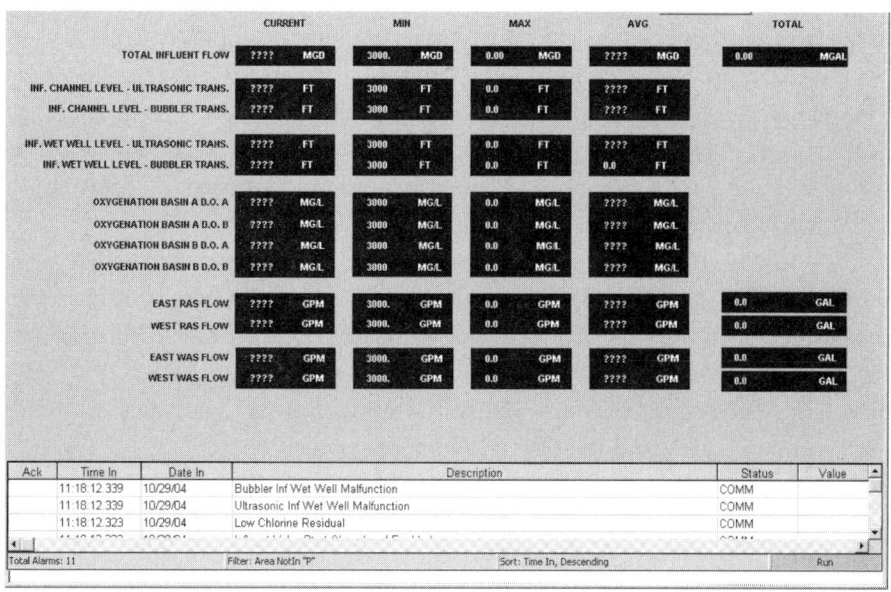

FIGURE 11.15 HMI historical trend and point displays.

When determining how much data to store in a local HMI station, design engineers should consider the data's *granularity*—the size of the data units. For example, if the data are collected once every minute, the station should store 72 hours' worth. If the data are collected once every 15 minutes, the station should store 40 days' worth. If the data are collected once a day, the station should store 13 months' worth.

In areas where continuous access to updated data could help improve process control, system designers should consider embedding appropriate real-time trends in the HMI screens. Most HMI-station packages permit this feature, so operators can track a recent process trend (e.g., rising tank level or falling dissolved oxygen) without having to access historical data, which can be time-consuming.

Historian. Historical trends use data that typically have been compressed and stored in the historian or another database. The historian is a dedicated workstation (or stations) in a PCS used to store and administer all historical data. These data are compressed so users can easily access and process weeks, months, and years of historical information.

A historical trend requires a defined time interval and compressed or averaged data. So, these trends are better for evaluating one or more process variables over a long time (e.g., shift, week, or year) than for reviewing current data. They also work well in reflecting gradually changing variables (e.g., levels in large tanks, temperatures in incinerators, or plant flows).

To ensure data availability for regulatory purposes, designers should store permit-related process-variable data (e.g., flow, dissolved oxygen, chlorine residual, ammonia, or phosphorus) in the historian whenever possible.

UPDATE TIMES AND SAMPLING INTERVALS. Engineers should configure the data-collection intervals properly. For example, process variables that change slowly and will be observed for hours or days should be sampled on the order of minutes (e.g., once a minute for a 24-hour trend). Process variables that change rapidly should be sampled more often, in fractions of a second.

If a process variable may not change significantly for several minutes or if its signal may be subject to noise, engineers can compensate via the following procedures (Figure 11.16):

Deadband. For gradually changing variables, engineers can configure the data-collection system to only update the trend point when the variable changes by a certain percentage. This saves processing power and allows more data to be acquired.

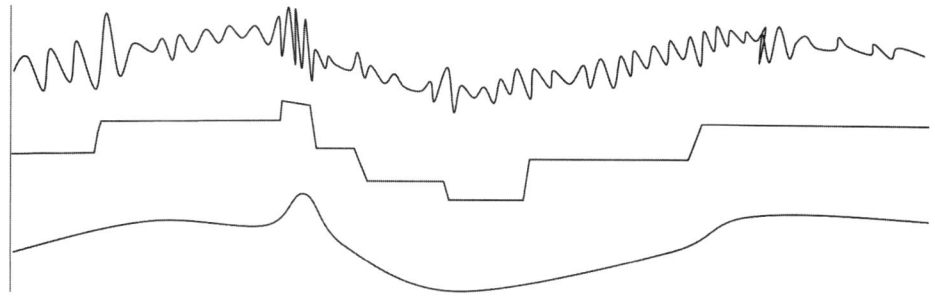

A. Raw data
B. Filtering using deadband, data compression
C. Smoothing using a moving average function

FIGURE 11.16 Data filtering.

Smoothing. For noise interference, engineers can add a moving-average (smoothing) algorithm to the process code, so an averaged value will be recorded in the historian. For example, the trending system may sample an input every 0.1 seconds but display 10-second averages of these data to users.

CONTROL ROOMS AND HUMAN ENGINEERING

The central control room is the hub of the HMI (Figure 11.17). More than any other location, the equipment in this room must be conducive to human interaction with the PCS and help them make informed decisions. Its location should be chosen based on an analysis of how the PCS should be operated. The analysis should answer such questions as

- Will operators be controlling equipment remotely or observing it directly?
- Should operators be roving or stationary?
- How accessible and visible should the control room be?

386 Automation of Wastewater Treatment Facilities

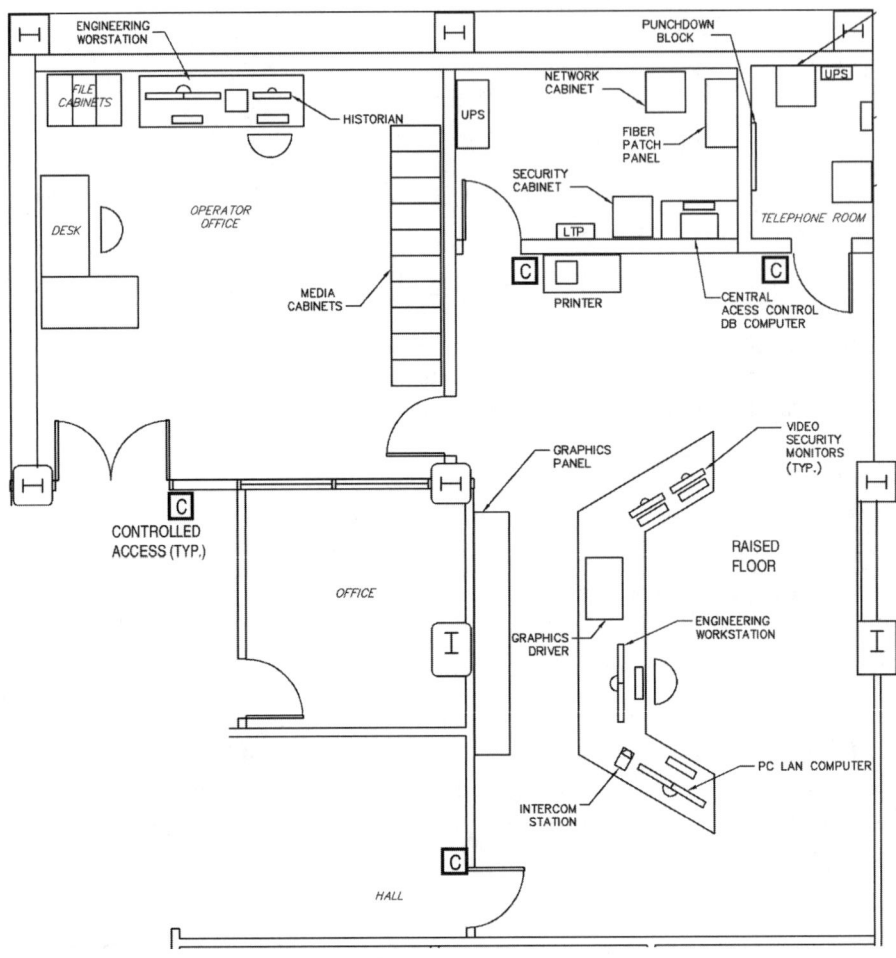

FIGURE 11.17   An example of a control room layout.

- How often should operators pass through it?
- How close should it be to equipment?

The control room must conform to all applicable engineering codes and regulations. Designers also should consult the ISA's Recommended Practices Series RP-60 and consider consulting with an architect, fire-protection expert, and mechanical and electrical engineers.

Because operators will spend a lot of time at the consoles and displays in this room, human-engineering aspects (e.g., console and display ergonomics, layout, ambient lighting, and personnel safety and security) are important. A good control-

TABLE 11.3 Equipment in the main control area.

Control Room / Console Area
• Plant control system console with PCS workstations and monitors
• Large screen display for plant processes
• Graphics driver(s) for large displays
• Video monitoring equipment (for process monitoring)
• Security system console (could also be in a separate area)
• Normal/emergency communication equipment
• Historian equipped with printer
• Media storage cabinets
• Remote pump station/collection system monitoring and telemetry hardware
Network/Server Room
• Communication connections
• Patch panels and media converters
• Network cabinet with network switches and servers
• Uninterruptible power supplies
• Security equipment

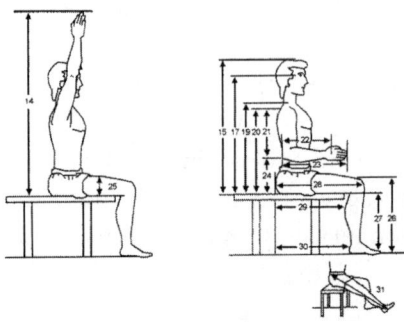

	Percentile values in centimeters					
	5th percentile			95th percentile		
Seated body dimensions	Ground troops	Aviators	Women	Ground troops	Aviators	Women
14 Vertical arm reach, sitting	128.6	134.0	117.4	147.8	153.2	139.4
15 Sitting height, erect	83.5	85.7	79.0	96.9	98.6	90.9
16 Sitting height, relaxed	81.5	83.6	77.5	94.8	96.5	89.7
17 Eye height, sitting erect	72.0	73.6	67.7	84.6	86.1	79.1
18 Eye height, sitting relaxed	70.0	71.6	66.2	82.5	84.0	77.9
19 Mid-shoulder height	56.6	58.3	53.7	67.7	69.2	62.5
20 Shoulder height, sitting	54.2	54.6	49.9	65.4	65.9	60.3
21 Shoulder-elbow length	33.3	33.2	30.8	40.2	39.7	36.6
22 Elbow-grip length	31.7	32.6	29.6	38.3	37.9	35.4
23 Elbow-fingertip length	43.8	44.7	40.0	52.0	51.7	47.5
24 Elbow rest height	17.5	18.7	16.1	28.0	29.5	26.9
25 Thigh clearance height		12.4	10.4		18.8	17.5
26 Knee height, sitting	49.7	48.9	46.9	60.2	59.9	55.5
27 Popliteal height	39.7	38.4	38.0	50.0	47.7	45.7
28 Buttock-knee length	54.9	55.9	53.1	65.8	65.5	63.2
29 Buttock-popliteal length	45.8	44.9	43.4	54.5	54.6	52.6
30 Buttock-heel length		46.7			56.4	
31 Functional leg length	110.6	103.9	99.6	127.7	120.4	118.6
	Percentile values in inches					
Seated body dimensions						
14 Vertical arm reach, sitting	50.6	52.8	46.2	58.2	60.3	54.9
15 Sitting height, erect	32.9	33.7	31.1	38.2	38.8	35.8
16 Sitting height, relaxed	32.1	32.9	30.5	37.3	38.0	35.3
17 Eye height, sitting erect	28.3	30.0	26.6	33.3	33.9	31.2
18 Eye height, sitting relaxed	27.6	28.2	26.1	32.5	33.1	30.7
19 Mid-shoulder height	22.3	23.0	21.2	26.7	27.3	24.6
20 Shoulder height, sitting	21.3	21.5	19.6	25.7	25.9	23.7
21 Shoulder-elbow length	13.1	13.1	12.1	15.8	15.6	14.4
22 Elbow-grip length	12.5	12.8	11.6	15.1	14.9	14.0
23 Elbow-fingertip length	17.3	17.6	15.7	20.5	20.4	18.7
24 Elbow rest height	6.9	7.4	6.4	11.0	11.6	10.6
25 Thigh clearance height		4.9	4.1		7.4	6.9
26 Knee height, sitting	19.6	19.3	18.5	23.7	23.6	21.8
27 Popliteal height	15.6	15.1	15.0	19.7	18.8	18.0
28 Buttock-knee length	21.6	22.0	20.9	25.9	25.8	24.9
29 Buttock-popliteal length	17.9	17.7	17.1	21.5	21.5	20.7
30 Buttock-heel length		18.4			22.2	
31 Functional leg length	43.5	40.9	39.2	50.3	47.4	46.7

FIGURE 11.18 Human static anthropometric data (sitting). (Copyright ISA 1985. All rights reserved. Reproduced and distributed with permission of ISA.)

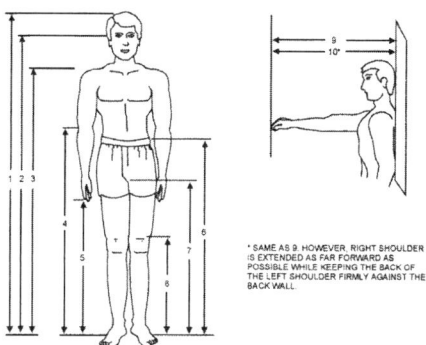

	Percentile values in centimeters					
	5th percentile			95th percentile		
	Ground troops	Aviators	Women	Ground troops	Aviators	Women
Weight (kg)	55.5	60.4	46.4	91.6	96.0	74.5
Standing body dimensions						
1 Stature	162.8	164.2	152.4	185.6	187.7	174.1
2 Eye height (standing)	151.1	152.1	140.9	173.3	175.2	162.2
3 Shoulder (acromiale) height	133.6	133.3	123.0	154.2	154.8	143.7
4 Elbow (radiale) height	101.0	104.8	94.9	117.8	120.0	110.7
5 Fingertip (dactylion) height		61.5			73.2	
6 Waist height	96.6	97.6	93.1	115.2	115.1	110.3
7 Crotch height	76.3	74.7	68.1	91.8	92.0	83.9
8 Kneecap height	47.5	46.3	43.8	58.6	57.8	52.5
9 Functional reach	72.6	73.1	64.0	90.9	87.0	80.4
10 Functional reach, extended	84.2	82.3	73.5	101.2	97.3	92.7
	Percentile values in inches					
Weight (lb)	122.4	133.1	102.3	201.9	211.6	164.3
Standing body dimensions						
1 Stature	64.1	64.6	60.0	73.11	73.9	68.5
2 Eye height (standing)	59.5	59.9	55.5	68.2	69.0	63.9
3 Shoulder (acromiale) height	52.6	52.5	48.4	60.7	60.9	56.6
4 Elbow (radiale) height	39.8	41.3	37.4	46.4	47.2	43.6
5 Fingertip (dactylion) height		24.2			28.8	
6 Waist height	38.0	38.4	36.6	45.3	45.3	43.4
7 Crotch height	30.0	29.4	26.8	36.1	36.2	33.0
8 Kneecap height	18.7	18.4	17.2	23.1	22.8	20.7
9 Functional reach	28.6	28.8	25.2	35.8	34.3	31.7
10 Functional reach, extended	33.2	32.4	28.0	39.8	38.3	36.5

FIGURE 11.19 Human static anthropometric data (standing). (Copyright ISA 1985. All rights reserved. Reproduced and distributed with permission of ISA.)

room design must address the following details: ergonomics, lighting, wiring (a raised floor may help), fire-protection systems, air conditioning and ventilation, personnel safety (e.g., no process lines in this room), security systems, and communication systems for both normal and emergency situations.

ERGONOMICS. The console and display design should accommodate human static and dynamic anthropometric data, as indicated in ISA RP-60.3 (ISA, 1985). The equipment layout should accommodate standing or sitting personnel and should be designed for frequent use without undue user strain. The ergonomics data shown in Figure 11.18 and Figure 11.19 are based on years of military research. These data are averages and should be tailored to the operators at the site.

EQUIPMENT LAYOUT. The main control room should be divided into two basic areas: the actual control console area and a restricted area for network, server, and plant-security equipment. Depending on the plant's size and complexity, designers should allocate space for all the equipment in Table 11.3.

PHYSICAL SECURITY. Physical security should be an inherent part of control-room design. For example, designers should include automatic exit doors that can be configured to default to "secure" (locked) or "safe" (open) if a power failure occurs. If a door must fail "secure", designers should provide a mechanical means of access (e.g., a key).

The building layout itself should limit access to control rooms by, for example, not requiring people to pass through control or electrical-equipment rooms to get to restrooms. However, access should not be too burdensome, or staff will circumvent it out of annoyance. Designers should consider where and how various personnel must travel during their shifts when laying out floor plans.

Basically, access controls should conform to the utility's standards, while restricting access to the appropriate individuals.

FIRE PROTECTION. To ensure that fire-protection measures conform to local regulations, control-system designers should consult with fire-protection experts. For example, paper storage should be limited in the main control room. The control room should have sufficient space for fire-protection control panels and containers of dispensing fluid, which cannot be hazardous to humans. Also, manual fire extinguishers should be available at the room's exits. If a raised floor is used, smoke detectors and fire protection are generally required for the space under the floor as well.

HVAC. Because of the equipment involved, control rooms often are hotter than other administrative areas. So, control-system designers should provide appropriate heat-dissipation calculations to mechanical engineers, so they can size the air-conditioning units properly. The air-conditioning system should maintain a temperature of 24°C (75°F) and 50% relative humidity (Battika, 2004). Backup power or redundant units should be considered if overheated equipment would compromise plant operations. The control room also should have external fresh-air intakes and self-closing doors to help maintain its ambient temperature. To avoid excessive ambient noise, the air conditioning units should be put in an adjacent mechanical, network, or server room.

LIGHTING AND ELECTRICITY. Most work areas typically should have 226 to 323 lx (21 to 30 ft-c) of illumination; however, control rooms should have 430 to 538 lx (40 to 50 ft-c) at 1.5 m (5 ft) above the floor (ISA, 1990). Control-room lighting should be diffuse and indirect to avoid excessive glare on console displays and computer monitors. Lights typically should be low near displays and monitors, and higher in printer and storage areas. Dimmers should be used wherever possible, so operators can adjust ambient and console lighting, as needed. There should be three independent lighting circuits:

- Display and console lighting,
- General room lighting, and
- Emergency lighting.

Electrical noise should be minimized. Communication and power cables underneath the raised floor should be kept at least 0.9 m (3 ft) apart and should cross at right angles.

Console displays and critical network equipment should have uninterruptible power supplies in the server or network area.

CONSOLES AND FURNITURE. Control rooms are not offices; they must accommodate round-the-clock staffing, heightened security, large projection screens or system displays, and the needs of specialists making critical operating decisions. Such characteristics must be reflected in an appropriate console design. Consoles typically consist of two parts: the mounting subsystem, which hides computer processors and cabling from view, and the console surface and appurtenances.

Console Subsystem. The equipment-mounting subsystem typically consists of fixed or slide-out shelves and interface hardware (e.g., brackets). Such equipment is designed to support the specified electronic equipment so it is ergonomically correct for the console operator. The computer monitor support should be able to adjust both the monitor height and its angle in the console.

Console Surface and Appurtenances. The exterior console panels include the work surface and the front, back, and top panels. Staff should be able to easily remove these panels and access the equipment mounted inside.

Design engineers also should take the following ergonomic considerations into account when creating consoles:

- *Corner Modules.* Corner modules can enable designers to arrange equipment for optimal workflow in the control room.
- *Linear Task Lighting.* Installing adjustable fluorescent task lights in the valence of each console module allows operators to augment the room's ambient lighting, as needed. The fixture should be designed to light the console work surface without introducing glare on monitor displays or the console face.
- *Glare-Reducing Glass.* The monitor surrounds or facings on the front of the console can be fitted with optical-quality glass panes to reduce eyestrain related to glare and reflections.
- *Footrests.* To ease operator fatigue, each console should include a footrest with antiskid feet and a rubber surface.
- *Integrated Power Bars and Cable Wireways.* To keep wires and power cords out of the operators' way, each console section should include integrated power bars and cable wireways.
- *Undercounter Keyboard Tray.* Each console counter section should be fitted with a keyboard tray that slides out from under the countertop. The low-profile tray includes a palm rest and typically accommodates keyboards up to 51 cm (20 in.) wide. It also is designed to maximize knee space under the counter.

Chairs. Each chair should have castors, a high back with lumbar support, and a swiveling seat with a pneumatic or hydraulic height adjuster. Designers should consider average anthropometric data when selecting chairs. The chairs also should be tested for stamina per the Business and Institutional Furniture Manufacturers Association's standards and approved by ANSI.

SECURITY

Personnel and PCS security should be an integral part of the design of any instrumentation and control (I&C) system for a wastewater treatment plant. The loss, damage, corruption, or improper operation of the PCS could disrupt plant operations and cause other serious problems. Plant managers should establish security policies and train all users to safeguard PCSs from accidental misuse or cyber attacks.

SECURITY VULNERABILITIES. Typically, a treatment plant's information technology (IT) system (used for e-mail, Internet use, billing and financial systems, etc.) has up-to-date firewall and antiviral-protection measures, but its PCS does not because utilities tend to segregate the two systems and not make IT personnel responsible for maintaining PCS infrastructure. This was not a problem when access to the PCS network was limited to the confines of the physical plant. However, as Ethernet-based protocols are introduced to PCSs and PCS information is more accessible to the IT network and databases, PCS vulnerability is becoming an issue.

During an intensive assessment of U.S. infrastructure—including wastewater treatment plants—the President's Critical Infrastructure Protection Board and the U.S. Department of Energy (2003) found that the

> "performance, reliability, flexibility, and safety of distributed-control/SCADA [Supervisory Control and Data Acquisition] systems are robust, while the security of these systems is often weak. This makes some SCADA networks potentially vulnerable to disruption of service, process redirection, or manipulation of operational data that could result in public safety concerns and/or serious disruptions to the nation's critical infrastructure."

Although design engineers have long planned for redundancy and reliability, cyber security has not been an integral part of the overall PCS architecture and design. However, they now need to secure PCSs from both physical and cyber attacks (Figure 11.20). Potential cyber threats range from amateur hackers inadvertently damaging a PCS to state-sponsored terrorists intent on mayhem.

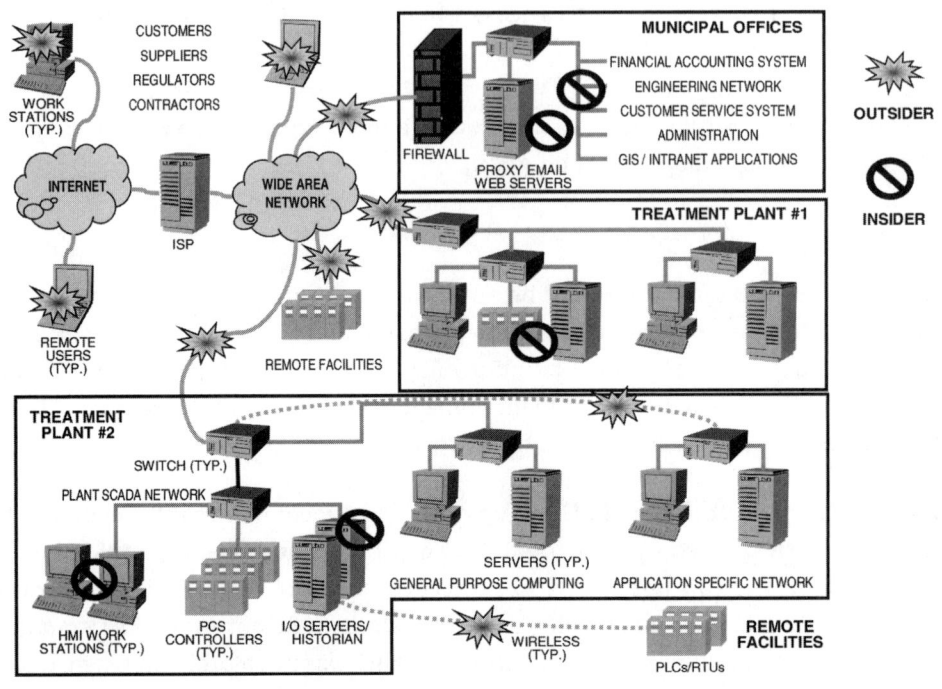

FIGURE 11.20 Common control system cyber vulnerabilities.

The challenge for designers is protecting the PCS while meeting the utility's accessibility requirements. For example, the PCS may need to be connected to the county's or city's LAN or wide area network (WAN) so information can be shared with various database and reporting systems [e.g., computerized maintenance management system (CMMS) or laboratory information management system (LIMS)]. Likewise, Internet or dial-up connections may be required so staff can access the PCS from remote facility sites or offsite.

It is also imperative that utilities establish a PCS security policy outlining who is permitted remote access and under what circumstances. Otherwise, PCS security

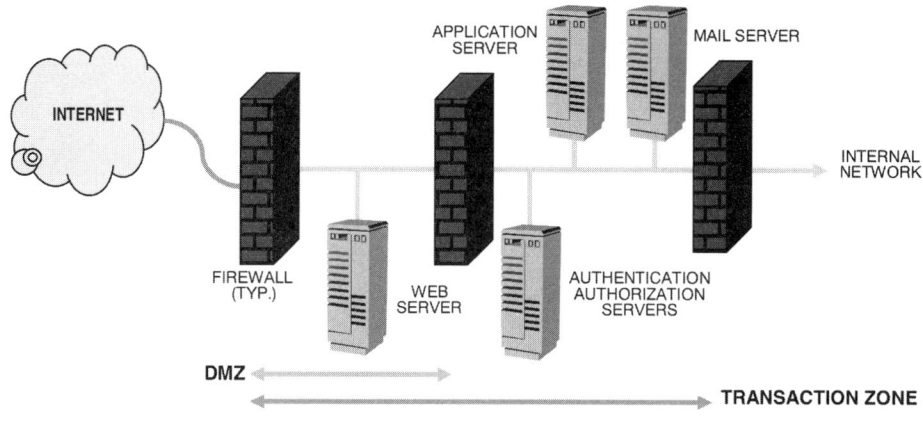

FIGURE 11.21 Firewall implementation.

could be easily compromised by someone posing as a telephone company representative, consultant, or contractor requiring remote access to the system. Security also could be compromised by someone with "network sniffing" software and knowledge of the PCS's wireless connection points.

NETWORK/PCS SECURITY. To better protect the PCS, design engineers should consider the U.S. Department of Energy's *21 Steps to Improve Cyber Security of SCADA Networks* (2003; Figure 11.21 and Table 11.4) when creating the system. The following steps can help safeguard a utility's PCS.

Separate the SCADA System from Other Networks. Design engineers should isolate the SCADA network from the county's or city's overall WAN. If the SCADA system must be connected to the WAN because of data transactions, etc., designers should put firewalls in the gateway server. They also should consider using switches or routers to separate the SCADA system from other networks.

TABLE 11.4 Twenty-one steps to improve the cyber security of SCADA networks. Source: U.S. Department of Energy, 2003.

1	Identify all connections to SCADA networks
2	Disconnect unnecessary connections to the SCADA network
3	Evaluate and strengthen the security of any remaining connections to the SCADA network
4	Harden SCADA networks by removing or disabling unnecessary services
5	Do not rely on proprietary protocols to protect your system
6	Implement the security features provided by device and system vendors
7	Establish strong controls over any medium that is used as a backdoor into the SCADA network
8	Implement internal and external intrusion detection systems and establish 24-hour-a-day incident monitoring
9	Perform technical audits of SCADA devices and networks, and any other connected networks, to identify security concerns
10	Conduct physical security surveys and assess all remote sites connected to the SCADA network to evaluate their security
11	Establish SCADA "Red Teams" to identify and evaluate possible attack scenarios
12	Clearly define cyber security roles, responsibilities, and authorities for managers, system administrators, and users
13	Document network architecture and identify systems that serve critical functions or contain sensitive information that require additional levels of protection
14	Establish a rigorous, ongoing risk management process
15	Establish a network protection strategy based on the principle of defense-in-depth
16	Clearly identify cyber security requirements
17	Establish effective configuration management processes
18	Conduct routine self-assessments
19	Establish system backups and disaster recovery plans
20	Senior organizational leadership should establish expectations for cyber security performance and hold individuals accountable for their performance
21	Establish policies and conduct training to minimize the likelihood that organizational personnel will inadvertently disclose sensitive information regarding SCADA system design, operations, or security controls

Separate SCADA and Security Functions. Design engineers should consider separating the SCADA and security networks. Although SCADA (or PCS) alarms could be security-system inputs, designers should avoid using the PCS as the security or intruder-detection system. The two systems' requirements are different, and most of the people who need access to the PCS are not the same as those who need access to the security system.

Control Access to SCADA Equipment. Design engineers should put all distributed control units (DCUs), PLCs, and communications cabinets and equipment in locked rooms with intruder-detection alarms. Access to network server and control rooms should be restricted to authorized personnel via key cards, access codes, etc. Remote terminal unit (RTU) cabinets should have doors with locks and tamper-resistant contact alarms, and the RTUs should be locked in "run" mode to prevent tampering with and sabotage of PCS programming. Design engineers also should provide locking doors on any cabinet where people could connect to the PCS via a laptop and a communications cable.

Secure Remote Connections. Treatment plant staff are not the only people who will need PCS access; other professionals will occasionally need to use the PCS to do the work that the utility has hired them to perform. So, utilities should establish access policies for "outsiders" (e.g., subcontractors and consultants), and design engineers should keep such policies in mind when creating the system. For example, designers could limit remote access connections to a designated server [e.g., a virtual private network (VPN) with at least 128-bit encryption] and create a system for monitoring those using the server. Designers also could give the PCS administrator the ability to create and monitor separate access levels for vendors. (They should not be given unlimited access.) Also, the lists of who has remote access and how much access they have also should be strictly controlled.

Implement Firewall Protection. A treatment plant's information technology network typically has firewalls between it and the Internet; so should its PCS (Figure 11.21). Design engineers should protect the PCS from hackers by including a robust firewall that monitors all access attempts from outsiders. The firewall should be sophisticated enough to detect a *"denial of service" (DoS) attack*—communication attempts at a frequency sufficient to disrupt proper operation of the system or to prevent its use by authorized staff.

Leverage All Logging Features. To help trace hostile actors, design engineers should maximize use of all the PCS software's logging features. The log should note the time, date, and username related to each system change (e.g., alarms, alarm acknowledgements, and setpoint changes).

Install Virus-Protection Software. Data corruption should be avoided at all costs. So, utilities should install virus-scanning software on the PCS, and periodically update the virus definitions. To avoid corrupted data by a virus unknowingly introduced via an employee's floppy disk, all floppy disk drives at all PCS workstations should be removed or disabled.

Implement the Highest Operating-System Security. Utilities should implement full operating-system security on all PCS workstations and update security patches on operating-system software whenever available. Treatment plant IT staff should eliminate the anonymous file-transfer protocol (FTP) account and limit software installation and file-system access to the PCS administrator.

Implement Backup and Recovery Procedures. Design engineers should provide procedures for operating-system-configuration backups and PCS-configuration recovery in case a system failure occurs. These procedures should complement the plant's overall disaster recovery plan. The backups should include both PCS software baseline configurations and subsequent modifications. Treatment plant staff should regularly back up relevant data and store it offsite with all PCS and operating-system backups.

HMI ACCESS. There are several ways to control onsite access to the PCS via its HMI. For example, the network administrator can set up passwords for each user. A keycard or proximity system could also be used. Either system would authenticate the user based on information residing on an access card, which is read by a swipe or proximity card reader connected to the PCS.

Remote access can be more challenging to secure. One commonly used method is a token or *smart card*—a portable device that displays a constantly changing password. This "one-time" password or code may either depend on the sequence number (i.e. each new code is different from the previous) or on the time it is generated [i.e., the code is only valid for a specified period of time from generation, (e.g., 2 minutes)]. In addition, the smart card contains a chip able to perform cryptographic operations. When logging in remotely, users must submit the currently displayed password or have the smart card read by a card reader. For this security system to be effective, staff issued smart cards (or tokens) must retain possession of them at all times.

In addition to limiting overall access to the PCS, utilities can further secure the system by assigning users various levels of access to system contents depending on job requirements. Each access level has its own environment—a set of display screens, menus, software resources, and accessible data—that users can navigate and control. Access rights are tied to each user's "username" and password, which are input via a workstation's login screen. Typical access levels are listed below:

Level I: View Only. This is the default environment when the PCS is turned on or a previous user has logged out. It only allows users to view overview and process screens, and to log into other access levels. No control, monitoring of details, or ability to acknowledge alarms is available.

Level II: Shift Supervisor/Control Room Operator. Staff granted access to this level's environment can view all process displays and initiate any operator action (e.g., changing setpoints or starting and stopping equipment).

Level III: Process Engineer. Staff granted access to this level's environment can view all process displays, initiate any operator action, access all PCS programs, and shut down the system.

Level IV: Software Engineer. Staff granted access to this level's environment can view all process displays and access the historian and all configuration software.

Level V: Administrator. Staff granted access to this level's environment have unlimited access to the PCS and can change other personnel's passwords. This level should be limited to the plant superintendent, deputy superintendent, plant manager, etc.

If tighter security is desired, these levels could be subdivided according to location or treatment process. For example, Level II could be subdivided into such environments as

Level IIA: Secondary Systems Operator. Staff granted access to this sublevel's environment can view all process displays and initiate operator actions associated with secondary treatment systems (e.g., blowers and aeration tanks).

Level IIB: Solids Handling Building Operator. Staff granted access to this level's environment can view all process displays and initiate operator actions associated with the gravity thickeners, thickening centrifuges, and wasting and return pumps.

SECURITY POLICY. Each utility should develop a policy on how to handle HMI-accessible information that an insider (or an outsider colluding with an insider) could use to exploit or damage a wastewater treatment facility. This policy should

- Outline the organization's security-related hardware, software, and activities;
- Heighten personnel's awareness of the need for security;
- Demonstrate the utility's commitment to protecting its vital information; and
- Be updated, as needed, based on emerging cyber and SCADA threats and changing system architectures.

The security policy should be posted to continually remind staff how sensitive some PCS information and controls are. For example, the HMI computer's login screen could furnish important policy guidelines.

REFERENCES

Battika, N. E. (2004) *The Condensed Handbook of Measurement and Control*, 2nd Ed.; Instrumentation, Systems, and Automation Society: Research Triangle Park, North Carolina

U.S. General Services Administration (1994) Federal Standard FS595B, www.gsa.gov.

Instrumentation, Systems, and Automation Society (1979) *Annunciator Sequences and Specifications*, Standard ANSI/ISA-S18.1-1979; Instrumentation, Systems, and Automation Society: Research Triangle Park, North Carolina.

Instrumentation, Systems, and Automation Society (1985) *Human Engineering for Control Centers*, Recommended Practices RP60.3-1985; Instrumentation, Systems, and Automation Society: Research Triangle Park, North Carolina.

Instrumentation, Systems, and Automation Society (1990) *Control Center Facilities*, Recommended Practices RP60.1-1990; Instrumentation, Systems, and Automation Society: Research Triangle Park, North Carolina.

Instrumentation, Systems, and Automation Society (1995) *Control Center Design Guide and Terminology*, Recommended Practices RP60.2-1995; Instrumentation, Systems, and Automation Society: Research Triangle Park, North Carolina.

President's Critical Infrastructure Protection Board (2003) *21 Steps to Improve Cyber Security of SCADA Networks*; U.S. Department of Energy: Washington, D.C.

U.S. Department of Energy (2003) *Common Vulnerabilities in Critical Infrastructure Control Systems*, 2nd Ed.; www.ea.doe.gov; Standard released in 1994. U.S. Department of Energy: Washington, D.C.

Chapter 12

Process Controllers

Applications	404	*Programming*	411
Collection Systems	404	*Soft PLCs*	412
Wastewater Treatment Plants	407	Remote Terminal Units	412
Types of Controllers	407	*Hardware*	412
Programmable Logic Controllers	407	*Software*	412
Components	408	*Diagnostics*	415
Input Relays (Coils)	408	*Basic Operations*	415
Internal Utility Relays (Contacts)	408	*Size*	415
Counters	409	*Standards*	416
Timers	409	*Recommended Specifications*	416
Output Relays (Contacts)	409	*PLCs versus RTUs*	417
Data Storage	409	Distributed Control Units	419
Analog I/O, Monitoring, and Control Capability	409	Single-Loop Controllers	420
Operations	410	Embedded Controllers	420
Step 1: Check Input Status	410	Input/Output Modules	420
Step 2: Execute Program	410	*Adapter*	420
Step 3: Update Output Status	411	*Discrete*	421
		Analog	421
		Specialty	421
		Communications	421

403

Ethernet	421	I/O Capacity and Program	
Wireless	422	Memory	423
Design Guidelines	422	Environmental Constraints	424
Market Position and Reputation	422	Programming Languages and Program Maintenance Costs	424
Maintenance Costs and Local Support	423	Security Features	425
		References	425

There are basically two types of process-control systems (PCSs) at wastewater utilities: in-plant and remote-facility. The chief differences between these systems are the communications methods and controller hardware involved. In-plant systems have historically used copper or fiber-optic communications media, while remote-facility systems used leased phone lines or wireless communications. (For more information on communications methods and their properties, see Chapter 9.)

As for controller hardware, remote-facility systems typically rely on supervisory control and data acquisition (SCADA) systems, which consist of process controllers, a communications network (see Chapter 9), and a human–machine interface (HMI; see Chapter 11; Figure 12.1; Irrinki, 1998). In-plant systems may rely on the same SCADA system, a distributed control system (DCS; Figure 12.2), or a plant control system.

APPLICATIONS

In the wastewater treatment industry, PCSs typically are divided into those used for collection systems and those used for wastewater treatment plants.

COLLECTION SYSTEMS. In collection systems, a SCADA system typically connects all remote wastewater lift stations to a central station for monitoring and control. The lift-station monitoring points may include

- Wet well level,
- Wet well high and low level alarms,

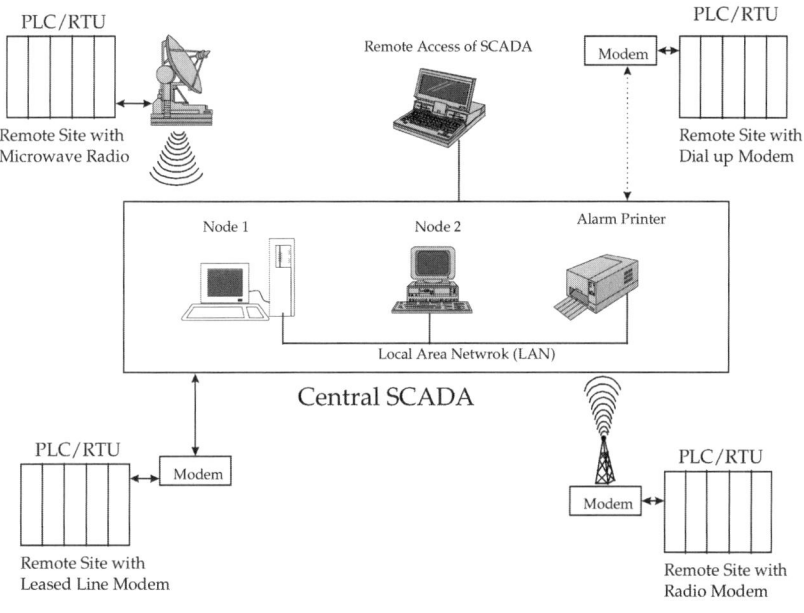

FIGURE 12.1 Schematic of a SCADA system.

- Pump run status (on or off),
- Pump run hours,
- Pump speed,
- Pump failure,
- Utility power failure,
- Emergency generator status,
- Intrusion alarm (security),
- Variable-frequency drive failure,
- Communications failure, and
- Low battery alarm.

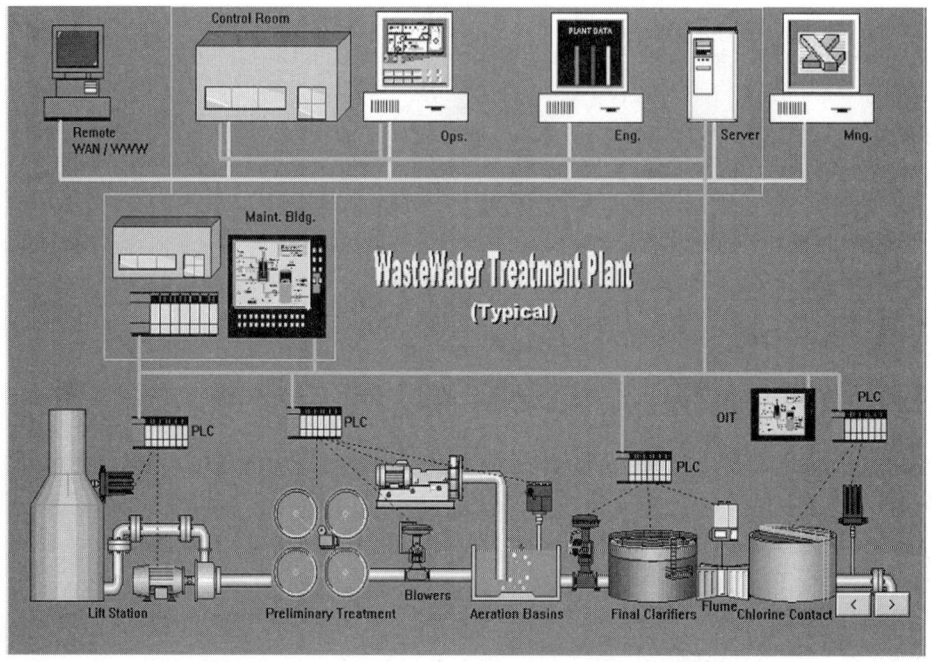

FIGURE 12.2 Schematic of a distributed control system.

The SCADA system can help collection-system personnel respond more quickly in emergencies. For example, if the power fails at a lift station without an emergency generator, the SCADA system can page on-call personnel and inform them where the power loss occurred. Then they can bring a portable generator or make alternate arrangements to run the lift station and avoid flooding local areas with wastewater.

Besides monitoring lift stations, the SCADA system can be used as an effective tool for infiltration and inflow (I/I) studies. Historical logs of pump running hours, when plotted against rainfall data, can show the effects of I/I on the collection system and lift station capacity. If the pumps are running longer during rainfall season or at high groundwater elevations than in the dry season, the collection system may have I/I. After studying the effects, staff can implement an I/I reduction plan.

WASTEWATER TREATMENT PLANTS. Wastewater treatment plants can control each unit process (e.g., preliminary treatment, aeration, disinfection, return and waste activated sludge, sludge dewatering, odor control, and chemical feed) with a programmable logic controller (PLC), DCS, or remote terminal unit (RTU). If these controllers are networked, staff can monitor and control the entire treatment plant from a central PCS workstation.

The system can help staff optimize each treatment process by tracking process variables (e.g., food-to-mass ratio, return activated sludge, waste activated sludge, mixed-liquor volatile suspended solids, sludge volume index, sludge age, and dissolved oxygen) and observing shifts in normal operating trends. With this information, staff can adjust each process as needed to meet National Pollutant Discharge Elimination System (NPDES) limits. The system also can be set up to print monthly reports of plant operating data for submission to regulators.

A process-control system also can help treatment plant staff save money. For example, it can monitor dissolved oxygen levels in the aeration tanks and adjust aeration rates to minimize excess aeration, resulting in power savings.

TYPES OF CONTROLLERS

The four most common controllers used in wastewater treatment facilities are

- PLCs,
- RTUs,
- Distributed controller units (DCUs), and
- Single-loop controllers (SLCs).

Each controller was originally developed to address a specific need. Over time, all have evolved into more sophisticated devices with more uses.

PROGRAMMABLE LOGIC CONTROLLERS. Programmable logic controllers were introduced in the late 1960s as a less-expensive alternative to complicated relay-based machine-control systems. The relay-based systems had to be changed whenever production requirements changed, and frequent changes were expensive. In addition, the numerous relays involved made troubleshooting tedious. Also, relays are electro-mechanical devices with a limited lifetime and a strict maintenance schedule.

To be an effective alternative, PLCs had to be rugged, long-lived, easily programmed by maintenance and plant engineers, and easily modified. So, PLC designers replaced mechanical parts with solid-state ones and used the familiar ladder-logic programming technique. The first commercial PLC was the MODICON 084.

During the 1970s, the dominant PLC technologies were sequencer-state machines and the bit-slice-based central processing unit (CPU). At first, conventional microprocessors lacked the power to quickly solve PLC logic, except in the smallest PLCs, but as microprocessors evolved, larger and larger PLCs were built. Meanwhile, Modbus was introduced, enabling a PLC to communicate with other PLCs within a couple of hundred feet of the machine it was controlling. Also, PLCs now could send and receive varying voltages, so they could handle analog inputs and outputs (I/O). The AMD 2901 and 2903 were widely used in this decade.

In the 1980s, General Motors' manufacturing automation protocol attempted to standardize PLC communications. Several PLC protocols were introduced in this decade. Programmable logic controllers also began shrinking and became programmable via a PC's symbolic programming rather than dedicated programming terminals or handheld programmers. Today, the world's smallest PLC is about the size of one control relay.

Fewer new PLC protocols were introduced in the 1990s, and popular ones were modernized. The latest standard (IEC 61131-3) attempts to merge PLC programming languages into one international standard, because PLCs can now be programmed in function-block diagrams, instruction lists, C, and structured text simultaneously. Meanwhile, PCs are replacing PLCs in some applications. For example, the company that originally commissioned the MODICON 084 now uses a PC-based control system.

Components. A programmable logic controller mainly consists of a CPU, memory, a power supply, and appropriate circuits for I/O data. Basically, in discrete control applications, a PLC is the equivalent of hundreds or thousands of relays, counters, timers, and data-storage locations. It simulates internal relays through bit locations in registers. Each part functions as follows:

- *Input Relays (Coils).* Typically transistors, these "relays" receive signals from switches and sensors.

- *Internal Utility Relays (Contacts).* These simulated relays enable a PLC to eliminate external relays. Some are always on, some are always off, some are dedicated to one task, and some are only used when the power is turned on to initialize stored data.

- *Counters.* These simulated counters can be programmed to count pulses (up, down, or both; Figure 12.3). Because they are simulated, their counting speed is limited. Some manufacturers use high-speed counter hardware instead.
- *Timers.* These simulated timers are available in many varieties and increments. The most common is an on-delay type. Others include off-delay, retentive, and non-retentive types. Increments vary from 1 millisecond through 1 second.
- *Output Relays (Contacts).* These transistors, relays, or triacs send on–off signals to solenoids, lights, etc. The type used depends on the PLC model chosen.
- *Data Storage.* Typically, assigned registers store data—usually temporary storage for math or data manipulation. They also can store data when the power is off and will still have this information when the power is turned back on. Both convenient and necessary!
- *Analog I/O Monitoring and Control Capability.* To manage analog I/O, modern PLCs primarily use manufacturer-specific function blocks embedded in the discrete control ladder logic. Although PLC analog monitoring and control capabilities are approaching those of DCSs, the embedded blocks make analyzing and troubleshooting these functions more difficult in PLCs than in DCSs.

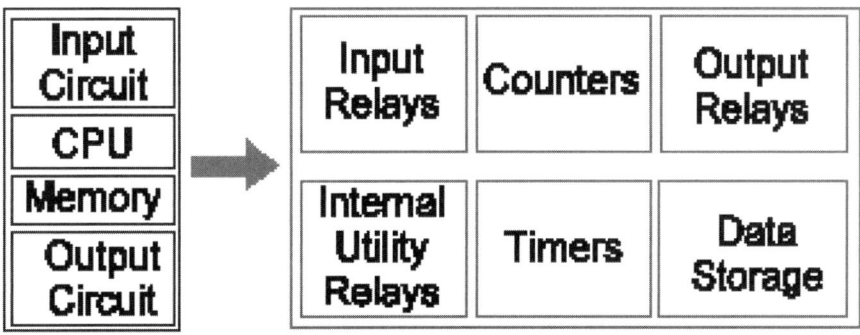

FIGURE 12.3 A simulated counter in a PLC.

Operations. A PLC operates in a continuous loop that basically consists of the following three steps (Figure 12.4).

Step 1: Check Input Status. First, the PLC looks at each input to determine whether it is on or off. In other words, is the sensor connected to the first input on? Is the second? Is the third? It records these data in its memory for use during the next step.

Step 2: Execute Program. Next, the PLC executes its program one instruction at a time, using the stored data. (If actual input data were used in this step, logic problems could occur if the inputs changed mid-execution.) For example, if the program said, "if the first input is on, then turn on the first output," the PLC notes whether the first output should be turned on and stores this information for use during the next step. It does this for each line of the program.

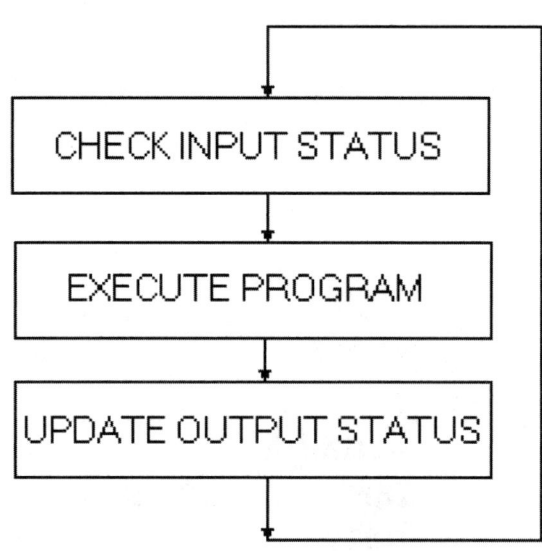

FIGURE 12.4 The scan cycle of a PLC.

Step 3: Update Output Status. Finally, the PLC updates the status of each output based on the information collected in the first and second steps. In other words, it will turn each discrete output on or off and adjust each analog output, as appropriate.

After the third step, the PLC goes back to step one and repeats the process. One *scan time* is the time it takes to execute all three steps (typically measured in milliseconds).

The PLC's hardware and programming are optimized to execute these three steps reliably, accurately, and efficiently. However, as with any computer, this is not instantaneous. When first turned on, a PLC first does a quick system status check to ensure that its hardware is working properly. If not, the PLC will halt and indicate the error. For example, if the PLC's backup battery is low and power was lost, the memory will be corrupt.

If the system-status check is successful, the PLC will begin scanning, and the loop continues indefinitely. Unlike personal computers (PCs), the PLC runs the entire program every scan.

Programming. The first PLCs were programmed via a ladder-logic technique that was based on relay-logic wiring schematics. This eliminated the need to teach electricians, technicians, and engineers how to program a PLC. It is still the most common technique for programming PLCs.

Other PLC programming techniques include instruction lists, function blocks, sequential function charts, and structured text. Instruction List is a low-level programming language. Its programs are lists of instructions that are executed sequentially, top to bottom. Function blocks, which originated in DCSs and SLCs used for continuous analog process control, are basically logic blocks organized by function (e.g., pump alternator, pumping station, or secondary clarifier). Sequential function charts graphically access, manage, and deconstruct logic blocks. Structured text is similar to a modern high-level computer programming language (e.g., BASIC or C).

The IEC 61131-3 standard is the first international attempt to standardize PLCs. It includes all five programming techniques mentioned above and is gradually being adopted by PLC and other process-controller manufacturers. Its two major advantages are the user-definable function block and the sequential function chart. The function block can combine programming in several languages to define a single block, or object (e.g., a valve, transmitter, or adjustable frequency drive). Each object can be maintained in an object library and combined into more complex function blocks for pumping stations, clarifiers, filters, etc. The complex function blocks also can be maintained in an object library and further combined into more complex facility objects. The blocks make programming standards easy to establish and maintain, as well as making the resulting programs easy for anyone to logically examine and understand.

The sequential-function chart language organizes the user-defined function blocks and controls their execution. It also navigates the program for troubleshooting, modifications, training, and batch programming, using the same logical relationships employed to execute the program. In addition, it can graphically deconstruct each function block on the screen to show its contents—down to the most basic level. All interfaces and relationships among the components are clearly depicted.

Soft PLCs. Many PLC manufacturers now offer soft PLCs—software versions of their controllers (Table 12.1). Soft PLCs typically allow an industrial computer (x86 based) to execute PLC programs and interface to PLC I/Os. They combine a PLC's process identifier (PID), discrete, and analog I/O control with a computer's high-performance data-handling, computational, and networking capabilities.

Soft PLCs typically incorporate the features, functionality, reliability, speed, and troubleshooting characteristics of proprietary "hard" PLCs into an open architecture system. They consist of an instruction set, a user program, and data table memory; a fast, deterministic program scan time; and an open architecture platform for connection to various I/O systems, other devices, and networks. They typically are configured as an embedded system and do not require a hard drive, monitor, or keyboard (Figure 12.5). Also, they do not require a Microsoft Windows operating system, so Windows-based stability and security concerns are irrelevant.

REMOTE TERMINAL UNITS. A remote terminal unit is a small, rugged computer. It originally was designed to handle communications for a SCADA system by collecting data from field devices and sending the data to the SCADA master on command (Figure 12.6).

Hardware. A remote terminal unit has a CPU, volatile memory, and nonvolatile memory for processing and storing programs and data. It can communicate with other devices via either serial port(s) or an onboard modem, with I/O interfaces for analog and digital I/O. It includes a secure power supply, with a backup battery and electrical protection against spikes. It also includes a real-time clock and a watchdog timer to ensure that it restarts.

Software. A remote terminal unit typically has a real-time operating system (RTOS), a communications driver, I/O system drivers, a SCADA application, and a method

TABLE 12.1 Characteristics of soft programmable logic controllers.

Hardware and operations	• Runs as an embedded 32-bit, real-time multitasking "kernel" on Pentium-compatible CPUs
	• Is independent of hardware bus (e.g., ISA, PCI, PC/104, and VME)
	• Minimal hardware requirements (e.g., rotating drives or monitor), so reliable and less expensive
	• Supports digital, analog, and specialty I/O
	• Supports COM ports to serial devices for ASCII communications
	• Supports user-configurable communication channels (network, data paths) for data or program logic access from other computer applications or PLCs
Software	• Operates ladder logic and user-developed functions written in C, C++, or Java
	• Online run mode program changes
	• Includes range of development, documentation, and online troubleshooting features
	• Runs imported or converted PLC programs
Special features	• Java Virtual Machine and Embedded Web Server provide a number of functions for data sharing, manipulation, and remote monitoring and maintenance
	• Built-in FTP server provides for remote maintenance, automated backups, and more
System requirements	• Pentium-compatible CPU
	• 8 MB RAM memory (16 to 32 MB for larger applications, or if using Web server)
	• 8 MB disk—typically FLASH, but other drives work, too (16 MB or more for larger applications, or if using Web server)
	• Ethernet port
	• Parallel port
	• User-specified I/O interface card(s) or port(s)
	• Other user-specified communications port(s) (e.g., COM ports)

414 Automation of Wastewater Treatment Facilities

FIGURE 12.5 A soft PLC.

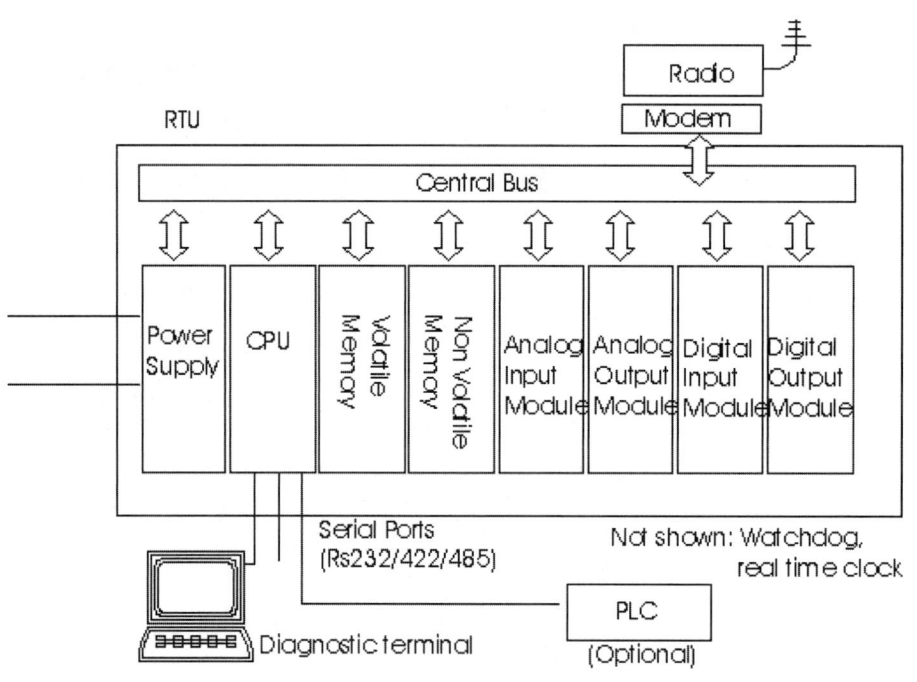

FIGURE 12.6 A remote terminal unit for a SCADA system.

for configuring user applications. These functions may be intermingled. The operating system may be a modern RTOS with monitoring and control functionality, or it may be code that began as one big loop scanning inputs and monitoring communications ports. The drivers link the unit to both the field devices and the SCADA master. The supervisory control and data acquisition application scans inputs, processes and stores data, and responding to requests from the SCADA master. User applications are generally used to provide local control capability.

Most modern RTUs also have monitoring and control capabilities similar to those of PLCs, but their monitoring-and-control programming languages typically do not use ladder logic. Many use scripting languages similar to BASIC. However, there is no standard programming language for RTUs.

Diagnostics. RTUs typically perform continuous or routine self-diagnostics. Some diagnostic results data are accessible by the master to provide limited remote diagnostic monitoring.

Basic Operations. Most RTUs rapidly scan inputs. They also may do some processing, such as changing an I/O's state; "time stamping" changes; sounding alarms, when appropriate; and storing data until the SCADA master requests them. Some RTUs even initiate reports to the SCADA master. Typically, though, the SCADA master asks the RTU for changes, and the RTU must respond to the request. A SCADA master's request may be as simple as "give me all your data" or as complex as "execute this control function".

Size. Size matters for RTUs. Tiny RTUs with eight-bit processors and minimal memory can be used for basic data collection and very limited control, whereas large, sophisticated RTUs can time-stamp data with millisecond accuracy, process hundreds of inputs, and even control small RTUs. Obviously, the larger ones are more expensive. Following are examples of the available range of RTUs:

- Tiny standalone RTUs often use single-chip processors with minimal memory and can run off batteries for a year or longer. They log data into EPROM or FLASH ROM and download them when physically accessed by operators. They may not be able to handle a sophisticated communications protocol.

- Small standalone RTUs can power up periodically and apply power to sensors (or radios) to measure or report. They usually run off solar batteries, which can maintain operations for at least 4 months during winter in the far north.

- Medium RTUs are dedicated single-board industrial computers, such as a desktop PC or industrial computer (e.g., VME, MultiBus, STD bus, and PC104).
- Large RTUs provide complete plant control, with all the bells and whistles. They sometimes are called DCSs because of their size and I/O capacities. Some even have enough redundancy to rival an actual DCS.

Standards. Because RTUs are specialty devices, there originally were no standards for them and RTUs from one supplier could not be mixed with those from another supplier. However, the industry has begun to develop protocol converters and emulators so different RTUs can "talk" to each other, and some RTU standards have begun to emerge: DNP3 and IEC870 for RTU communications and IEC 61131-3 for programming RTUs.

Recommended Specifications. When specifying RTUs, design engineers should consider the following:

- Physical size (make sure it will fit);
- Power consumption;
- Temperature ratings for the application (e.g., -10 to 65 °C);
- Relative humidity (e.g., 0 to 95% noncondensing);
- Dust, vibration, rain, salt, and fog protection;
- Electrical noise immunity;
- Hardware and software compatibility;
- SCADA master compatibility (e.g., analog data stamps and communications protocols);
- I/O capability and capacity [always allow some spare (e.g., 10 to 20%), do not ask for analog output unless needed, and check analog accuracy and signal type];
- Analog input filters;
- Communications capability [e.g., support for radio, public switched telephone network (PSTN) landline, microwave, satellite, or X.25; if supporting PSTN, RTU must time stamp and store the data, and the SCADA master must accept the data backlog and backfill its database and trend files accordingly];

- Communications protocols [consider standard protocols (e.g., DNP3, IEC870, and MMS) rather than proprietary ones];
- Maximum addressability (e.g., 255 RTUs);
- Peer-to-peer communications support [e.g., data storage and forwarding if communications are difficult (especially radio)];
- Data transfer rate (e.g., 1200 baud FSK or 9600 baud radio);
- Additional serial ports (e.g., PLC interfaces);
- Programming features (e.g., time stamps, processor, and memory capacity);
- Time stamp accuracy {the accuracy standard in the electricity industry seems to be 1 millisecond, which is not achievable without fast processors and an accurate time signal [e.g., from a global positioning system (GPS)]};
- Ease of programming and configuration (see IEC61131-3);
- Local and remote diagnostics;
- Local diagnostics indicators; and
- Error log with remote access.

Table 12.2 shows typical specifications for RTUs.

PLCs versus RTUs. The inputs and outputs of PLCs and RTUs are similar, but one originally specialized in control programming and the other in communications. A PLC is a small industrial computer designed to replace relay logic. It was programmed to repeatedly scan all connected data inputs and respond with the appropriate outputs. At first, PLCs could not communicate with other devices, but as they began to be used in situations where communications were desirable, PLC communications modules were developed to support Ethernet (for use in DCSs) and Modbus communications protocols [for use over dedicated (wire) links].

RTUs, on the other hand, were designed to handle communications for SCADA systems. They collected information from field devices and sent that information to the SCADA master, doing only what the master told them to do. As SCADA systems became more sophisticated, however, the desire to program RTUs increased, and manufacturers responded accordingly.

Today, more RTUs and PLCs are handling both tasks, and the distinctions have nearly disappeared.

TABLE 12.2 Comparison of RTU systems.

Evaluation criterion	System type					
	Bristol Babcock	Data Flow Systems	Allen Bradley	PLC Direct	Square D 6370 AS	Motorola
Processor, base, and I/O	RTU 3305	TAC II with PLC module	SLC 5/03	DL450	Modicon Compact	MOSCAD RTU
Communications ports	3 RS232 1 S232/RS485	Radio port 1 RS232 port	1 DH485 1 RS232	2 RS232: 2 support Modbus master/slave	2 RS232	1 RS232 1RS232/485 1 Modulated or RS 232
Removable field wiring terminal strips	Yes	No	Yes	Yes	Yes	Yes
I/O capability and expansion	Fixed; up to 4 AI, 2 AO, 8 DI, 2 DO, 6 DI/DO configurable Can add local RTU 3301 as I/O	Modular plug-in cards	Modular plug-in cards	Modular plug-in cards	Modular plug-in cards	Modular plug-in cards
Programming	Function block	Industrial basic	Ladder logic	Ladder and flowchart	Ladder logic	Ladder logic
Program integrity	EEPROM for program Battery-backed RAM for data	EEPROM for program Battery-backed RAM for data	EEPROM for program Battery-backed RAM for data	EEPROM for program Battery-backed RAM for data	EEPROM for program Battery-backed RAM for data	EEPROM for program Battery-backed RAM for data
Memory	512-kb program 512-kb data	32-kb program 20 kb data	12-k program 4-k data	15-k program 15-kdata	16-k program 4-k data	128-k user
Local MMI connectivity	Yes	No	Yes	Yes	Yes	Yes
Protocols	DF1, Modbus, BSAP, and more	TAC II	DF1	With unsolicited: Modbus Other: DirectNet	Modbus	With unsolicited: MDLC Others: Modbus, DF1, and more

EEPROM = electrically erasable programmable read-only memory
k = kilo
kb = kilobytes
RAM = random-access memory

DISTRIBUTED CONTROL UNITS. Distributed control units typically function much like PLCs and RTUs, with the following exceptions (Table 12.3):

- They are an integrated part of a DCS;
- They traditionally have been programmed using function-block or scripting languages rather than ladder logic;
- They are very redundant, including duplicate controllers and I/O modules; and
- They have one global database for I/O points that everyone can see.

The global database allows I/O points to be tagged and mapped to a device on the plant floor or to some internal DCS variable with a name indicative of its function. Then, whenever the variable is needed, engineers or operators simply ask for it by name without worrying about where it is coming from or what type of device it is.

The redundancies, safeties, and diagnostics built into DCUs make them both reliable and as available as possible. Manufacturers of PLCs and RTUs are beginning to include this feature in their controllers because it drastically reduces programming costs.

TABLE 12.3 Differences between DCUs and PLCs.

DCU	PLC
Process control	Discrete
Expensive	Economical
High degree of redundancy	Limited redundancy
Process state-driven	Event-driven
For large systems with lots of analog data (more than 1000 I/O)	For small systems with less than 1000 I/O

There are three basic types of DCUs: SLCs, embedded controllers, and I/O modules.

SINGLE-LOOP CONTROLLERS. Single-loop controllers provide DCU functionality for one analog loop. Many are made by DCU manufacturers. A typical SLC would have both discrete and analog I/O supporting a hybrid analog control strategy that could be adjusted based on discrete status inputs. Depending on the manufacturer, SLCs may be programmed via instruction lists, function-block languages, or other programming languages. Many SLCs also have networking capabilities.

Historically, SLCs were used with PLCs to provide single-loop integrity for important analog loops. To do this, one SLC was assigned to each analog output, thereby reducing the probability that one hardware failure would affect more than one analog loop. However, because of the expense and complexity involved, SLCs are rarely used for single-loop integrity today.

EMBEDDED CONTROLLERS. An *embedded controller* is any controller embedded in a process-measurement or final-control device—not an internal microprocessor used for signal processing, diagnostics, and calibration. However, it is a short leap from signal processing, diagnostics and calibration to process control, so embedded controllers could easily proliferate if supported by market demand.

Embedded controllers have long been available in complex instruments (e.g., analyzers and ultrasonic measuring devices) and final-control elements (e.g., control valves). However, the availability of inexpensive, rugged, and powerful processors developed for other industries (e.g., automotive) and the adoption of Fieldbus and programming language standards may provide the increased functionality and communications needed for widespread use in the wastewater treatment field.

INPUT/OUTPUT MODULES. Single-loop controllers and smaller PLCs and RTUs all have integrated I/O modules—I/O signals connected directly to the SLC, PLC, or RTU that limit their I/O capacity. However, DCSs and larger PLCs and RTUs have *modular I/O*—I/O modules installed in their chassis that can be tailored to the specific application. Modular I/O also allows one PLC, RTU, or DCU to accommodate thousands of I/O signals.

Adapter. Input/output adapter modules are typically installed in each chassis to provide a communications link between I/O modules in that chassis and the remote processor.

Discrete. Discrete modules accept discrete signals (e.g., on–off) from pushbuttons, limit switches and other devices, and produce discrete signals to control motor starters, pilot lights, annunciators, and other actuators. The outputs are directly controlled by the state of corresponding bits in the PLC data table, and the inputs directly control the state of corresponding bits in the PLC data table.

Design engineers may want to isolate discrete inputs to reduce the probability of physical damage to the module if the field wiring becomes exposed to unusually high voltages. Similarly, discrete outputs can be optically isolated so the field signal is not directly connected to the module's internal logic. Alternatively, interposing relays can be added.

Analog. Analog modules accept analog signals from pressure transmitters, analyzers and other devices; produce analog signals to control modulating valves, adjustable frequency drives, and other devices; and convert signals from analog to digital and digital to analog, as appropriate, to directly interface with PLC data table values using up to 16-bit resolution. Analog modules also can be user-configured for a desired fault-response state if I/O communication is disrupted. This feature limits the extent of faults and provides for safe, predictable responses to them.

Isolating analog signals is usually a good idea to protect them from electrical surges. Non-isolated analog inputs share a common ground or return connection. Isolated analog inputs have separate connections for each pair of analog input signals. The isolation levels offered by analog modules vary considerably. The isolation provided by differential input modules, for example, only equals the module's power supply voltage [usually less than ±15 V direct current (DC)]. Isolated input modules, on the other hand, usually provide more isolation than either optical or transformer isolation.

Specialty. Several specialty modules (e.g., process-control modules, positioning modules, high-speed logic-controller modules, and configurable flow-meter modules) are available to further enhance control systems. Each provides a unique, easy-to-use interface between digital or analog I/O circuits and the PLC processor.

Communications. Communications modules allow PLC processors to communicate with, monitor, and control individual instruments or actuators.

Ethernet. An Ethernet I/O module uses inexpensive Ethernet connections to a local area network to connect thousands of I/O measurements and control points to the processor. It can be used to both collect remote data and control machines and industrial processes. These modules can

- Pre-empt relay and digital outputs via interlock and emergency stop signals;
- Display status information on comprehensive diagnostic light-emitting diodes (LEDs);
- Allow field wiring to connect directly to I/O boards; and
- Allow I/O modules to connect to the server module via inexpensive Category 3 cables.

Wireless. Wireless I/O modules basically collect raw discrete or 4- to 20-mA analog signals from I/O points and transmit them via radio to a central processing device (e.g., a PC or PLC). This method does not involve Fieldbuses or wireless local area networks; it is slower and proprietary but simpler to implement. Most wireless I/O systems are simply used to acquire data, but some closed-loop control applications have been successfully deployed. The main limitation of wireless I/O modules is effective bandwidth, which dictates the number of signals that can be transmitted at once. They typically have process-loop reaction frequencies of 1 to 20 Hz, which is enough for many control applications.

DESIGN GUIDELINES

Design engineers typically should consider the following when comparing and selecting controllers:

- Market position and reputation,
- Hardware maintenance costs and quality of local support,
- I/O capacity and program memory,
- Environmental constraints,
- Communications requirements (see Chapter 9),
- Programming languages and program maintenance costs, and
- Security features.

MARKET POSITION AND REPUTATION. Controllers are commodities commonly available with a wide range of features from a number of manufacturers. However, the automation industry is competitive and driven by rapid, continuous technological advances, and there is no guarantee that any manufacturer or product line will be available well into the future. So, design engineers should choose one or more market-leading manufacturers with reputable product lines well-suited to wastewater treatment applications. A market leader, by definition, has a large customer base that promotes a product's continued existence and increases the likelihood that a graceful transition strategy will be available when it becomes obsolete.

MAINTENANCE COSTS AND LOCAL SUPPORT. Because most of today's commercial controllers are commodities with similar life expectancies and competitive prices, controlling maintenance costs is now less of an issue. Basically, design engineers should avoid selecting a niche product only available from one manufacturer.

Modern controllers also are available with features that ease maintenance, including

- Removable terminal blocks that allow the I/O module to be replaced without removing wiring;
- "Hot swappable" modules, which can be replaced while the controller is powered and operating;
- "Hot swappable" redundant power supplies, controllers, and I/O modules that continue operating after a module has failed and while it is being replaced; and
- Custom operator-interface modules that function as inexpensive, easy-to-use workstations.

The availability and reputation of local maintenance support, however, is more variable and important. Nevertheless, design engineers should be able to find suitable products and acceptable local support from several leading manufacturers in most areas.

I/O CAPACITY AND PROGRAM MEMORY. Although a controller's I/O capacity and program memory typically are well-matched, this is not always true. So, design engineers should establish requirements for both features when selecting a controller.

Also, when comparing controllers, design engineers should account for the effect of each controller's programming architecture on program-memory requirements. The amount of program memory required for each application depends on the complexity of the program and the programming language used. Controllers using ladder logic, older scripting languages, and older internal data communications methods typically will use less memory per function than controllers using more modern, flexible languages (e.g., those complying with IEC 61131-3) and more modern internal data communications methods. Ladder-logic PLCs typically have 4 to 64 kilobytes of memory. Soft PLCs, DCSs, and larger devices programmed via newer languages will have memory capacities in the megabytes.

ENVIRONMENTAL CONSTRAINTS. Important environmental constraints include electricity, temperature, humidity, and potential exposure to corrosive or hazardous substances. Protection from the corrosive and hazardous environment is especially important for both safety and maintenance reasons. Most industrial controllers are environmentally hardened and well-suited for wastewater utilities. However, aftermarket communications modules typically are not. Designed for office use, these modules have become more common as business network protocols, such as Ethernet, have proliferated in wastewater treatment plants. Using such communications interface modules compromises the controllers' reliability. Also, they often require 120 V alternating current (AC) power and uninterruptible power supplies, rather than 24 V DC batteries. Fortunately, third-party communications modules that use 24 V DC and are designed specifically for industrial environments are becoming more common and should be used when available.

Most critical applications should have backup power to maintain monitoring and control during power outages. The backup-power duration depends on the application. Plants with standby generators that are on-line minutes after power loss only need a few minutes of local-controller backup power. Critical remote critical facilities without generators, however, may need hours or days of local-controller backup power to continue functioning during an extended power outage.

PROGRAMMING LANGUAGES AND PROGRAM MAINTENANCE COSTS. Historically, programming languages were hardware-specific: RTUs typically used proprietary scripting languages, PLCs used proprietary ladder-logic programs, and DCUs used proprietary function-block languages. However, as more RTU, PLC, and DCU manufacturers adopt the IEC 61131-3 PLC programming-language standard for their products, the functional separation among their languages is shrinking. Although the underlying programs are still proprietary and hardware-specific, the user interface is becoming more standardized, making programming language less of a concern when choosing controllers.

Another important language-related consideration is program maintenance costs. Historically, program files were maintained on one or more laptop computers using ad hoc backup procedures, with variable and unexpected results. It is not uncommon for program files to be rewritten from scratch after a hardware failure destroys the working copy and the backup copy cannot be found or is corrupted.

Emerging configuration-management systems can automate and simplify program configuration, as well as provide better security by authenticating users and limiting access. When combined with IEC 61131-3 programming languages and well-maintained function-block and object-model standards and libraries, configuration management systems can significantly reduce long-term program maintenance costs by providing modification logs that provide a historical record of programming changes, enforcing change management policies, and making it easier for someone unfamiliar with a specific program to understand and troubleshoot the program.

SECURITY FEATURES. To date, controllers have few security features. However, as more emphasis is placed on security and protection against cyber attacks, security-related modifications are beginning to emerge. These modifications should be considered when choosing controllers. In particular, engineers should look for devices with features that

- Support encrypted operations and programming communications,
- Require proper identification and authentication for programming and configuration access, and
- Reduce susceptibility to tampering and sabotage.

REFERENCES

Allen Bradley Home Page. http://www.ab.com (accessed November 2003).

Foxboro Home Page. http://www.foxboro.com (accessed November 2003).

Irrinki, S. (1998) *Basics of Supervisory Control and Data Acquisition (SCADA), Proceedings of OWEA Annual Conference*; Tulsa, Oklahoma, May 13.

PLCopen Home Page. http://www.plcopen.org (accessed September 2004).

Chapter 13

Process Control Narratives

Purpose of Process Control Narratives	428	PLC Functions	437
		Remote Manual Controls	437
Typical Process Control Narrative Components	428	Remote Automatic Controls	437
		Alarms	438
Equipment Tag Numbers	428	System Components	438
Process and Instrumentation Diagrams	437	Other Methods of Conveying Process Descriptions	438
Process and Equipment Descriptions	437	Animations	438
Local Controls	437	Static Process Models	439
Motor-Control-Center Functions	437	Interactive Process Models	439

PURPOSE OF PROCESS CONTROL NARRATIVES

Process control narratives are written for a number of reasons. They reveal the design engineer's overall control philosophy and objectives. They should simply and clearly state how the control systems are expected to function, so everyone—suppliers, contractors, and operators—will understand.

A coherent process control narrative helps coordinate the various disciplines involved in the design. It enables process, electrical, and instrumentation and controls (I&C) engineers to work together effectively. Periodically reviewing the process control narratives with each discipline will significantly improve design coordination.

Process control narratives help the construction team understand the intent of the contract documents. They should make the requirements clear to the general contractor, I&C subcontractor, electrical subcontractor, and related suppliers. To encourage these people to read the descriptions during the bidding process, the engineer should refer to them as "related documents" in Part 1 of each appropriate specification.

Process control narratives also provide a framework for operations staff once construction has been completed and the facilities are in service. The narratives should answer the operators' questions about how the processes are supposed to work. They also can be an operator-training tool.

Finally, the process control narratives are a source of institutional knowledge, which can be especially important at older treatment plants (those more than 10 to 15 years old) where staff familiar with its design and operations have left. There, a set of well-prepared process control narratives is invaluable. The process control narratives are the one place where all of the interactions among various facility components are documented and explained.

Tables 13.1, 13.2, 13.3, and 13.4 are examples of process descriptions.

TYPICAL PROCESS CONTROL NARRATIVE COMPONENTS

Below are the major components of a process control narrative. Some of these items may be unnecessary for simple I&C systems, but all should be included in a design for a relatively complex facility.

EQUIPMENT TAG NUMBERS. Unique equipment tag numbers are essential to avoid the significant process errors that can result if the incorrect equipment and/or

TABLE 13.1 An example of a process description for a headworks lift station.

Item	Description
Equipment tag numbers included	Pump 160; Pump 170; Pump 180; Pump 190 LIT 161, LC162, LAH 163, LAHH 164
Related P&ID number	Sheet E28
Process description	This is the first phase of a five-phase wastewater treatment plant development. This process is the influent lift station to the first phase of the new treatment plant. In this phase, most of the controls are local, but the facility interfaces with the utility's SCADA system. The lift station consists of four pumps that are actuated based on the level in the wet well. The pumps are alternated.
Local controls	HOA (hands-off-automatic) switch for each pump Indicator lights for each pump: Red = Off; Green = Run Small PLC in vendor panel controls pump operation.
MCC functions	Manual = Off = Remote operation Manual = On and Off at MCC Off = Off Remote = Control from the panel at the lift station. This will permit operation at the lift station, or remotely via the SCADA system.
PLC functions	No plantwide PLC for this phase of project. PLC at lift station (LC 262) provides for pump alternation controls. It turns pumps on and off at predetermined elevations, based on a signal from LT 161. Lead pump On = 506.10 ft MSL (mean sea level) Second Pump On = 507.5 ft MSL Third Pump On = 508.1 ft. MSL Third Pump Off = 507.7 ft. MSL Second Pump Off = 507.0 ft. MSL Lead Pump Off = 505.8 ft. MSL
Remote manual controls	When HOA switch is set to "AUTO" at the lift station, the pumps can be controlled manually via the SCADA system.
Remote automatic controls	None
Alarms	High-level alarm (LAH 163) and very-high-level alarm (LAHH 164) are based on level switches set in wet well and provide visual and aural indication of alarm. Both alarms are equipped with a push-to-test button. LAH = 508.0 ft. MSL; LAHH = 508.5 ft. MSL
Components	Manufacturer-supplied control panel containing LC 262 for local control.

TABLE 13.2 An example of a process description of an effluent pumping station.

Item	Description
Equipment tag numbers included	Pump 210; Pump 220; Pump 230; Pump 240 LIT 206, LQC 001 LSLL 201, LSL 202, LSH 203, LSHH 204, LQC 002 FCV 110, ZS 111, ZS 112 FCV 120, ZS 121, ZS 122 LIT 401
Related P&ID number	Sheet E8

Process description	This pumping station is constructed on a parallel 36-in. main that normally discharges to a river. When the river rises above flood stage, this pumping station is designed to close the sluice gates on a flow-control structure and pump into the river.
	Pumps 210, 220, and 230 are the main pumps. Pump 240 functions as a sump pump. The water level is monitored in a downstream manhole by LIT 401.
	   1. When the water surface elevation reaches 606.3 feet MSL in the downstream manhole, the main PLC (LQC 001) closes the two sluice gates on the 36-in. lines (FCVs 110 and 120).
	   2. Water level in the wet well is monitored by LIT 206, which sends a signal to the PLC.
	   3. The PLC then controls when pumps start, stop, and alternate.
	When the signal from the ultrasonic level transmitter is re-established, LQC 001 resumes control of the station.
	The SCADA system can monitor lift-station operations and assume control of the station, but only if the controls at the station are set to "remote." Under this mode of operation, all of the SCADA pump control is manual.

Local controls	HOA switch for each pump
	Indicator lights for each pump:
	   Red = Off
	   Green = Run
	Small PLC in vendor panel controls pump operation if the signal from the ultrasonic level transmitter (LIT 206) is lost. In this situation, the manual floats provide level information to a backup PLC (LQC 002), which then assumes control of the lift station.

MCC functions	*This is actually at the vendor-supplied panel at the lift station.*
	   Manual = Off = Remote operation
	   Manual = On and Off at MCC
	   Off = Off
	   Remote = Control from the panel at the lift station. This will permit operation remotely via the SCADA system.

TABLE 13.2 An example of a process description of an effluent pumping station. *(continued)*

Item	Description
PLC functions	This lift station has two PLCs: LQC 001 and LQC 002.
	The primary PLC (LQC 001) provides for pump-alternation controls to turn pumps on and off at predetermined elevations, based on a signal from LIT 206. Lead Pump On = 418.0 ft. MSL Lag Pump On = 424.0 ft. MSL Lag Pump Off = 424.0 ft. MSL All Pumps Off = 408.5 ft. MSL
	If the signal from LIT 206 fails, then control reverts to LQC 002, which is actuated off of four level switches (201 through 204).
Remote manual controls	When HOA switch is set to "AUTO" at the lift station, the pumps can be controlled manually via the SCADA system.
Remote automatic controls	No remote automatic operation is possible over the SCADA system.

Alarms	High-level alarm (LAH 163) and very-high-level alarm (LAHH 164) are based on level switches set in the wet well and provide visual and aural indication of alarm. Both alarms are equipped with a push-to-test button.
	   LAH = 421.0 ft. MSL
	   LAHH = 425.5 ft. MSL

Components	1. Manufacturer-supplied control panel for lift station. 2. Manufacturer-supplied control panel for sluice gates at the gate chamber. 3. Manufacturer-supplied panel at DS manhole for the ultrasonic level transmitter (LIT 401).

valves are operated. Additionally, unique tag numbers eliminate confusion in the operations and maintenance manual. The Instrumentation, Systems, and Automation Society (ISA) has developed a widely used method for assigning tag numbers, but some utilities prefer to use their site-specific tagging methods. The method used is less important then the result: every piece of equipment has a unique tag number, thereby avoiding uncertainty. The tag numbers should also be used in the process descriptions.

TABLE 13.3 An example of a process description of an oxidation ditch with MLSS-recirculation pumps.

Item	Description

Equipment tag numbers included	Oxidation Ditch No. 1
	  04-PMP-101, 04-pmp-102
	  04AE-189A, 04TE-189B
	  04FIT-103, 04-FCV-101
	Oxidation Ditch No. 2
	  04-PMP-201, 04-pmp-202
	  04AE-289A, 04TE-289B
	  04FIT-203, 04-FCV-102
	Oxidation Ditch No. 3
	  04-PMP-301, 04-pmp-302
	  04AE-389A, 04TE-389B
	  04FIT-303, 04-FCV-103

Related P&ID numbers	4N02A, 4N02B, 4N03A, 4N03B, 4N04A, 4N004B
Process description	The MLSS-recirculation pumps pump MLSS from the "C" channel to the "A" channel for biological nutrient control.
Local controls	Provide HAND/OFF/AUTO local manual control for the pumps. When the switch is placed in the HAND position, START the pump. When the switch is placed in OFF position, STOP the pump. When the switch is placed in the AUTO position, the PLC will START and STOP the pump. Provide AUTO signal to PLC. Provide LOCKOUT/STOP. Provide LOCAL/REMOTE selector switch for modulating flow-control valve. When the switch is in the LOCAL position, the modulating valve is controlled locally with the OPEN/STOP/CLOSE pushbuttons. When switch is in the REMOTE position, the modulating valve is controlled by a position signal from the PLC. When in REMOTE, provide AUTO indication to PLC. Provide modulating valve position indication and send signal to PLC. Annunciate FAIL indication upon modulating-valve failure and provide this signal to the PLC. Provide POWER ON indication when modulating-valve actuator is powered.
MCC functions	Annunciate pump RUN status and provide signal to the PLC. Provide pump fail (thermal OVERLOAD) signal to the PLC.
PLC functions	The PLC will control the pumps and flow-control valve based on either flow, flow pacing, or a manual code. In manual mode, the SCADA operator enters a valve position that is output to the valve.

TABLE 13.3 An example of a process description of an oxidation ditch with MLSS-recirculation pumps. *(continued)*

Item	Description
PLC functions	In flow-control mode, the flow-control valve modulates to maintain a flow setpoint entered by the SCADA operator. Use a PID controller that runs every 1 minute.
	In ratio control, the flow-control valve modulates via a PID algorithm to maintain a flow ratio entered by the SCADA operator using the following formulae: (1) N = determined by infl. gate position (2) Q_{inf} = MAX (01F1700A, 01F1700B)/N (3) Q_{recir} = R x Q_{inf} where N = number of oxidation ditches in operation, Q_{inf} = influent flow to each oxidation ditch, R = recirculation ratio setpoint (0.0 - 2.0, default 0.5), and Qrecir = recirculation flow rate setpoint to PID controller.
	Always run at least one (lead) recirculation pump. When in the flow- or ratio-control modes, START second (lag) pump when the flow setpoint is above 3.0 mgd. STOP a pump when the flow setpoint is below 2.8 mgd.
	Recirculation pumps shall be rotated on a FIRST ON/FIRST OFF basis. Rotate pumps at least once daily. Do not rotate a pump to the lead position if it is not in AUTO control mode.
SCADA manual control	SCADA operator may manually START or STOP a pump. SCADA operator may select either manual, flow, or ratio control. SCADA operator may select a valve position, flow setpoint, or a flow ratio setpoint, depending on the control mode.
SCADA automatic control	Accumulate, indicate, and store running times for the pumps.
Alarms	FAIL TO RUN—SET alarm bit if pump fails to start after commanded for a preset time period (initially set to 10 seconds). RESET alarm if pump RUNS, local control is set to HAND or OFF, or control mode is set to SCADA MANUAL.
	FAIL TO STOP—SET alarm bit if pump fails to stop after commanded for a preset time period (initially set to 10 seconds). RESET alarm if pump STOPS, local control is set to HAND or OFF, or control mode is set to SCADA MANUAL.
Components	MiniCAS unit is supplied by pump manufacturer.

TABLE 13.4 An example of a five-step biological nutrient removal process.

Item	Description

Equipment tag numbers included	210 Anaerobic Basin
	    Mixer 211 and SC 212
	    AE212 (ORP) and AIT 213
	220 First Anoxic Basin
	    AE 222 (ORP) and AIT 223
	    FCV 224, FE 225, and FI 226 (airflow)
	230 First Aerobic Basin
	    AE 232 (DO) and AIT 233
	    FCV 237, FE 238, and FI 239 (airflow)
	    FE 234, FE 235, and FI 236 (airflow)
	    261 mixed liquor internal recycle
	240 Second Anoxic Basin
	    AE 242 (ORP) and AIT 243
	    FCV 244, FI 245, and FI 246 (airflow)
	250 Second Aerobic Basin
	    AE 252 (DO) and AIT 253
	    FCV 254, FE 255, and FI 256
	310 Final Clarifier
	    310 sludge collector drive
	    WAH 311 and WAHH 312
	    360 Scum Pump
	    LSL 363 and LSH 364
	    370 RAS pump and SC 371
	320 Final Clarifier
	    320 sludge collector drive
	    WAH 321 and WAHH 322
	    350 Scum Pump
	    LSL 353 and LSH 354
	    380 RAS pump and SC 381
	FE 390 and FIT 391 RAS flow meter and secondary
	292 and 293 motor operated valves for RAS and WAS
	Aeration blowers
	    Blower 970
	    Blower 980
	    Blower 990

Related P&ID numbers	Sheet E31
Process description	This process is the second phase of a wastewater treatment plant with multiple expansions planned during the next 25 years. The plant has both phosphorus and ammonia limits. At this point in the development of the facility, most of the process-control functions are performed man-

TABLE 13.4   An example of a five-step biological nutrient removal process. *(continued)*

Item	Description
	ually. However, the facility is on the utility's SCADA system and limited manual control can be implemented in the remote manual mode.
	This treatment process is a five-stage biological process designed to remove both phosphorus and nitrogen.
	Stage 1—Anaerobic Basin 210 Maintenance of anaerobic conditions is critical for phosphorus release and subsequent uptake in the following basins. So, the mixer speed is adjusted to maintain an ORP value that is indicative of true anaerobic conditions. However, mixer speed cannot be reduced too much or solids will settle in the basin.
	Stage 2—Anoxic Basin 220 (1st anoxic) This basin functions as a selector basin and the first stage of the phosphorus removal process, as well as providing substrate removal. The basin should be maintained in anoxic conditions, so ORP control is critical to operations. Airflow is adjusted to maintain the ORP at the desired levels. One of the process objectives is to use the nitrate in the RAS as an oxygen source (reduce nitrate to nitrogen gas) and reduce aeration requirements.
	Stage 3—Aerobic Basin 230 (1st aerobic) This basin is operated to be truly aerobic, and a dissolved oxygen (DO) concentration of 1.0 to 2.0 is desirable. This basin provides both phosphorus and organic substrate removal. Another process objective is to nitrify any remaining ammonia nitrogen to nitrate nitrogen. Control is accomplished by manually varying the airflow and monitoring the basin's DO concentration. Mixed liquor is recycled from this basin to the second stage to optimize nitrogen removal.
	Stage 4—Anoxic Basin 240 (2nd anoxic) Like Basin 220, this basin is operated in an anoxic mode, so maintaining the appropriate ORP level is important for process performance. Control is manual control of airflow to maintain correct ORP. Process objective is to denitrify and convert any nitrate to nitrogen gas, thus removing additional nitrogen from the process.
	Stage 5—Aerobic Basin 250 (2nd aerobic) This is a truly aerobic basin, and control is manual airflow control based on the basin's DO concentrations. Process objectives are to nitrify any residual ammonia, air strip any remaining nitrogen gas bubbles from the mixed liquor, and increase the DO to prevent possible denitrification in the final clarifier.
	Final Clarifiers 310 and 320 Flow-paced RAS control and timer-controlled WAS flows.

TABLE 13.4 An example of a five-step biological nutrient removal process. *(continued)*

Item	Description
Local controls	Stage 1—Anaerobic Basin 210 Speed control for basin mixer and process ORP measurement. Stage 2—Anoxic Basin 220 (1st anoxic) Flow-control valve for basin airflow control and process ORP measurement. Stage 3—Aerobic Basin 230 (1st aerobic) Two flow-control valves for airflow control, DO probes and indicators, and mixed-liquor recycle-flow-control valve for internal recycle. Stage 4—Anoxic Basin 240 (2nd anoxic) Airflow-control valves and process ORP measurement. Stage 5—Aerobic Basin 250 (2nd aerobic) Airflow control and process DO measurement. Final Clarifiers

MCC functions	Clarifier drives
	    On and Off, with indicator lights
	RAS pumps
	    On and Off, with indicator lights
	Aeration blowers
	    On and Off, with indicator lights
	Mixer 211
	    On and Off, with indicator lights

PLC functions	There is no PLC used to control the biological process at this phase of the plant's development.

Remote manual controls	The utility SCADA system can be used for the following control functions:
	1. Aeration blowers
	    Start and stop
	2. Clarifier drives
	    Start and stop
	3. RAS pumps
	    Start, stop, and speed control via the VFDs
	4. Mixer 211
	    Start, stop, and speed control via the VFD
	5. Sludge wasting
	    Manual control

Remote automatic controls	No remote automatic operation is possible over the SCADA system.
Alarms	Clarifier drive torque alarms. Low aeration-air pressure.
Components	1. Manufacturer-supplied control panel for each clarifier. 2. Manufacturer-supplied control panel for Mixer 211. 3. Manufacturer-supplied panels for RAS pumps. 4. Manufacturer-supplied panel for each blower.

PROCESS AND INSTRUMENTATION DIAGRAMS. Process and instrumentation diagrams (P&IDs) are the basis of the overall process-control strategy, so they must be properly prepared. These diagrams are also the one place where every piece of equipment—and its tag number—can be found.

PROCESS AND EQUIPMENT DESCRIPTIONS. These descriptions should detail which types of equipment are needed, how they all fits together, and how each process should function.

LOCAL CONTROLS. Design engineers should describe any local equipment controls and detail their relationship and interactions with the programmable logic controller (PLC) or supervisory control and data acquisition (SCADA) system. This information is especially important to ensure compatibility among all the engineering disciplines involved.

MOTOR-CONTROL-CENTER FUNCTIONS. Design engineers should detail all the functions performed by and at the motor control center. In particular, any work required by the electrical engineering group and the electrical subcontractor and supplier should be clearly identified.

PLC FUNCTIONS. Design engineers should detail the PLC's (or other controller's) role in both the specific process being described and the overall process-control architecture.

REMOTE MANUAL CONTROLS. Design engineers should describe the manual equipment controls available to operators at remote locations. Although the controls generally will be implemented by a SCADA system, operations staff should know how the SCADA system functions. Also, the SCADA interfacing requirements will be important during facility design.

REMOTE AUTOMATIC CONTROLS. Design engineers should describe the automated equipment controls available to operators at remote locations. Although the controls generally will be implemented by a SCADA system, operations staff should know how the SCADA system functions. Also, the SCADA interfacing requirements will be important during facility design.

ALARMS. Design engineers should describe when and how local and remote alarms will be generated. The description should provide enough detail so programmers can create the appropriate software and operators will understand which responses are needed.

SYSTEM COMPONENTS. Design engineers should describe all of the control components that will be supplied by vendors (rather than the instrumentation subcontractor) to avoid conflicts and potential change orders.

OTHER METHODS OF CONVEYING PROCESS DESCRIPTIONS

It can be difficult to describe a treatment process or facility so everyone understands the designer's or facility manager's intent—especially facilities with complex treatment schemes. Many of the larger, older treatment plants have been expanded many times, and the interconnections between expansion phases can be difficult to follow. Also, people who do not work in the wastewater treatment industry may not immediately understand the underlying biochemistry or chemistry of each treatment process.

So, to avoid losing readers, discussions of process descriptions should begin simply, with the "big picture", and become progressively more detailed. The following tools may help.

ANIMATIONS. Animations help eliminate confusion and clearly depict how a particular process or plant is intended to work. If "a picture is worth 1000 words", then an animated picture is worth at least a million words! Animations of various treatment processes (e.g., conventional activated sludge systems, sequencing batch reactors, five-stage biological nutrient removal processes, and complex hydraulic systems) have helped people better understand them.

With the tools currently available to digital graphic artists, animation development is relatively straightforward and not too costly. The effort and costs are more than justified by the ease with which viewers understand the project design.

Animation also can be invaluable during the construction phase. The transportation industry often uses animation at public meetings to illustrate a project at various stages of construction, and attendees typically feel much better-informed as a result.

STATIC PROCESS MODELS. Static process models are valuable teaching tools: they allow users to change one process variable and see how it affects process results. To be effective, the interface should be user-friendly, and the format should enable the user to easily see the process responding to changes. Successful models typically are relatively simple spreadsheets, frequently developed in Microsoft Excel. Basically, if you can draw a graph of the process, you can develop a useful model.

These models have been developed for many applications (e.g., chemical-feed controls; anaerobic- or aerobic-digester loads; and sludge age, mean cell residence time, and food-to-microorganism ratio calculations). One of the most useful models was developed for use in the water treatment industry to predict changes in water chemistry because of different chemical treatments.

INTERACTIVE PROCESS MODELS. Interactive process models allow users to become familiar with the rate at which a process change occurs. These models are usually spreadsheets with interactive bar charts (sliders) displaying the important process variables. Users can change any process parameter and watch other variables respond. The basic difference between the interactive and static models is the "real time" response to process variable changes.

Chapter 14

Advanced Applications for Wastewater Treatment

Energy Management	442	*Decision Support*	454
Conservation	442	A Case Study	455
Real-Time Power Monitoring	442	Biological Nutrient Removal	457
Reducing On-Peak Energy Demand	443	Why Is BNR Control Needed?	457
Capital Improvements	443	Selecting Setpoints for a BNR Plant	460
Artificial Intelligence	444	Ammonia Control	460
Expert Systems	446	Denitrification Control	462
The Knowledge Base	446	Respirometry	463
The Inference Engine	446	Intermittent Aeration	464
Case-Based Reasoning Systems	449	Sequencing Batch Reactors	464
Fuzzy Logic	451	IWA/COST Benchmark Simulation Environment for Control Strategy Comparison	465
Artificial Neural Networks	452		
Intelligent Decision-Support Systems	453	References	466
Data Gathering	453		
Diagnosis	454		

This chapter describes the use of control systems for energy management, operations decision support, and optimization of nutrient removal.

In addition to these advanced control applications, modern control systems now can provide automated reports; plan, schedule, make, or support business decisions; and make commitments (e.g., procurement). Other advanced applications include integration of supervisory control and data acquisition (SCADA) systems with various computer systems (e.g., computerized maintenance management systems).

ENERGY MANAGEMENT

Energy efficiency is important in today's competitive wastewater treatment industry. Energy costs typically consume 15 to 30% of a treatment plant's operations and maintenance (O&M) budget and are the second-highest controllable cost (behind labor).

An integrated energy-management program leverages both supply-side and demand-side opportunities. Supply-side opportunities include selecting the best rate structure, negotiating new power rates, and participating in power "wheeling". These opportunities are beyond the scope of this manual. [For more information on energy management in treatment plant design and operations, see *Energy Conservation in Wastewater Treatment Facilities* (Water Environment Federation, 1997)].

Demand-side opportunities include conservation, power monitoring, process modifications, aeration control, power-demand shifts, energy-efficient equipment upgrades, and power-factor correction.

CONSERVATION. Simple conservation (e.g., turning off lights in unpopulated rooms or lowering the thermostat) can reduce electrical costs. Promoting conservation is a simple, effective first step in any energy management program.

REAL-TIME POWER MONITORING. Another simple step is monitoring power demand in real time. Such monitoring shows treatment plant personnel how their actions affect power costs and help them minimize these costs.

Real-time power monitoring involves connecting an electric power meter—the existing billing meter or an analog power meter—to the instrumentation and control (I&C) system. Adding extra contacts to the billing meter is inexpensive and enables staff to see the same signal that the electric power provider is using to bill the plant. It also requires a software program to average and total the power readings (often in 15-minute blocks) and sort them into on-peak, mid-peak, and off-peak time periods.

Most modern programmable logic controllers (PLCs) include a real-time clock and calendar and have sufficient programming capabilities to do this.

The control system then can use this information to graphically display trends in demand, current billing demand, maximum monthly demands (including date and time of each), and the total kilowatt-hours used for each period. It can also calculate the monthly power bill to date and provide an estimate for the entire month.

REDUCING ON-PEAK ENERGY DEMAND. Shifting power usage from on-peak to off-peak demand times (usually evenings and weekends) whenever possible can lower energy bills. For example, utilities could use the control system to concentrate pumping during off-peak times, move solids-processing activities to evenings, use chemical precipitants to reduce oxygen loading, and lower dissolved oxygen setpoints during on-peak hours.

CAPITAL IMPROVEMENTS. Numerous capital improvements can reduce a utility's power costs. One strategy is replacing older motors with high-efficiency models. Another is installing power-factor-correcting equipment.

The power factor is the ratio between actual load power (kW) and apparent load power (kVA). It measures how effectively the current is being converted into useful work output, and is a good indicator of the load current's effect on the supply system's efficiency. A power factor of 1.0 is the most efficient loading of the supply system, while a power factor of 0.5 indicates much higher losses.

All current causes losses in the power supply system. A poor power factor, however, could be the result of a significant phase difference between voltage and current at the load terminals—typically caused by an inductive load (e.g., an induction motor, power transformer, lighting ballasts, welder, or induction furnace). It also could be due to a high harmonic content or a distorted or discontinuous current waveform—typically caused by a rectifier, variable speed drive, switched mode power supply, discharge lighting, or other electronic load. If caused by an inductive load, a poor power factor can be improved by adding a power-factor correction. If caused by a distorted current waveform, however, a poor power factor could only be improved via expensive harmonic filters or a change in equipment design.

A power-factor correction is achieved by adding capacitors in parallel with connected motor circuits. The resulting capacitive current is leading current used to cancel the lagging inductive current flowing from the supply. A controller monitors the power factor and turns capacitor banks on and off to keep the power factor above a preset limit (typically 0.95).

ARTIFICIAL INTELLIGENCE

During wastewater treatment plant upsets or abnormal conditions, success depends on plant staff's ability to quickly identify the problem, diagnose it properly, and initiate recovery actions. Such effective troubleshooting depends, in turn, on the optimal use of an ever-increasing amount of data from multiple heterogeneous sources (Figure 14.1). Troubleshooting is beyond the scope of conventional control systems. Although today's I&C systems can monitor plant conditions and control essential operations via a PLC network (Jepsson et al., 2002), even the most advanced conventional systems have trouble solving problems that require qualitative information and heuristic reasoning. Intelligent control systems, however, could be the answer.

Intelligent control systems are those with the following features (Stephanopoulos and Han, 1996):

- They use logic, sequencing, reasoning, or heuristics in addition to numerical algorithms;
- They are essentially nonlinear controllers with more autonomy than conventional controllers; and
- They rely on representational forms and decision-making procedures that emulate human or biological systems.

Basically, intelligent control systems use artificial intelligence techniques to better control complex processes. There are two types of artificial intelligence techniques: knowledge-based approaches and soft computing approaches. Knowledge-based approaches use heuristic knowledge and human experience to reason about the problem. Knowledge-based systems include both expert or rule-based systems (Jackson, 1990) and case-based reasoning systems (Kolodner, 1993; Aamodt and Plaza, 1994).

Soft computing approaches include several theories, methods, and techniques to solve nonlinear problems (e.g., pattern recognition, systems control, prediction, and optimization). The most common are artificial neural networks (Hecht-Nielsen, 1990; Hertz et al., 1991), fuzzy logic (Klir and Folger, 1988), genetic algorithms (Goldberg,

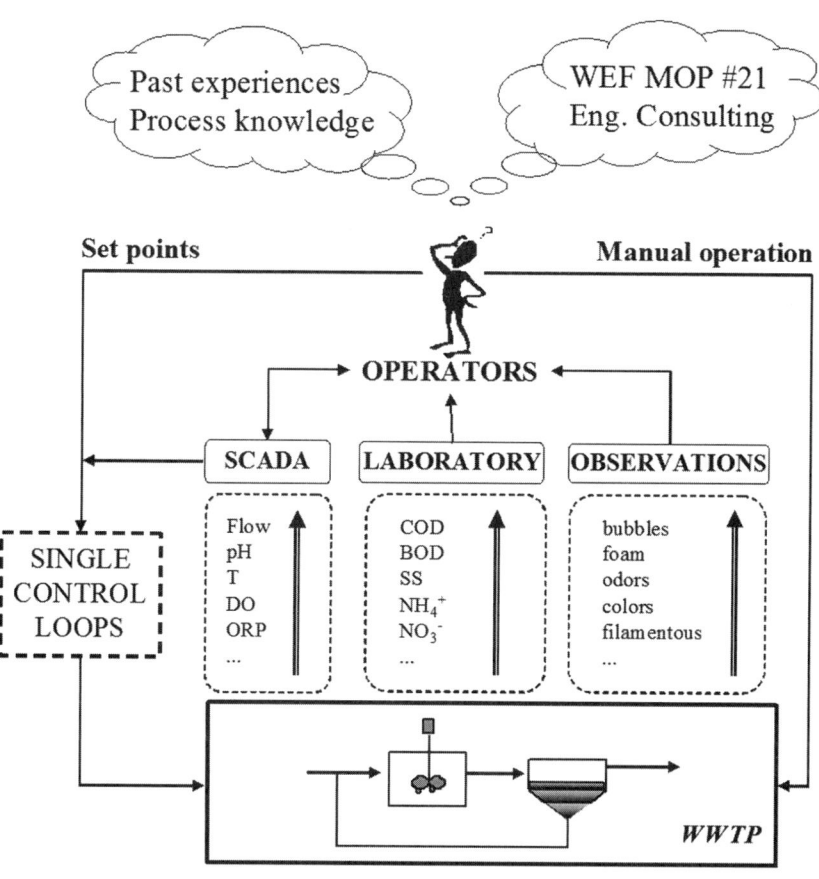

FIGURE 14.1 Data, knowledge, and information available to help operators make optimal decisions.

1989), probabilistic reasoning (Pearl, 1987), and rough set theory (Pawlak, 1991). Of these intelligent control techniques, expert systems, case-based reasoning systems, fuzzy logic, and artificial neural networks have been tested in wastewater treatment plants with promising results.

EXPERT SYSTEMS. Expert systems are computer programs that try to emulate the way experts reason and make decisions to solve problems. They use linguistic rules or conditional statements elicited from human experts to encode expertise, which when chained in logical sequences, can explore a situation and reach some conclusions [e.g., if the food-to-microorganism (F:M) ratio is decreasing and the sludge volume index (SVI) is increasing, then the risk of filamentous bulking is true].

Expert systems were introduced to the wastewater treatment industry in the late 1970s and 1980s. The first prototypes were based on simplified decision trees, which only included general knowledge available in textbooks, and used a long, tedious algorithm based on a series of questions and answers between the computer and the operator to infer a solution.

In the 1990s, more sophisticated and realistic expert systems were created to diagnose process conditions; select, design, and optimize processes; and help make decisions. These applications never really succeeded, though, because they were too complex and existing knowledge could not be captured in reliable models and advisory systems. In fact, these expert systems were never installed in real treatment plants; instead, they were tested using hypothetical problems, simplified case studies, or specific experiments at pilot plants.

Expert systems have two main modules:

- *The Knowledge Base.* The knowledge base includes a set of decision trees (Figure 14.2) with the overall knowledge of the process. Each branch of the decision tree is an expert rule (e.g., if problems are observed in primary settlers and water color is green, then chrome waste is possible). A sophisticated example of a decision tree developed to supervise deflocculation in activated sludge processes can be found in Comas et al. (2003).

- *The Inference Engine.* The inference engine is the software that controls the reasoning operation by scanning the knowledge in the knowledge base (several user-friendly commercial software packages, including G2, Prolog, and OPS5, have an inference engine).

Expert systems also may have connections to a process database and external applications, and can provide explanations or justifications via a graphical user interface.

The core—and bottleneck—of expert system development is acquiring the knowledge to build decision trees. It involves eliciting, analyzing, and interpreting the knowledge that experts use to solve a particular problem. The challenge is that,

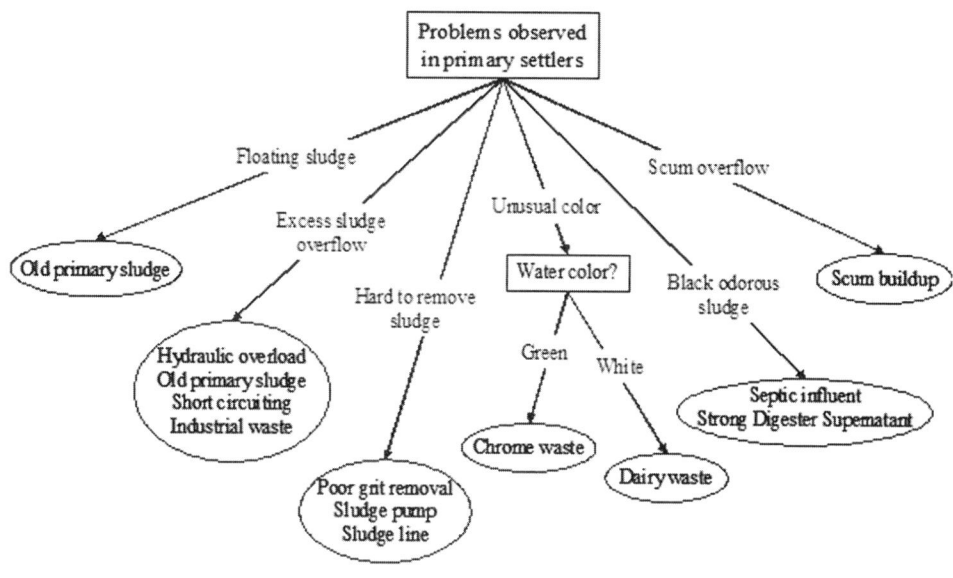

FIGURE 14.2 A simplified example of an expert system's decision tree to solve primary settler problems.
Source: Patry and Chapman, 1989.

because of the complexity involved, experts cannot always explain precisely how they solved a particular problem.

Of course, computers themselves can be used to acquire the knowledge for problem-solving. As several authors have recently noted, there is a difference between "data rich" and "information rich", and computers can be used to "mine" large volumes of data for useful information. Data-mining techniques—statistical techniques and machine-learning methods (Langley, 1996; Mitchell, 1997)—automate the detection or discovery of relevant data, meaningful patterns, and usable knowledge in a database, thereby helping to increase productivity.

Machine-learning methods determine correlations [e.g., the statistical relationship(s) between two or more variables] and recognize patterns (e.g., a process' systematic behavior) by classifying groups of measurements or observations in an appropriate multidimensional space. The methods can be supervised—capable of learning from examples by a process of generalization—or unsupervised—the system establishes the classes itself based on the pattern's statistical regularities.

Several data-mining techniques (e.g., time-series analysis, cluster analysis and k-means, principal component analysis, wavelet time and frequency domain analysis, rule induction, and decision tree induction) have been used in the wastewater treatment field (Comas et al., 2001). These techniques define and identify deviations from regular operating conditions, but become even more powerful when integrated into an expert system framework, where they can be used to establish operating rules.

Figure 14.3 shows the two-phase methodology developed by Rodríguez-Roda et al. to acquire and fix knowledge from complex processes (e.g., the activated sludge process). The first phase consists of literature reviews and site interviews with operators and process experts. The second phase involves using data-mining tools to handle, classify, interpret, and codify data to find relevant information. The two-phase method also lets users explore the database(s) to discover new pieces of knowledge. The overall goal is to ease the knowledge-acquisition process while taking advantage of existing information technology.

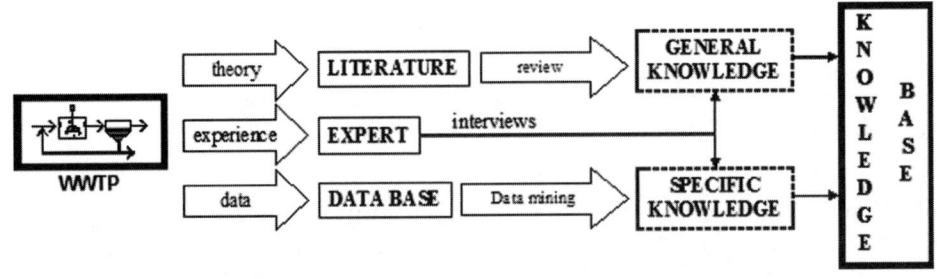

FIGURE 14.3 Knowledge acquisition method used to build an exhaustive knowledge base.
Source: Rodríguez-Roda et al., 2001.

CASE-BASED REASONING SYSTEMS. Case-based reasoning systems are computer programs that use past experiences to solve new problems. Basically, they automatically compare the current situation to previous ones, identify similarities, and reuse the approach of the closest match to solve the current problem. So, they require less effort than control systems that must always "start from scratch" to solve problems. Case-based reasoning systems have been used for fault diagnosis, medical detection, equipment selection, information retrieval from historical meteorological databases, forest firefighting plans, and process engineering designs. They also have been proposed as decision-support tools in the activated sludge process and could help newer operators draw on others' expertise when solving problems.

The basic idea behind case-based reasoning systems is learning from experience. Humans often recall previous solutions and mistakes when solving a new problem; why not computers? With this idea in mind, a case-based reasoning system is designed to cycle through the following steps (Figure 14.4):

- Gathering and processing available data to define the current situation (case);
- Searching for and retrieving similar experiences from the case library (using an algorithm based on a numerical similarity comparison among the cases);
- If no perfect match exists, adapting the most similar solution of a previous problem to the existing situation using, for example, a simple parameter-adjustment interpolation;
- Applying the adapted solution to the process;
- Evaluating its consequences; and
- Adding new information to the case library.

Both successes and failures must be included in the library because both are useful in solving problems.

Each case (Figure 14.5) is composed of the following elements:

- A situation [diagnosis of the process (e.g., storm, foaming)];
- The values of its key variables (e.g., flow, temperature, filamentous abundance);
- The chosen control strategy [e.g., "increase the sludge retention time (SRT) setpoint by 15% for 1 week, unless scum is excessive in aeration tanks"]; and

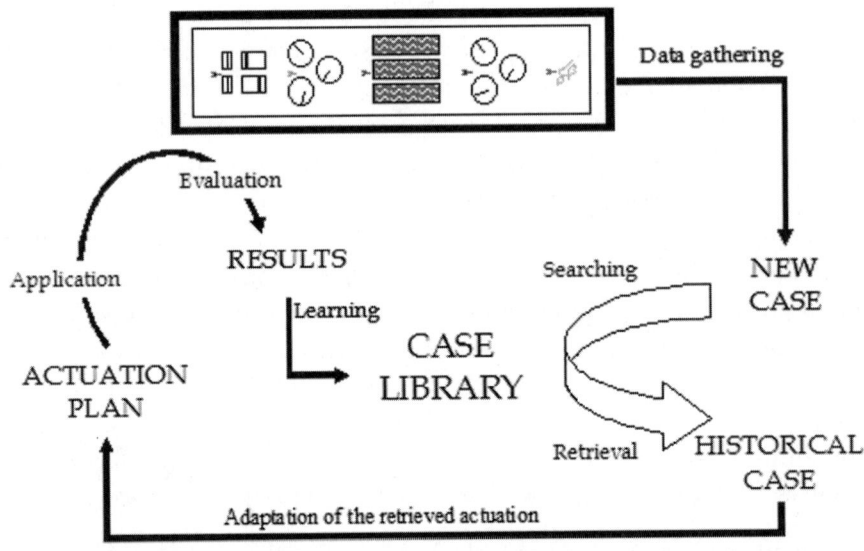

FIGURE 14.4 The case-based reasoning system's working cycle. Source: Rodríguez-Roda et al., 1999.

identifier	Case-07/23/1998		
Diagnosis	Foaming		
Process data	Inflow COD-influent Ammonia-influent Filamentous abundance Turbidity-effluent ...	6.35 455 45 Category 5 High ...	MGD mg O_2/L mg N/L - -
Control action	increase 15% SRT set point monitor TSS-effluent examine mixed liquor ...		
Evaluation	Success in 5 days. Nocardioform foam mostly disappears.		

FIGURE 14.5 Hypothetical example of a case for the wastewater treatment plant domain.

- Its results (e.g., "SRT was kept high, even though Nocardioform foam covered 40% of the aeration tank surface; problem was successfully solved in 5 days").

The structure and organization of the case library is different from that of a traditional database, where a full match of variables is required. Hierarchical structures instead of classical flat memories are common to ensure the system's efficiency both in response time and in successfully finding suitable comparisons.

To function most effectively, a case-based reasoning system should be initialized with a set of common experiences (cases) obtained from technical books, historical data, and process operators and experts. Then, the case library will be updated automatically as a natural byproduct of problem-solving, and the case-based reasoning system will become more and more accurate and sophisticated over time.

FUZZY LOGIC. Fuzzy logic was the first soft-computing approach to be used in industrial control systems, and it is now considered one of the leading advanced control techniques in industry (Manesis et al., 1998). Its use has significantly improved product quality, productivity, and energy consumption.

Fuzzy logic was introduced by L.A. Zadeh, who extended classical Boolean logic to real numbers. In Boolean algebra, 1 = "true" and 0 = "false"; in fuzzy logic, all the values between zero and one are also included to indicate partial truth.

Fuzzy logic grew out of a need to quantify rule-based systems. Rule-based reasoning is grounded in qualitative knowledge, and fuzzy logic blends a quantitative approach with such knowledge. Basically, it enables propositions to be assigned degrees of truthfulness or falsehood by quantifying such qualifiers as approximately, often, rarely, several, few, and very. For example, the statement "the reactor is foaming" might be 100% true if the reactor is fully covered with foam, 80% true if it is mostly covered, 50% true if it is somewhat covered, and 0% true if it has no foam at all.

Fuzzy logic is not a substitute for statistics; rather, it is used when statistical reasoning is inappropriate. Statistics (e.g., variance, standard deviation, and confidence intervals) express the extent of specific knowledge about a value. Fuzzy logic distinguishes among data when no sharp, well-defined boundaries exist.

Fuzzy logic begins with measurements made by either sensors or people. These measurements are then appropriately labeled (e.g., very low, low, normal, high, or very high) based on predetermined definitions of each label by means of membership functions. This is called *data fuzzification*. Then, a typical fuzzy logic controller compares each labeled measurement to a predetermined set of rules (e.g. "if input A is low and input B is low, then output is high") and acts accordingly. The action(s) related to each rule are first translated into a crisp control signal (*defuzzified*), which is then transmitted to the appropriate instruments or equipment.

Fuzzy-logic-based control systems have been successfully used in a number of cases; thorough surveys can be found in Sugeno (1985) and Hirota (1993). They have several potential applications in wastewater treatment facilities, including identifying better sludge and aeration control strategies (e.g., Ferrer et al., 1998), controlling wastewater neutralization, and predicting sludge-settling behavior. [For more information on fuzzy control, see Driankov et al. (1993).]

ARTIFICIAL NEURAL NETWORKS. Artificial neural networks were inspired by the human brain's amazing parallel-processing abilities. The idea is to re-create these abilities by constructing artificial models of biological neurons and their networks. The power of biological neural structures stems from their enormous number of highly interconnected simple units. However, once their complex electrochemical processes are abstracted, the resulting computation turns out to be conceptually simple.

Artificial neural networks are information-processing structures without a global or shared memory, in which each computing element operates only when all of its incoming information is available. Basically, the network is made up of a number of simple, highly interconnected nodes with internal and adjustable parameters that process information and modify themselves in response to external inputs. Each node has a weight expressing the strength with which two involved nodes are connected.

The data used to implement an artificial neural network are usually split into training, validation, and test sets, and used for model construction, parameter selection, and model assessment. The nodes' weights are initialized either randomly or based on existing knowledge. Each node weight typically is determined during training, by minimizing some form of error that is a function of the difference between actual and expected outputs. The goal is to determine a set of weights that will minimize errors, while generalizing well in the presence of new data.

One of the challenges of using an artificial neural network to model biological systems is specifying the network architecture. Implementing a standard artificial neural network to deal with a specific problem typically involves much trial and error concerning learning-parameter specifications, the model's architecture, the training algorithm, or even the assessment procedure (Aitkenhead et al., 2003). However, many advances have been made in automating these tasks.

Artificial neural networks are designed to find satisfactory solutions for problems about which little or nothing is known. They are known to be resilient to noise and effective in capturing the nonlinear relationships that might exist among variables in complex systems.

Artificial neural networks have been widely and successfully used in a variety of environmental applications since the mid-1980s. For example, they have been used to predict the occurrence of future bulking episodes in wastewater treatment plants (Belanche et al., 2000). They also have been used in groundwater modeling, water quality assessment, streamflow prediction, and other hydrologic applications.

One often-quoted limitation of artificial neural network models is their intrinsic "black box" behavior (i.e., users have difficulty finding an intuitive explanation of the results), and some experts fear that this will hinder their widespread acceptance in wastewater treatment control (Olsson et al., 1998). Nevertheless, many approaches as feature selection, feature relevance determination, or rule extraction have been put forward to alleviate this shortcoming.

INTELLIGENT DECISION-SUPPORT SYSTEMS. A new generation of intelligent control systems is emerging that is based on combinations of artificial-intelligence tools. Called intelligent decision-support systems, these computer programs reduce decision-making time, improve decision-making consistency and quality, and help ensure effective process performance. They have been used for such applications as controlling air quality, supporting storm and weather forecasting; assessing environmental effects; managing forests, forest fires, and chemical and environmental emergencies; and supporting water and wastewater management.

Building an intelligent decision-support system involves six steps: analyzing the problem(s); gathering relevant data and knowledge; choosing the most appropriate reasoning models; integrating the models into a flexible, modular architecture; implementing the system; and validating it. [For details, see Poch et al. (2004).] Although each system's structure depends on the type of environmental problem addressed, all typically include the following three levels:

- ***Data Gathering.*** The first level includes all the tasks involved in gathering and registering data into databases. Because of various measurement and data-recording problems, recorded data are sometimes erroneous and must be filtered before they can be understood and interpreted via a database.

- *Diagnosis.* The second level includes the reasoning models (e.g., statistical, numerical, and artificial intelligence) used to infer the state of the process so a reasonable action can be proposed.

- *Decision Support.* The third level includes all the tasks involved in gathering and merging the conclusions derived at the diagnosis level and predicting the results of various proposed actions. It also includes an interactive, graphical user–machine interface.

Intelligent decision-support systems are tools designed to cope with complex, multidisciplinary problems, like those in the wastewater treatment field, where wrong management decisions may have disastrous social, economic, and ecological consequences. They are not only efficient mechanisms for finding optimal solutions, but also for making the entire decision-making process more open and transparent. They can be used for such tasks as

- Managing process knowledge;
- Finding out what information is needed and the best way to capture, present, and share it;
- Training new operators;
- Supervising the SCADA system;
- Acquiring knowledge more easily; and
- Systematically registering the decisions made and their consequences.

When measuring the success of an intelligent decision-support system used in an environmental application (e.g., the activated sludge process), users should base their evaluation on what the system knows, how it uses what it knows, how fast it learns something new, how reliable it is, and how well it performs overall.

Future intelligent decision-support system research is expected to be focused on the following issues (Poch et al., 2004):

- Integrating several sources of data and knowledge,
- Improving knowledge-acquisition methods,
- Better sharing and reusing knowledge,
- Involving system users in intelligent decision-support system development, and

- Developing benchmarks for validating intelligent decision-support systems.

A CASE STUDY. Recent results in Europe show that intelligent decision-support systems can be successfully implemented and evaluated in full-scale activated sludge facilities (Rodríguez-Roda et al., 2002). The project's goal was to develop a real-time intelligent decision-support system to improve operations at the 30 280-m^3/d (8-mgd) Granollers Wastewater Treatment Plant.

The plant treats a mix of urban and industrial wastewater. The process consists of preliminary, primary, and secondary treatment to remove organic matter, suspended solids and, under some conditions, nitrogen. The secondary treatment process is based on a modified Ludzack-Ettinger configuration for BNR. The sludge treatment process includes thickening, anaerobic digestion, and dewatering.

Figure 14.6 shows the architecture of the intelligent decision-support system implemented at Granollers. The system's first level gathered all the available process information and structured it into a database. Some of the data were fed into the classical control system, a simple control loop to regulate dissolved oxygen in the bioreactor compartments and the recirculation flow rate.

The first level gathered 220 digital and 20 analog data (e.g., pH; air, water, and sludge flowrates, equipment status for valves and pumps, and dissolved oxygen) in real-time from the SCADA system. Measurements from 158 monitored variables [e.g., chemical oxygen demand (COD); biochemical oxygen demand (BOD); volatile suspended solids (VSS); nitrogen; phosphorus; temperature; metal contents; and fats, oils, and grease] at different sampling locations were added daily to a Microsoft Excel database, along with qualitative data (e.g., floc characterization, filamentous bacteria, predominant protozoa identification, presence of bubbles, and settling observations). These data were automatically validated to filter out noise and identify typos, outliers, and missing values.

Meanwhile, the modules in the second level used the database to diagnose the state of the process. This level included two artificial-intelligence systems: an expert system and a case-based reasoning system. The expert system consisted of about 700 rules and 200 procedures. It was programmed to address eight problems related to mechanical or electrical failures (e.g., clogged pumps and air-system failure); seven primary and 17 secondary treatment problems (e.g., storms, foaming, low pretreatment efficiency, rising sludge, dispersed growth, pinpoint floc, scum, overloading, and underloading); and six transition states to intermediate alarms (e.g., filamentous alarm, foaming alarm, and rising sludge alarm).

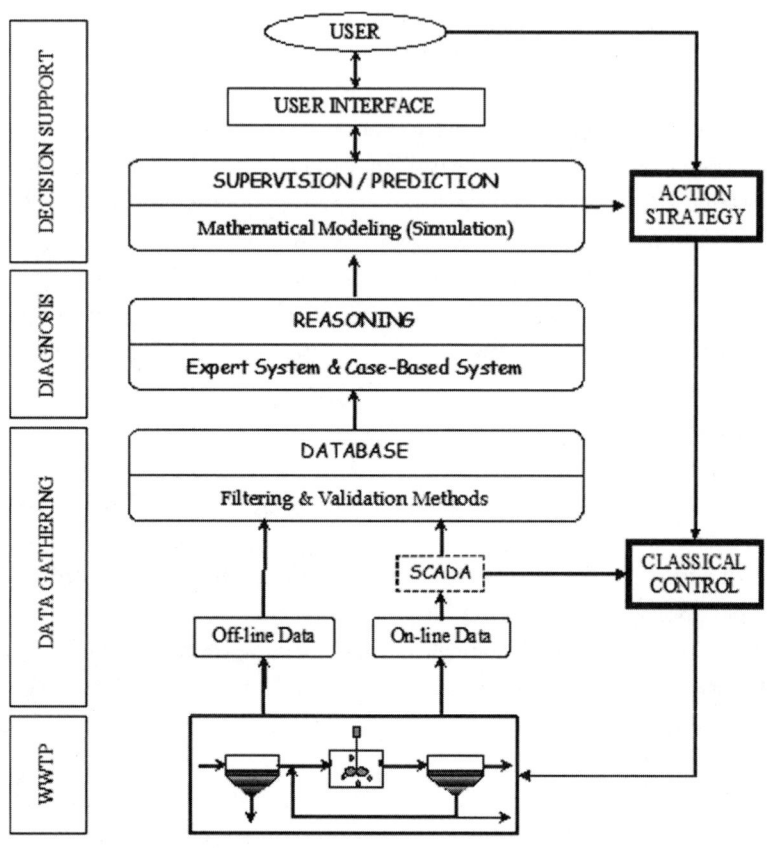

FIGURE 14.6 Multilevel architecture of the intelligent decision-support system implemented in the Granollers Wastewater Treatment Plant.
Source: Rodríguez-Roda et al., 2002.

The case-based reasoning system was programmed to evaluate the treatment process according to 17 variables: influent flow rate, COD, suspended solids, and nitrogen; primary effluent COD and suspended solids; effluent COD, suspended solids, and nitrogen; and mixed-liquor suspended solids, SVI, F:M, waste flow, dissolved oxygen 1, dissolved oxygen 2, settleability aspect, and predominant filamentous organism. The example distance criterion (Sànchez et al., 1996) was used to compare the similarity among cases. The cases were stored in a floating memory database, and every reasoning cycle retrieved the three most similar cases.

The second-level's conclusions were sent to the third level, where the suggested actions could be evaluated via a predictive mechanistic process model. This model was built and calibrated using a mechanistic process simulator.

The conclusions were sent to the operator via the computer's graphical user interface (Figure 14.7) or, when possible, applied automatically to the plant, modifying the setpoints of the automatic control loops.

The intelligent decision-support system acted in supervisory cycles. The cycle was routinely initialized once a day, although it could also be started manually at any time. It also was initialized whenever any alarm symptom was fulfilled.

Researchers evaluated the intelligent decision-support system for 18 weeks. It identified 123 problems, including foaming, rising sludge, filamentous bulking, underloading, overloading, deflocculation (including possible toxic), hydraulic shock, mechanical fault, poor primary settling, nonbiological clarifier problems (sudden oscillations of upflow velocity and an excess of biomass concentration), and influent nitrogen/organic matter shock. Of all the problems that occurred during the 18 weeks, the system

- Successfully identified 79.7% (approximately one-third in advance and two-thirds on the same day);
- Incorrectly identified 8.1% (10 situations); and
- Did not identify 2.2% (15 situations).

BIOLOGICAL NUTRIENT REMOVAL

Biological nutrient removal (BNR) plants are designed to remove ammonia, nitrate, and sometimes phosphate, as well as traditional BOD and total suspended solids (TSS) pollutants from wastewater. They cost more and are often more sophisticated to operate than a conventional activated sludge process. Automation can make BNR plants less expensive and more reliable.

WHY IS BNR CONTROL NEEDED? Process control would be simple if external conditions (e.g., flow rate and ammonia variations) did not change. In real life, how-

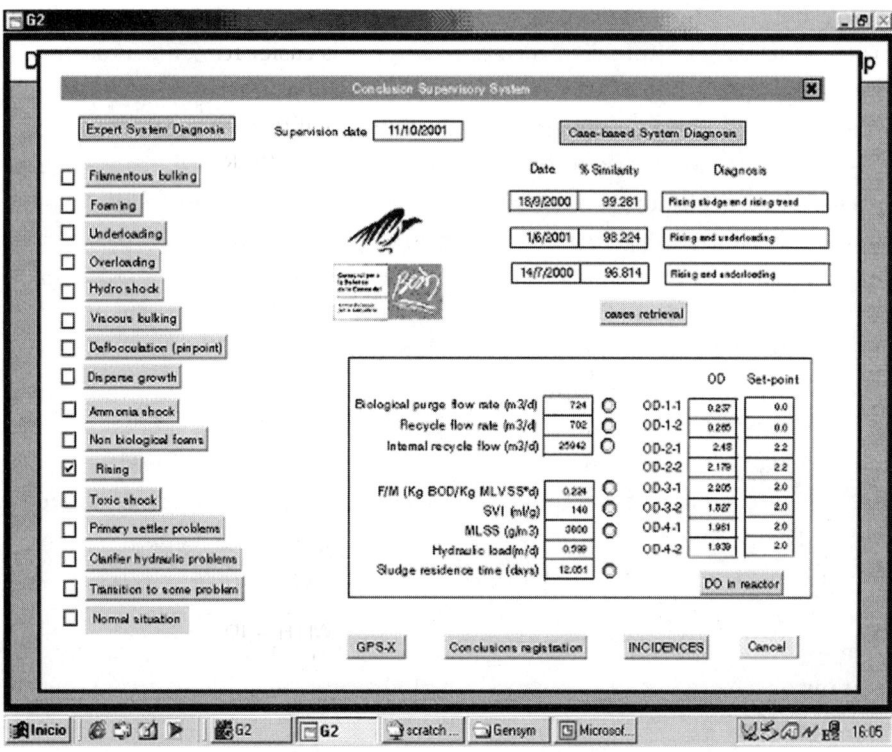

FIGURE 14.7 Operator interface of the intelligent decision-support system implemented in Granollers Wastewater Treatment Plant.

ever, the ratio of maximum to minimum oxygen demand (needed for both BOD and ammonia removal) typically varies from 3:1 to 5:1 between peak and off-peak hours. At small treatment plants, this ratio is sometimes as high as 16:1. The ratio is larger at BNR plants than at conventional plants because influent nitrogen levels vary more than influent BOD levels.

Equalizing the nitrogen load over 24 hours can substantially improve BNR operations (Figure 14.8). Fortunately, such equalization is easier for a BNR plant than for a conventional plant because it recycles about 30% of ammonia. By adding a small storage tank for flow recycled from the dewatering facility and using a timer so it fills during peak hours and empties during off-peak hours, a BNR plant can significantly reduce its influent ammonia variations. A more sophisticated option would be to use an ammonia analyzer and a controller to control recycled water flows and maintain a constant ammonia load. Similar control loops could be used to equalize the influent ammonia load via the influent equalization tank.

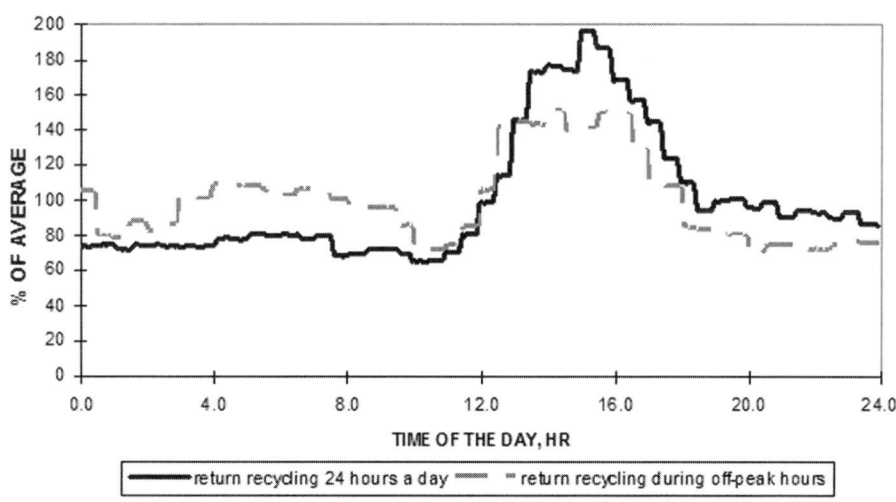

FIGURE 14.8 Influent ammonia concentration with and without ammonia equalization.

However, nothing will completely eliminate nitrogen and COD load variations. So, to maintain optimum process performance, BNR plant staff must compensate for these variations by adjusting airflow supply, waste flow, internal recycling flow, and other control parameters accordingly. Automated process control can help.

SELECTING SETPOINTS FOR A BNR PLANT. Selecting the optimum setpoint for each I&C system control loop is more challenging for BNR plants than for conventional plants (Table 14.1). Increasing SRT, for example, may cause foam, but reducing it may cause chlorine disinfection problems because of the nitrite in BNR effluent. According to Ekster and Rodríguez-Roda (2003), maintaining an optimum SRT using an SRT/MLSS automatic controller will save a 38 000-m^3/d (10-mgd) BNR plant between $10,000 and $25,000 a year.

AMMONIA CONTROL. Based on the Monod equation, ammonia removal is proportional to the dissolved oxygen concentration:

$$\text{Ammonia removal rate} = DO/(DO + K)*NH_3/(NH_3+K_1)$$

Where K = half saturation coefficient. The value of K is site-specific and depends on a number of factors (e.g., solids retention time and temperature). If K is unknown, the International Association of Water Quality (IAWQ) recommends using a default value of 1 mg/L. Maintaining a constant dissolved oxygen level will maintain almost a constant ammonia removal rate and will also contribute to stability of the denitrification process. Despite the critical importance of dissolved oxygen control for the BNR process, some BNR plants do not use it because of dissolved oxygen control reliability problems. Ekster and Wang (2005) showed at a full-scale facility that data mining algorithm used in an off-the-shelf dissolved oxygen control system (Ekster and Associates, Fremont, California) can resolve the dissolved oxygen control reliability problem. An operating experience of this control system also showed that reliable and effective dissolved oxygen control can resolve settleability problems and provide 15% energy savings (Moise and Norris, 2005). Further BNR process improvement can be achieved by an increase of the ammonia removal rate during peak times and rate reduction during off-peak times. This could be done by appropriately varying the dissolved oxygen setpoint. The challenges of this approach are the long delay time of ammonia changes after a change of air flow is introduced and the reliability and expense of on-line ammonia analyzers. To date, successful control methods include rule-based control (Krause et al., 2001; Pelletier and Sloan, 1999), cascade con-

TABLE 14.1 Issues to consider when selecting setpoints.

Process	Parameter	Potential problems if value is	
		Lower than optimum	Higher than optimum
Primary treatment	Sludge depth	Less VFA production, and so less influent denitrification potential	More effluent TSS, soluble and particulate BOD, and H_2S generation
		Subsequent sludge treatment processes are less efficient because of excess water in thin sludge discharged from primary clarifiers	Phosphate release if primary and BNR wasted sludge are co-thickened
			More floating solids because of gasification
			More mechanical stress on sludge- and scum-collection mechanism
BNR process	Solids retention time	Chlorination problems because of nitrite in the BNR effluent	Low F:M foaming and bulking
			More clarifier solids loading
		Inadequate removal of phosphate, ammonia, and nitrate	Phosphate release
			More energy needed to sustain endogenous respiration
		Deterioration of waste sludge thickening because of thin sludge discharged from secondary clarifiers	
	DO concentration	Chlorination problems because of nitrite in the BNR effluent	Inhibition of denitrification
			High energy cost because of excessive airflow supply
		Inadequate removal of ammonia	Breakup of floc, and so more effluent TSS
		Foaming problems caused by Microthrix	
		Gasification in the clarifier because of the presence of nitrogen	
	Internal recycle	Less nitrate to be denitirified	Lower dentirification rate because of more oxygen in the anoxic compartment
			Increased pumping costs

TABLE 14.1 Issues to consider when selecting setpoints. *(continued)*

Process	Parameter	Potential problems if value is	
		Lower than optimum	Higher than optimum
	Return florate from the clarifier	Improved denitrification in the clarifer	Lower dentirification rate in the clarifer
		Increased effluent TSS due to increased sludge depth in clarifier and solids floating due to denitrification if nitrate concentration is above 6 mg/L	Better dentirification in the first anoxic compartment
			Increased pumping costs
			Deterioration of waste sludge thickening because of thin sludge discharged from secondary clarifiers
	Methanol feed	Incomplete denitrification	Higher cost of methanol, aeration, and sludge processing
			More clarifier mass loading

trol using proportional-integral-derivative (PID) control algorithm (Ingildsen et al., 2001), and cascade control using the Monod equation (Liu et al., 2003). All listed methods require significant development efforts and may be impractical for some facilities.

An even simpler control method is to turn on and off the air supply to the *swinging compartment*—a compartment that can be used for either nitrification or denitrification—based on the ammonia concentration (Samuleson and Carlson, 2001).

DENITRIFICATION CONTROL. If the BNR plant includes internal mixed-liquor recycle (MLR) from the aerobic zone to the anoxic zone, denitrification can be controlled by the MLR flowrate—with or without adding methanol or other sources of COD. Increasing the MLR increases the extent of denitrification to a certain point; if the MLR is too high, too much dissolved oxygen can be returned to the anoxic zone, thereby inhibiting denitrification. For practical purposes, an MLR up to four to six times the influent flow is usually sufficient to improve denitrification.

MLR flow can be controlled in proportion to the influent or based on the nitrate concentration in the anoxic compartment using either a proportional algorithm (Yuan et al., 2001) or fuzzy logic (Serralta et al., 2001). Implementing these control loops can both improve nitrogen removal and significantly reduce power usage for recirculation. The control schemes are fairly easy to implement, but nitrate analyzer reliability, initial costs, and maintenance requirements should be considered before making a decision.

Chemical addition can be controlled via a flow-proportional dose system. More advanced options involve measuring nitrate and controlling chemical addition via a simple proportional control algorithm (Devissher et al., 2001) or fuzzy logic (Marsilli-Libelli and Giunti, 2001). Another control option involves a cascade PID algorithm and two nitrate analyzers (Cho et al., 2001). One analyzer is installed in the mixed-liquor channel and the other in the anoxic compartment. The analyzer in the mixed-liquor channel provides signals to the primary PID controller, which calculates the nitrate concentration target in the anoxic compartment. The secondary PID controller maintains this target based on data from the analyzer in the anoxic compartment. The drawbacks of this method are the difficulties in maintaining two analyzers and tuning the controllers.

Both ammonia and nitrate control schemes need to take into account the challenges of operating ammonia and nitrate analyzers. Even the most advanced control method can cause significant operating problems if it does not use fail-safe logic. Developing such logic sometimes is three to four times more effort than developing the control algorithm itself.

RESPIROMETRY. Respirometry has been widely used to characterize waste and determine kinetics (Chandran and Smets, 2001; Spanjers et al., 2002). Online respirometers have been used to survey wastewater treatment facilities, to develop diurnal load profiles, and to optimize treatment plant operations. Respirometric control could improve any activated sludge process by providing better nitrogen removal, increasing nitrifier viability by reducing the time that the biomass is endogenous, and reducing energy costs by matching aeration to nitrogenous oxygen demand. Several respirometric control approaches have been suggested, and oxidation ditches, completely mixed tanks, sequencing batch reactors (SBR), and plug-flow reactors are all suitable candidates (Barnard et al., 2003)

For example, respirometric control was shown to be the most effective method for improving nitrogen removal in an oxidation ditch in Beemster, Holland (Draaijer et al., 1997). In this application, an *in situ* respirometer was used to measure the ditch's oxygen load, and aerators were adjusted to match the demand.

Respirometric control is an obvious choice for SBRs, and some promising investigations have been done (e.g., Yoong et al., 2000, and Cohen et al., 2003). For example, Shaw and Watts (2002) proposed a method in which aeration is discontinued once the respirometric measurement indicates that the biomass is endogenous, and then the reactor is left idle for a period to promote endogenous denitrification, thereby improving overall nitrogen removal.

Plug-flow reactors are the greatest challenge for respirometric control, although they were the first considered for advanced control by John Watts, the foremost proponent of respirometric control (Watts and Garber, 1993). He proposed that anoxic or aerobic swing zones could be used with respirometric measurements to adjust the overall aerobic volume to match the influent load. Rapid respirogram generation is required to facilitate dynamic control, or alternatively, respirometry can be used to build a picture of typical load variations and this can be used to adjust anoxic or aerobic volumes.

INTERMITTENT AERATION. Intermittent aeration in a continuous-flow system has been proposed as an effective way to achieve nitrification and denitrification in one reactor. In this process, the biomass in the main reactor (or reactors) is subjected to alternating aerobic and anoxic conditions as the aeration system is turned on and off. To control this process, several I&C systems rely on oxidation–reduction potential (ORP; Mori et al., 1997; Caulet et al., 1997; Charpentier et al., 1998). Oxidation–reduction potential is used to determine when denitrification is complete under anoxic conditions and when nitrification is complete under aerobic conditions. Based on experiments at the Cisterna di Latina Wastewater Treatment Plant, however, Carucci et al. (1997) contend that pH is a more suitable control parameter to indicate the same endpoints in the denitrification and nitrification processes.

Another option is to use dissolved oxygen and timers to achieve treatment goals (Japan Society of Industrial Machinery Manufacturers, 2001). The aeration system would be run at 0.2 mg/L of dissolved oxygen for a set period to provide mixing but no nitrification. Then, the dissolved oxygen would be increased to 2.5 mg/L to provide fully aerobic conditions for nitrification.

SEQUENCING BATCH REACTORS. Sequencing batch reactors can be controlled by adjusting the duration of various treatment phases. Currently, there are several examples of I&C control of pilot-scale SBRs, but few of full-scale plants. This is partly because of the difficulty in synchronizing multiple-basin timings, particularly for SBR basins that share one blower system (typically not a concern in pilot-scale tests).

After comparing ORP, dissolved oxygen, and pH control options in a pilot-scale SBR, Cho et al. (2001) concluded that ORP provided the best total nitrogen removal. Andreotta et al. (2001) agreed. Likewise, Demoulin et al. (1997) used an ORP-based system for the cyclical activated sludge technology (CAST) system at the Großarl Wastewater Treatment Plant. It is not clear if this process is fully automated.

IWA/COST BENCHMARK SIMULATION ENVIRONMENT FOR CONTROL STRATEGY COMPARISON.

Simulation allows control strategies for activated sludge plants to be evaluated cost-effectively. To avoid unbiased comparison of the results, a standardized simulation benchmark was developed by the European Cooperation in the field of Scientific and Technical Research (COST) 682/624 Actions. This work is continued under the umbrella of an IWA Task Group (2005–2008).

The former version of the COST/IWA benchmark simulation environment, called the Benchmark Simulation Model No. 1 (BSM1; Copp, 2002), includes a plant layout, simulation models and parameters, a detailed description of influent disturbances (dry weather, storm, and rain events), as well as performance-evaluation criteria (e.g., effluent quality index, effluent violations, operating costs, and controller assessment) to determine the relative effectiveness of proposed control strategies. The benchmark plant layout consists of five completely mixed reactors, with a total volume of 5999 m^3, including a pre-denitrification section that occupies one-third of this volume, and a secondary settler (Figure 14.9). The ASM1 (Henze et al., 1987) models biological processes, while the Takács ten-layer model (Takács et al., 1991) describes settling processes. Kinetic and stoichiometric model parameters, as well as basic operating conditions, are also provided in the benchmark description.

All simulations and performance assessments in BSM1 must be done as specified in the benchmark protocol (http://www.benchmarkwwtp.org; i.e., perform a 150-day steady-state simulation to obtain adequate initial values, simulate the plant with the dry weather scenario for 14 days, and then apply the dry, rain, or storm influent file conditions for another 14 days). Only the last week of the simulation is used for plant performance evaluation.

The IWA Task Group on Benchmarking is currently working on extending the benchmark's goal to encompass other processes in a wastewater treatment system

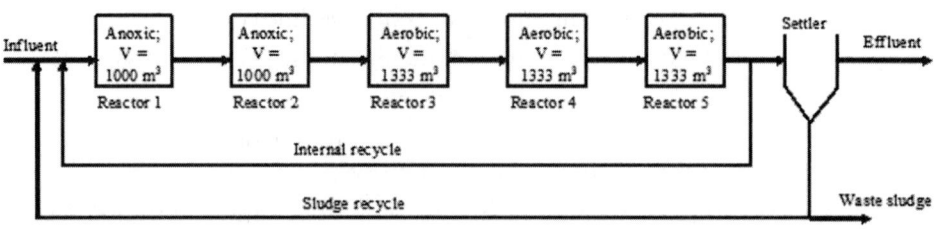

FIGURE 14.9 COST/IWA BSM1 plant layout.

(mainly primary clarification, anaerobic digestion, and sludge thickening; BSM2; Jeppsson et al., 2006), as well as tools that will enable users to evaluate long-term control strategies (BSM1_LT; Rosen et al., 2004). This extended benchmark simulation model will enable plantwide control evaluation when confronted with a two-year influent file with many perturbations and seasonal effects. Temperature's influence in the biological processes is also considered in BSM2. Finally, an extension module to the IWA/COST simulation benchmark for microbiology-related solids separation problems was recently presented (Comas et al., 2006). The aim of this knowledge-based model is not to diagnose filamentous bulking, foaming, or rising sludge with absolute certainty but to quantify in dynamic scenarios whether simulated control strategies lead the process to favorable conditions for them to arise or not.

REFERENCES

Aamodt, A.; Plaza, E. (1994) Case-based reasoning: fundamental issues, methodological variations and system approaches. *AI Communications*, 7 (1), 39–59.

Aitkenhead, M. J.; McDonald, A. J. S.; Dawson, J. J.; Couper, G.; Smart, R. P.; Billet, M.; Hope, D.; Palmer, S. (2003) A novel method for training neural networks for time-series prediction in environmental systems. *Ecological Modeling*, 162, 87–95.

Andreottola, G.; Foladori, P.; Ragazzi, M. (2001) On-line control of an SBR system for nitrogen removal from industrial wastewater. *Water Science & Technology*, 43 (3), 93–100.

Barnard, J.L.; Shaw, A.; Watts, J. B. (2003) Optimized Control of Nitrifying Suspended Growth Systems Using Respirometry. *Proceedings of Ozwater 2003 Convention & Exhibition, AWA 20th Convention*; Perth, Australia.

Belanche, L. L.; Valdés, J. J.; Comas, J.; Rodriguez-Roda, I.; Poch, M. (2000) Prediction of the bulking phenomenon in wastewater treatment plants. *Artificial Intelligence in Engineering*, 14 (4), 307–317.

Carucci, A.; Rolle, E.; Smurra, P. (1997) Experiences of On-Line Control at a Wastewater Treatment Plant for Nitrogen Removal. *Proceedings of the 7th International Workshop of the IAWQ Specialist Group on Instrumentation, Control, and Automation*; Brighton, United Kingdom, July.

Caulet, P.; Lefevre, F.; Bujon, B.; Reau, P.; Philippe, J. P.; Audic, J. M. (1997) Automated Aeration Management in Waste Water Treatment: Interest of the Application to Serial Basins Configuration. *Proceedings of the 7th International Workshop of the IAWQ Specialist Group on Instrumentation, Control, and Automation*; Brighton, United Kingdom, July.

Chandran, K.; Smets, B. F. (2001) Estimating biomass yield coefficients for autotrophic ammonia and nitrite oxidation from batch respirograms. *Water Research*, 35 (13), 3041–3282.

Charpentier, J.; Martin, G.; Wacheux, H.; Gilles, P. (1998) ORP regulation and activated sludge: 15 years of experience. *Water Science & Technology*, 38 (3), 197–208.

Cho, B. C.; Chang, C. N.; Liaw, S. L.; Huang, P. T. (2001) The feasible sequential control strategy of treating high strength organic nitrogen wastewater with sequencing batch biofilm reactor. *Water Science & Technology*, 43 (3), 115–122.

Cho, J.-H.; Sung S. W.; LeeI, B. (2001) *Cascade Control Strategy for External Carbon Dosage in Predenitrification Process*; Conference preprints of the First IWA Conference on Instrumentation, Control and Automation, Malmo, Sweden, 67–75.

Cohen, A.; Hegg, D.; de Michele, M.; Song, Q.; Kasabov, N. (2003) An intelligent controller for automated operation of sequencing batch reactors. *Water Science & Technology*, 47 (12), 57–63.

Comas, J.; Dzeroski, S.; Gibert, K.; Rodriguez-Roda, I.; Sànchez-Marrè, M. (2001) Knowledge discovery by means of inductive methods in wastewater treatment plant data. *AI Communications, 14* (1), 45–62.

Comas, J.; Rodríguez-Roda, I.; Sànchez-Marrè, M.; Cortés, U.; Freixó, A.; Arráez, J.; Poch, M. (2003) A knowledge-based approach to the deflocculation problem: integrating on-line, off-line, and heuristic information. *Water Research, 37*, 2377–2387.

Comas, J.; Rodríguez-Roda, I.; Poch, M.; Gernaey, K.; Rosen, C.; Jeppsson, U. (2006) Extension of the IWA/COST simulation benchmark to include expert reasoning for system performance evaluation. *Water Science & Technology,* 53 (4–5), 331–339.

Copp, J. B. (ed.) (2002) *The COST Simulation Benchmark: Description and Simulator Manual*; Office for Official Publications of the European Communities, Luxembourg, ISBN 92-894-1658-0.

Demoulin, G.; Goronszy, M. C.; Wutscher, K.; Forsthuber, E. (1997) Co-current nitrification/denitrification and biological P-removal in cyclic activated sludge plants by redox controlled cycle operation. *Water Science & Technology, 35* (1), 215.

Devisscher, M.; Bogaert, H.; Bixio, D. (2001) *Feasibility of Automatic Chemical Dosage Control—A Full-Scale Evaluation*; Conference preprints of the First IWA Conference on Instrumentation, Control, and Automation, Malmo, Sweden, 623–631.

Draaijer, H.; Buunen, A. H. M.; van Dijk, J. W. (1997) Full-Scale Respirometric Control of an Oxidation Ditch. *Proceedings of the 7th International Workshop of the IAWQ Specialist Group on Instrumentation, Control, and Automation*; Brighton, United Kingdom, July.

Driankov, D.; Hellendoorn, H.; Reinfrank, M. (1993) *An Introduction to Fuzzy Control*; Springer Verlag: New York.

Ekster, A.; Rodríguez-Roda, I. (2003) The Effect of Sludge Age on the Operations Cost of an Activated Sludge System. *Proceedings of the 76th Annual Water Environment Federation Technical Exhibition and Conference* [CD-ROM]; Los Angeles, California, Oct 11–15; Water Environment Federation: Alexandria, Virginia.

Ekster, A.; Wang, J.(2005) Reliable DO Control Is Available. *Water Environment & Technology*, 17 (2), 41–43.

Ferrer, J.; Rodrigo, M. A.; Seco, A.; Penya-Roja, J. M. (1998) Energy saving in the aeration process by fuzzy logic control. *Water Science & Technology*, 38 (3), 209–217.

Goldberg, D.E. (1989) *Genetic Algorithms for Search, Optimization, and Machine Learning*; Addison-Wesley: Reading, Massachusetts.

Hecht-Nielsen, R. (1990) *Neurocomputing*; Addison-Wesley: Reading, Massachusetts.

Henze, M.; Grady Jr., C. P. L.; Gujer, W.; Marais, G. v. R.; Matsuo, T. (1987) *Activated Sludge Model No. 1*, IAWPRC Scientific and Technical Report No. 1; International Association on Water Quality: London, United Kingdom.

Hertz, J.; Krogh, A.; Palmer, R. G. (1991) *Introduction to the Theory of Neural Computation*; Addison-Wesley: Reading, Massachusetts.

Hirota, K. (1993) *Industrial Applications of Fuzzy Technology*; Springer Verlag: New York.

Ingildsen, P.; Jeppsson, U.; Olsson, G. (2001) *Dissolved Oxygen Controller Based on On-Line Measurement of Ammonia Combining Feed-Forward and Feedback.* Conference preprints of the First IWA Conference on Instrumentation, Control, and Automation; Malmo, Sweden, 631–639.

Jackson, P. (1990) *Introduction to Expert Systems*, 2nd ed.; Addison-Wesley: Reading, Massachusetts.

Japan Society of Industrial Machinery Manufacturers (2001) *Introduction of Japanese Advanced Environmental Equipment*; Tokyo, Japan. nett21.gec.jp/JSIM_DATA/index.html (accessed June 2006).

Jeppsson, U.; Alex, J.; Pons, M. N.; Spanjers, H.; Vanrolleghem, P. A. (2002) Status and future trends of ICA in wastewater treatment—a European perspective. *Water Science & Technology*, 45 (4/5), 485–494.

Jeppsson, U.; Rosen, C.; Alex, J.; Copp, J.; Gernaey, K. V.; Pons, M.-N.; Vanrolleghem, P. (2006) Towards a benchmark simulation model for plant-wide control strategy performance evaluation of WWTPs. *Water Science & Technology*, 53 (1), 287–295.

Klir, G. J.; Folger, T. A. (1988) *Fuzzy Sets, Uncertainty, and Information*; Prentice Hall Int. Editions: New York.

Kolodner, J. (1993) *Case-Based Reasoning*; Morgan Kaufmann: San Francisco, California.

Krause, K.; Bocker, K.; Londong, J. (2001) *Simulation of a Nitrification Concept Considering Influent Ammonium Load*. Conference preprints of the First IWA Conference on Instrumentation, Control and Automation; Malmo, Sweden, 555–563.

Langley, P. (1996) *Elements of Machine Learning*; Morgan Kaufmann: San Francisco, California.

Liu, W.; Lee, G.; Goodley, J. (2003). Using On-line Ammonia and Nitrate Instruments to Control Modified Ludzack-Ettinger (MLE) Process. *Proceedings of the 76th Annual Water Environment Federation Technical Exhibition and Conference* [CD-ROM]; Los Angeles, California, Oct 11–15; Water Environment Federation: Alexandria, Virginia.

Manesis, S. A.; Sapidis, D. J.; King, R. E. (1998) Intelligent control of wastewater treatment plants. *Artificial Intelligence in Engineering*, 12, 275–281.

Marsili-Libelli S.; Giunti, L. (2001) *Fuzzy Predictive Control for Organic Carbon Dosing in Denitrification*. Conference preprints of the First IWA Conference on Instrumentation, Control and Automation; Malmo, Sweden, 35–43.

Mitchell, T. M. (1997) *Machine Learning*; McGraw-Hill: New York.

Moise, M.; Norris, M. (2005). Process Optimization and Automation Improves Reliability and Process Efficiency of Oxnard WWTP. *Proceedings of the 78th Annual Water Environment Federation Technical Exhibition and Conference* [CD-ROM]; Washington, D.C., Oct 29–Nov 2; Water Environment Federation: Alexandria, Virginia.

Mori, Y.; Sasaki, K.; Yamamoto, Y.; Tsumura, K.; Ouchi, S. (1997) Countermeasures for Hydraulic Load Variation in Intermittently Aerated 2-Tank Activated Sludge Process for Simultaneous Removal of Nitrogen and Phosphorus. *Proceedings of the 7th International Workshop of the IAWQ Specialist Group on Instrumentation, Control, and Automation*; Brighton, United Kingdom, July.

Olsson, G.; Aspegren, H.; Nielsen M. K. (1998) Operation and control of wastewater treatment—a Scandinavian perspective over 20 years. *Water Science & Technology*, 37 (12), 1–13.

Patry, G.; Chapman, D. (1989) *Dynamic Modeling and Expert Systems in Wastewater Engineering*; Lewis Publishers: Chelsea, Michigan.

Pawlak, Z. (1991) *Rough Sets: Theoretical Aspects of Reasoning about Data*; Kluwer Academic Publishers: Norwell, Massachusetts.

Pearl, J. (1987) *Probabilistic Reasoning in Intelligent Systems: Networks of Plausible Inference*; Morgan-Kaufmann, Inc.: San Francisco, California.

Pelletier, R.; Sloan, D. (1999) Orlando's Nitrogen Profile... a New Approach to the Activated Sludge Control. *Proceedings of the 72nd Annual Water Environment Federation Technical Exposition and Conference* [CD-ROM]; New Orleans, Louisiana, Oct 9–13; Water Environment Federation: Alexandria, Virginia.

Poch, M.; Comas, J.; Rodríguez-Roda, I.; Sànchez-Marrè, M.; Cortés, U. (2004) Designing and building real environmental decision support systems. *Environmental Modeling and Software*, 19 (9), 857–873.

R.-Roda, I.; Comas, J.; Poch, M.; Sànchez-Marrè, M.; Cortés, U. (2001) Automatic knowledge acquisition from complex processes for the development of knowledge based systems. *Industrial & Engineering Chemical Research*, 40 (15), 3353–3360.

R.-Roda, I.; Poch, M.; Sànchez-Marrè, M.; Cortés, U.; and Lafuente, J. (1999) Consider a case-based system for control of complex processes. *Chemical Engineering Progress*, 95 (6), 39–45.

Rodriguez-Roda, I.; Sànchez-Marrè, M.; Comas, J.; Baeza, J.; Colprim, J.; Lafuente, J.; Cortés, U.; and Poch, M. (2002) A hybrid supervisory system to support wastewater treatment plant operations. *Water Science & Technology*, 45 (4/5), 289–297.

Rosen, C.; Vanrolleghem, P.A.; Jeppsson, U. (2004) Towards a common benchmark for long term process control and monitoring evaluation. *Water Science & Technology*, 50 (11), 41–49.

Samuelsson, P.; Carlsson, B. (2001) *Control of Aeration Volume in an Activated Sludge for Nutrient Removal*. Conference preprints of the First IWA conference on Instrumentation, Control and Automation; Malmo, Sweden, 429–437.

Sànchez M.; Cortés U.; Lafuente J.; Rodriguez-Roda, I.; Poch, M. (1996) DAI-DEPUR: a integrated and distributed architecture for wastewater treatment plants supervision. *Artificial Intelligence in Engineering*, 10 (3), 275–285.

Serralta, J.; Ribes, J.; Seco, A.; Ferrer, J. (2001) *A Supervisory Control System for Optimizing Nitrogen Removal and Aeration Consumption in Wastewater Treatment Plants.* Conference preprints of the First IWA conference on Instrumentation, Control and Automation; Malmo, Sweden, 429–437.

Shaw, A.; Watts, J. (2002) The Use of Respirometry for the Control of Sequencing Batch Reactors—Principles and Practical Application. *Proceedings of the 75th Annual Water Environment Federation Technical Exhibition and Conference* [CD-ROM]; Chicago, Illinois, Sep 28–Oct 2; Water Environment Federation: Alexandria, Virginia.

Spanjers, H.; Patry, G. G.; Keesman, K. J. (2002) Respirometry-based on-line model parameter estimation at a full-scale WWTP. *Water Science & Technology*, 45 (4/5), 335–343.

Stephanopoulos, G.; Han, C. (1996) Intelligent systems in process engineering: a review. *Computers in Chemical Engineering*, 20 (6/7), 743–791.

Sugeno, M. (1985) *Industrial Application of Fuzzy Control*; North Holland: Amsterdam.

Takács, I.; Patry, G. G.; Nolasco, D. (1991) A dynamic model of the clarification thickening process. *Water Research*, 25, 1263–1271.

Water Environment Federation (1997) *Energy Conservation in Wastewater Treatment Facilities*, Manual of Practice No. FD-2; Water Environment Federation: Alexandria, Virginia.

Watts, J.; Garber, W. (1993) On-Line Respirometry: A Powerful Tool for ASP Operation and Design. *Proceedings of the 6th IAWQ—ICA of Water and Wastewater Treatment*, Hamilton, Canada.

Yoong, E. T.; Lant, P. A.; Greenfield, P. F. (2000) In Situ Respirometry in an SBR treating wastewater with high phenol concentrations. *Water Research*, 34 (1), 239–245.

Yuan, Z.; Oehmen, A.; Indildsen, P. (2001) *Control of Nitrate Recirculation Flow in Predenitrification Systems.* Conference preprints of the First IWA Conference on Instrumentation, Control and Automation, Malmo, Sweden, 27–35.

Chapter 15

Instrumentation and Control System Specifications

The Documentation Process	474	The Specifications Process	475
Process and Instrumentation Diagrams	474	Related Documents	477
		Organization	481
Control-System Block Diagrams	474	*General*	481
Physical Drawings	475	*Products*	483
Special Mounting Details	475	*Control System*	483
Signal I/O and Instrument Lists	475	*Instrumentation*	483
		Execution	484
Loop Drawings	475	Suggested Readings	486

As instrumentation and control (I&C) systems have become more complex, so have the related specifications. Specifying engineers should document all control-system requirements for both hardware and software, as well as every installation detail, from equipment procurement to startup and training. This includes all the types of communications (e.g., dedicated leased phone lines, licensed and unlicensed radio systems, satellites, and wireless WI-FI 802.11g) involved. The specifications should be clear to all interested parties (e.g., the owner, potential bidders, and regulators) and result in an effective bidding and construction process—and an I&C system that functions as intended.

THE DOCUMENTATION PROCESS

Design engineers should document every aspect of the I&C system. Control-system documentation typically begins with a control-system block diagram that illustrates every controller in the plant and at remote locations. It also notes all the types of communications media required. Then, design engineers should write up the hardware and software specifications for all control-system components. Then, in the mechanical and electrical drawings, they should show precisely where every control-system console and panel will be installed. Next, they should write process-control narratives detailing how the control system is supposed to function. Finally, they should produce console and panel layouts, fiber-optic patch panels, in-plant and remote site wiring, and mounting details, as needed.

Instrumentation documentation typically begins with a process and instrumentation diagram (P&ID). Then, design engineers should put together data sheets or specifications for all instruments. Then, in the mechanical and electrical drawings, they should show precisely where every instrument will be installed. Then, they should generate signal input/output (I/O) and instrument lists. Once the P&IDs are completed, they should make loop drawings, if required. Finally, they should produce panel layouts and special mounting details, as needed.

PROCESS AND INSTRUMENTATION DIAGRAMS. Design engineers should create P&IDs for every project. These diagrams "outline" the I&C system requirements and are useful communication tools. They should be distributed to all design team members so the process, electrical, and instrumentation disciplines can coordinate their efforts more effectively. (For more information on P&IDs, see Chapter 3 and Chapter 4.)

CONTROL-SYSTEM BLOCK DIAGRAMS. Design engineers also should develop a control-system block diagram for every project. These diagrams are "road maps" of all in-plant and remote control units connected to the control system. The remote control units can consist of either programmable logic controllers (PLCs), distributed control units (DCUs), remote I/O units (RIO), or remote terminal units (RTUs). The control system block diagrams should also show all interfaces to existing control systems and IT networks and note the type of communications that will be used with each type of control unit. (For more information on control-system block diagrams, see Chapter 3 and Chapter 4.)

PHYSICAL DRAWINGS. Mechanical or process drawings should show precisely where each field instrument will be mounted so I&C engineers can verify that installation requirements (e.g., piping, straight runs, and uninterruptible power source) will be met. Likewise, electrical system drawings should show every control-system console and control panel.

Electrical designers use the P&IDs, console and control-panel drawings, and mechanical or process drawings to develop the conduit, wire, and fiber-optic plans and schedules. They also use the information to develop elementary diagrams of the control and signal interlocking and interfacing logic requirements.

SPECIAL MOUNTING DETAILS. Special mounting details show contractors how to properly install I&C system components. Such details include console and panel layouts; fiber-optic patch panels; all piping and sizing requirements; and any special fittings, wiring, or mounting material fabrication requirements.

SIGNAL I/O AND INSTRUMENT LISTS. Some design engineers use P&ID software that automatically generates lists of instruments and signal I/Os. Others create the lists manually using Microsoft Excel or another spreadsheet program (Table 15.1).

LOOP DRAWINGS. Typically, system integrators create the detailed loop drawings, but sometimes they are the responsibility of the design engineer. In either case, the loop drawings should be made after the P&IDs are completed and before the electrical designer develops the conduit, wire, and fiber-optic plans and schedules.

THE SPECIFICATIONS PROCESS

The specifications process includes all the steps required to specify and procure a complete I&C system. Basically, this consists of specifications and drawings, bidding information, contract terms and conditions, and payment provisions.

Correct and detailed I&C system specifications are important. The specifications should include detailed hardware and software requirements, I/O signal listings, and control narratives that instruct the contractor how to configure the control-system software to meet the project's goals.

Design engineers typically rely on two key guides when developing detailed specifications: *The Project Resource Manual*, published by the Construction Specifica-

TABLE 15.1 I/O signal list.

Signal type	PLC number	Tag number	Equipment name	Signal function	Range on text	Units off text
DI	1	YS-100	Station door switch	Entry alarm	Closed	Open
AI	1	PT-100	City of Chicago water pressure	Station supply pressure	0 to 100	psi
AI	1	TT-101	Valve vault A	Vault temperature	0 to 100	°F
AI	1	TT-102	Valve vault B	Vault temperature	0 to 100	°F
DI	1	YS-103	Engine generator	Failure	Fail	Normal
DI	1	LSH-104	Pump room sump	High-level alarm	Alarm	Normal
DI	1	YS-105	Station fire	Fire alarm	Alarm	Normal
AI	1	AT-106	Chlorine residual analyzer	Chlorine residual	0 to 2	ppm
AI	1	LT-110	Reservoir level signal	Level	0 to 15	ft
DI	1	LS-110	Reservoir level low cutout signal	Low-level pump cutout	Cutout	Normal
AI	1	MI-111	Pump number 1	Amperes	0 to 150	amps
DI	1	QSA-111	Pump number 1 hand switch	In auto	Auto	---
DI	1	QSB-111	Pump number 1 hand switch	In off	Off	---
DI	1	QSC-111	Pump number 1 hand switch	In hand	Hand	---
DO	1	HS-111	Pump number 1 hand switch	Start/stop	Start	Stop
DI	1	MN-111	Pump number 1	Pump run	Running	Not running
DI	1	XS-111	Pump number 1	Future VFD Failure	Fail alarm	Normal
AO	1	SIC-111	Pump number 1	Future VFD speed control	0-100	%
AI	1	ST-111	Pump number 1	Future VFD speed	0-1800	rpm
AI	1	MI-112	Pump number 2	Amperes	0-200	amps

tions Institute (CSI), and *Specification Forms for Process Measurement and Control Instruments, Primary Elements and Control Valves* (Standard S20), published by the Instrumentation, Systems, and Automation Society (ISA). *The Project Resource Manual* is a master specification document intended to provide every content-related detail that design engineers would need to specify a complete I&C system.

Specification Forms for Process Measurement and Control Instruments, Primary Elements and Control Valves has standardized both the content and form of instrumentation data sheets for all types of I&C devices (Figure 15.1 and Figure 15.2). These forms are valuable checklists for specifying both field and panel-mounted instruments.

RELATED DOCUMENTS. For an I&C system, detailed specifications may need to be developed for

- Control panels;
- The control system;
- Human–machine interface (HMI) software;
- Instruments;
- Maintenance;
- Network requirements;
- Operations and maintenance (O&M) manuals;
- Process display requirements;
- Report requirements;
- Startup services;
- Supervisory control and data acquisition (SCADA) systems, including RTUs, DCUs, RIO units, and communications media (e.g., leased phone lines, fiber-optic cable, radio, and satellite);
- System documentation;
- System-integrator requirements;
- Telemetry systems;
- Transition plan;

		MAGNETIC FLOWMETERS				SHEET	OF
						SPEC. NO.	REV.
		NO	BY	DATE	REVISION		
						CONTRACT	DATE
						REQ. - P.O.	
						BY	CHK'D	APPR.

	1		Meter Tag No.				
	2		Service				
	3		Location				
METERING ELEMENT	4	CONN'S	Line Size, Sched.				
	5		Line Material				
	6		Connection Type				
	7		Connection Materials				
	8	METER	Tube Material				
	9		Liner Material				
	10		Electrode Type				
	11		Electrode Material				
	12		Meter Casing				
	13		Power Supply	Elec. Code			
	14		Grounding, Type & Material				
	15		Enclosure Class				
	16						
	17	FLUID	Fluid				
	18		Max. Flow, Units				
	19		Max. Velocity, Units				
	20		Norm. Flow	Min. Flow			
	21		Max. Temp.	Min. Temp.			
	22		Max. Press.	Min. Press.			
	23		Min. Fluid Conductivity				
	24		Vacuum Possibility				
	25						
ASSOCIATED INSTRUMENT	26		Instrument Tag Number				
	27		Function				
	28		Mounting				
	29		Enclosure Class				
	30		Length Signal Cable				
	31		Type Span Adjustment				
	32		Power Supply				
	33	TRANS.	Transmitter Output				
	34						
	35	DISPLAY	Scale Size	Range			
	36		Chart Drive	Speed			
	37		Chart Range	Chart Number			
	38		Integrator				
	39	CONTR.	Modes	Output			
	40		Action	Auto-Man.			
	41						
	42	ALARM	Contact No.	Form			
	43		Rating	Elec. Code			
	44		Action				
	45		Manufacturer				
	46		Meter Model Number				
	47		Instrument Model Number				

NOTES:

© 1981 ISA   ISA FORM S20.23

FIGURE 15.1   Form S20.23 magnetic flowmeters.

Magnetic flowmeters

Instructions for ISA Form S20.23

1) Tag number of meter only.
2) Refers to process application.
3) Show line number or identify associated vessel.
4) Give pipeline size and schedule. If reducers are used, so state.
5) Give material of pipe. If lined, plastic or otherwise non-conductive, so state.
6) Give connection type: FLANGED, DRESSER COUPLINGS, ETC.
7) Specify material of meter connections.
8) Select tube material. (Non-permeable material required if coils are outside tube).
9) Specify material of line.
10) Select electrode type: STD., BULLET NOSED, ULTRASONIC CLEANED, BURN OFF, etc.
11) Specify electrode material.
12) Describe casing: STD., SPLASH PROOF, SUBMERSIBLE, SUBMERGED OPERATION, etc.
13) Give ac voltage and frequency, along with application NEMA identification of the electrical enclosure.
14) State means for grounding to fluid: GROUNDING RINGS, STRAPS, etc.
15) State power supply and enclosure class to meet area electrical requirements.
17) State fluid by name or description.
18) Give maximum operating flow and units; usually same as maximum of instrument scale.
19) Give maximum operating velocity, usually in ft/s.
20) List normal and minimum flow rates.
21) List maximum and minimum fluid temperature °F.
22) List maximum and minimum fluid pressure.
23) List minimum (at lowest temp.) conductivity of fluid.
24) If a possibility of vacuum exists at meter, so state and give greatest value (highest vacuum).
26) List tag number of instrument used directly with meter.
27) Control loop function such as INDICATE, RECORD CONTROL, etc.
28) Mounting: FLUSH PANEL, SURFACE INTEGRAL WITH METER, etc.
29) Give NEMA identification of case type.
30) State cable length required between meter and instrument.

FIGURE 15.2 Instructions for ISA Form S20.23.

31) Span adjust: BLIND, ft/s DIAL, OTHER.

32) Give ac supply voltage and frequency.

33-34) If a transmitter, state analog output electrical or pneumatic range, or pulse train frequency for digital outputs, i.e., pulses per gallon.

35) List scale size and range.

36) Recorder chart drive — ELECT. HANDWIND, etc. and chart speed in time per revolution or inch per hour.

37) List chart range and number.

38) If integrator is used, state counts per hour, or value of smallest count; such as "10 GAL UNITS."

39) For control modes: (Per ANSI C85.1-1963, "Terminology for Automatic Control.") Write-in PI_f, I_f, PI_s, $PI_f D_f$, etc.

P = proportional (gain)

I = integral (auto reset)

D = derivative (rate)

Subscripts:

f = fast

s = slow

n = narrow

State output signal range, pneumatic or electronic.

40) Controller action in response to an increase in flowrate — INC. or DEC.

State auto-man. switch as NONE, SWITCH ONLY, BUMPLESS, etc.

42) Number of alarm lights in case. Give form of contacts; SPDT, SPST, etc.

43) Contact electrical load rating. Contact housing General Purpose, Class 1, Group D, etc., if not in the same enclosure described in line 29.

44) Action of alarms: HIGH, LOW, DEVIATION, etc.

45-47) Fill in manufacturer and model numbers for meters and instrument after selection.

FIGURE 15.2   Instructions for ISA Form S20.23. *(continued)*

- Testing;
- Training; and
- Warranty.

Typically, the following documents and drawings also should be provided with the above specifications:

- Process-control-system configuration and communication schematics,
- P&IDs,
- Console and control-panel layouts,
- RTU panel layouts,
- Fiber-optic-patch panel layouts,
- Instrumentation installation details,
- Mechanical and electrical drawings noting the location of I&C system components,
- Process-control narratives,
- Instrument lists,
- I/O lists,
- Conduit and wire schedules, and
- Loop diagrams.

ORGANIZATION. Specifications should be organized logically, so anyone can find the information they need. Design engineers typically use CSI's format, which divides specifications into the following three topics: general, products, and execution.

General. Information that is applicable to the entire I&C system may include the following:

- An outline of the overall work and requirements for the equipment and services to be provided.
- Related work or requirements specified elsewhere (e.g., a new treatment plant's I&C system components may be provided by multiple parties, or a flow meter may be supplied by one party, installed by a second, and electrically wired by a third). References to related work help to describe the equipment's final installed condition.
- Submittal documentation required from the contractor, including

- A bill of materials;
- Manufacturers' catalog sheets;
- Manufacturers' data sheets showing options selected and calibration data;
- Loop diagrams showing each loop, with all terminations shown (Instrument Loop S5.4, 1981);
- Wiring schematics with point-to-point wiring tables;
- Panel and console drawings showing the layout of front panel-mounted and internal panel-mounted devices;
- Diagrams and lists of software programs developed for programmed devices (e.g., programmable logic controllers or computers);
- Software submittals documenting the implementation of control-system narratives;
- Requirements for equipment samples and shop drawing submittals;
- Manufacturers' calibration certifications or test reports; and
- As-built drawings and wiring tables.
- O&M manuals furnished by the contractor, including
  - Record drawings consisting of final as-built contract drawings showing where the actual installed work varies from the work as originally shown;
  - Copies of approved submittal documentation;
  - Copies of certifications and test reports;
  - Copies of manufacturers' instrumentation and technical bulletins, installation instructions, O&M instructions, troubleshooting instructions, repair instructions, and spare parts lists;
  - An instrument schedule, including instrument number, type, service, location, calibrated range, and settings;
  - Software flow charts and annotated logic diagrams; and
  - Electrical interconnection diagrams.

- Spare parts
- Special tools

Products. Required data for each specified control-system device, control-system narrative, and instrument includes

Control System

- Control-system hardware and software;
- Control-system narratives;
- Communications equipment;
- Networking equipment;
- Database-configuration requirements;
- Process-display requirements;
- Report requirements; and
- Historical data collection requirements.

Instrumentation

- Instrument type;
- Pipe size (for in-line instruments);
- Minimum and maximum operating data (e.g., flow, pressure, temperature, and pH);
- Environmental requirements (e.g., temperature, humidity, and corrosive or explosive atmosphere);
- Method of connection to physical system (e.g., flange);
- Special installation hardware required;
- Special tagging requirements;
- Range of measurement;
- Accuracy;
- Precision;

- Enclosure;
- Electrical requirements;
- Signal output;
- Indication;
- Diagnostics;
- Application vendor options;
- Maintenance and calibration requirements;
- Input signal required;
- Type of setpoint adjustment (e.g., fixed or adjustable), if applicable;
- Alarm switch with limit settings and the amount of deadband required when switches return to normal;
- Construction materials;
- References to standards [e.g., ISA, Underwriters' Laboratory (UL), and National Electrical Code (NEC)]; and
- Appurtenances (e.g., mounting hardware, special tools, or calibration equipment).

Execution. Before a utility accepts a new I&C system, it should have passed three performance tests: a factory acceptance test (FAT), a system availability test (SAT), and a site demonstration test (SDT). The factory acceptance test proves that the control system is built to the project's requirements, configured correctly, and ready for shipment to the job site. The system availability test typically duplicates the FAT at the job site, using the installed instruments to completely test the control system, wiring, and communications media.

Instruments may be tested at the factory or in the field and are witnessed by the owner or design engineer. Both individual instruments (e.g., flow meters) and groups of devices (e.g., control panels) can be tested. After field installation, instrumentation testing should be integrated into the SDT.

For all these tests, the design engineer develops the test criteria, but the system integrator develops the detailed test procedures because each control system is unique. The tests are conducted by the system integrator and witnessed by the owner or design engineer.

Instrumentation and Control System Specifications

During startup of the I&C system, the contractor may have to tune the control loops or demonstrate that the instrument works properly. For some devices (e.g., magnetic flowmeter), a representative of the instrument manufacturer should start up the equipment.

Training specifications may include a minimum number of hours, require trainers' resumes, and require the contractor to submit a detailed outline of each course. Training courses should cover I&C system hardware, software, and instrumentation.

The specifications also should require the contractor(s) to provide the following control-system documentation:

- Corrected as-built drawings,
- Shop drawings,
- Panel drawings,
- Fiber-optic patch panel drawings,
- Loop drawings (if not done by the designer),
- Control-system hardware and software,
- Control-system interconnection drawings,
- Communications equipment hardware and software,
- Manufacturers' data sheets for all instruments, and
- O&M manuals.

These documents should be furnished in an electronic format as part of the final as-built documentation. These electronically formatted data can then be incorporated into the facility's documentation or maintenance management systems.

The specifications also should require the following types of schedules:

- Instrument lists,
- Annunciator window engraving,
- Panel devices,
- Graphical screens,
- Reports, and
- I/O signal interface.

SUGGESTED READINGS

American Water Works Association; American Society of Civil Engineers (1990) *Water Treatment Plant Design*, 3rd Ed.; McGraw-Hill: New York.

American Water Works Association (2001) *Instrumentation and Control*, Manual of Water Supply Practices—M2. American Water Works Association: Denver, Colorado.

Construction Specification Institute (2004) *The Project Resource Manual—Manual of Practice.* Construction Specification Institute: Alexandria, Virginia.

Instrumentation, Systems, and Automation Society (1991) *Instrument Loop Diagrams*, Standard ANSI/ISA-S5.4. Instrumentation, Systems, and Automation Society: Research Triangle Park, North Carolina.

Instrumentation, Systems, and Automation Society (1981) *Specification Forms for Process Measurement and Control Instruments, Primary Elements, and Control Valves*, Standard ISA-S20. Instrumentation, Systems, and Automation Society: Research Triangle Park, North Carolina.

Liptak, B.; Venczel, K. (1985) *Instrument Engineers' Handbook*, Revised Ed.; Chilton Book Co.: Radnor, Pennsylvania.

Water Environment Federation (1993) *Design of Wastewater and Stormwater Pumping Stations*, 2nd Ed.; Manual of Practice No. FD-4; Water Environment Federation: Alexandria, Virginia.

Whitt, M. D. (2004) *Successful Instrumentation and Control Systems Design*; Instrumentation, Systems, and Automation Society: Research Triangle Park, North Carolina.

Chapter 16

Instrumentation Maintenance

Maintenance Management System Components	488	Implementation Methods	502
		Textbook	502
Computerized Maintenance Management Systems	489	Modular	505
		Build as You Go	506
Verification of Instrument Performance	498	Turnkey	506
Quality Control Procedures	498	Reference	507
Control Charts	499	Suggested Readings	507
Example	500		

There are basically two types of maintenance work: repetitive (preventive) and non-repetitive (corrective or unscheduled). Preventive maintenance extends the lifespan and efficiency of equipment; an effective preventive maintenance program retards equipment deterioration via regular inspections, routine maintenance, parts replacement, calibration, and statistical performance-monitoring techniques. Corrective maintenance simply fixes broken components. Initially, as much as 80% of a maintenance program's labor hours are spent on corrective maintenance, but as utilities focus on maximizing equipment use, corrective maintenance can drop to as little as 50% of labor hours (Plant Maintenance, 1987).

Instrument maintenance is not just a matter of keeping the device functioning but also of periodically validating its readings. Proper instrument calibration and verification depends on statistical quality-control (QC) techniques, and the statistical data must be collected and stored with other instrument data.

Traditionally, less maintenance was performed than was needed because of

- The lack of funds or management support,
- The perception that maintenance is more labor-intensive than analytical,
- Managers' focus on capital projects,
- Internal resource requirements, and
- The lack of immediate results.

However, maintenance has been gaining importance as government agencies have begun requiring utilities to be accountable for their assets. The agencies' goal is to ensure that all government-funded equipment, including instrumentation, has as long and useful a life as possible. Both maintenance and instrument validation are major components of an asset management program.

MAINTENANCE MANAGEMENT SYSTEM COMPONENTS

A maintenance management system basically consists of the equipment to be maintained, the maintenance instructions and tools, the replacement parts, and a comprehensive schedule of maintenance tasks.

A successful maintenance management program depends on an effective work-order system. The work order authorizes the specific maintenance task to be done, detailing what, where, when, who, and how. It also provides necessary information for inventory, purchasing, and personnel files. Work order records typically are filed with other equipment documentation so the equipment's history can be easily recalled. These records also verify instrument upkeep, needed to meet warranty requirements.

Successful maintenance programs also require an effective inventory management system. A well-stocked inventory of replacement instruments, parts, and materials can keep a facility running smoothly. An overstocked warehouse, however, is a misuse of resources. The inventory management system should establish the facility's history of actual parts usage, so maintenance staff can create a realistic list of parts to keep on hand. It also should include methods for verifying the contents of the warehouse and ordering fresh stock, as needed.

The maintenance program's purchasing system should focus on three tasks: generating purchase requisitions when inventory drops to reorder levels, issuing purchase orders based on the requisitions, and expediting issued orders. Purchasing data should be available from a history file by requisition number, purchase order number, part number, and costs.

Maintenance personnel's data also must be tracked to calculate labor costs, classify crafts, establish task labor standards, and generate history records. Much of this information is available in work orders and employee time cards.

COMPUTERIZED MAINTENANCE MANAGEMENT SYSTEMS

Maintenance management is a complex system. It involves large numbers of equipment, parts, tasks, instructions, and schedules. It also involves substantial data-entry, -storage, -management, and -distribution requirements. Related communications cross departmental lines, and operations and maintenance (O&M) objectives often conflict. So, maintenance management is well-suited for computer software.

Currently available maintenance-management software establishes the program structure, provides documentation, distributes information, posts changes from one area to another, maintains a central data repository, manipulates the data to meet user needs, and makes the process more understandable. Reasonably priced computerized maintenance management systems (CMMSs) typically have a payback period of one or two years. (Procedures for calculating the return on investment are readily available.)

A well-designed CMMS automatically issues work orders for preventive maintenance tasks at regularly scheduled intervals. It automatically adjusts inventory records as parts are used and reorders parts as needed. It makes analyzing the facility's maintenance history a practical, speedy process. Also, smart transmitters and Fieldbus-compliant hardware make it easier to diagnose process and field instrumentation problems via the instrumentation and control (I&C) system.

To better understand how a CMMS functions, consider the following example, which involves the replacement of a membrane in a dissolved oxygen meter. First, the dissolved oxygen meter is added to the CMMS's equipment records (Figure 16.1).

```
Equipment Data Sheet

    Equipment No.        Description           BHPID                 Parent
           131.2         DO Sensor             ST:INS-AE-131.2       ST:INS

General Specifications

        Eq.Service:     DO Measurement
          Location:     Aeration Tank 2       Contract Number:   260
      Manufacturer:     Rosemount             Spec Reference:    13300
     Serial Number:     910                   O&M Doc Ref. :     Vol 6, Sec. 7
       Part Number:     0012-50-786

Financial Data

      Asset Number:     20737
       Cost Center:     Secondary             Initial Startup Date: 9/14/2000

   General Leger No. :  124                   Purchase Date:     Contract

     Original Cost:     $2,750.00             Warranty Exp Date: 9/14/2002
  Replacement Cost:     $3,250.00             Life Expectancy:   120 Mo
   Total Labor Cost:    $375.00
Total Material Cost:    $2,100.00
```

FIGURE 16.1 An example of an equipment record.

This record includes information on the sensor's location, service status, general specifications, and links to O&M and financial information. Based on the data in this record, the CMMS will generate appropriate preventive-maintenance work orders and maintain a history of work done, parts used, and labor hours involved. It also will notify staff when related warranties or service contracts are about to expire and produce work orders with instructions to renew or cancel them, as appropriate. The system will also track the meter's "total labor cost" and "total material cost" throughout its lifespan to help employees determine when it should be replaced.

```
Employee

        Last Name:   XXXXXXXXXX           Employye ID:  XXXXXX
        First Name:  XXXXXXXXXX                   SSN:  XXX-XX-XXXX

    Street Address:  XXXXXXXXXX
         Zip Code:   XXXXX
            Phone:   XXX-XXX-XXXX

             Craft:  Inst Tech              Hire Date:   6/14/1998
             Class:  4                   Review Period:  Annual
             Shift:  1                  Last Review Date 4/30/2002
       Hourly Wage:  $21.85              Sick Time Rate: 10
                                          Vacation Time: 10
  Accum Sick Time:  33
   Accum Vacation:  15

  Notes:  Employee is a self starter and actively pursues training opportunities.
```

FIGURE 16.2 An example of an employee record.

Employee data are also entered into the CMMS (Figure 16.2) so labor costs can be calculated based on actual-wage or overhead rates.

To keep the dissolved oxygen meter operating properly and cost-effectively, the CMMS will establish a schedule of related preventive-maintenance tasks, such as replacing the meter's membrane (Figure 16.3). This schedule is based on the manufacturer's recommendations. However, because actual replacement needs depend on the equipment's environment, the CMMS will initially schedule preventive maintenance tasks more frequently than the manufacturer recommends and gradually increase the interval between tasks until an optimal interval is determined.

In the work order associated with each preventive maintenance task (Figure 16.4), the CMMS also indicates the tools and parts typically required, and it notifies the warehouse to reserve them for the maintenance crew. In addition, the system

```
Task Record

Equipment No. :    131
Task No.           0001
Desc:              Quarterly replacement of DO sensor Membrane

Work Order Type:   PM              In-sevice Task [Y/N] :   Y
Task Priorrity:    2                       Next PM date:   9/15/2003
Perform every:     91    days    Calc'd From last [C/P] :   C

Crafts

        Craft       Resources    Task
                    Req'd        Duration (hrs)
        Inst Tech   1            0.5

Required Parts / Special Tools

        Inventory No.    Description            No. Req'd
        01-3458-25       DO Sensor Membrane     1
        03-3892-73       DO Sensor O-ring       1
        01-3458-25       Sensor Electrolyte     1
```

FIGURE 16.3 An example of a task form.

forecasts the labor hours involved, and this estimate can be compared with the actual time spent on the task to evaluate a maintenance crew's efficiency or the CMMS's accuracy. (Labor, tools, and parts estimations are based on task form information.) The work order also will include a set of instructions that the maintenance crew can take with them in the field. These instructions—basically, a checklist of "things to do"—are particularly effective for training new personnel.

When the task is completed and the work order "closed", the CMMS deletes the used parts from the warehouse inventory and charges the parts and actual labor to the appropriate cost center. So, if a sensor membrane were replaced, the parts inven-

```
Work Order Report

    Work Order Number:    P381031
          Description:    PM on DO Sensor Tank 2

     Work Start Date:     #######         CP Num. :        260
 Work Completion Date:    #######         O&M Doc. :   Vol 6 / Sec 7
          Status Code:    Ready           Spec Section :    13300
          Report Date:    #######

 Equipment No.           Description         BHPID
    131.2                DO Sensor           ST:INS-AE-131.2

 Tech ID(s)   Work Type    Priority    Cost Cntr    Requestor
   INST 2        PM           2        Secondary      Smith

              Labor Est. :    0.5                      Spare Part / Tools
              Labor Act. :    0.75                         [Y/N]      No.


 Task No:    0001
 Job Operations:
     10     Make sure the O-ring has not been damaged
     20     O-ring damaged, replace O-ring                          [Y]   03-3892-73
     30     Replace membrane
     31     Unscrew membrane cap and replace membrane.  Replace cap
     31     and hand tighten.  So not use tools to tighten!  Once cap is tightened,
     32     do not loosen!  Even if membrane cap is partially loosened and then
     33     re-tightened, the sensor will not operate properly.     [Y]   01-3458-25
     40     Once the membrane has been replaced, dry the tip and hold sensor
     41     with tip facing down.  After short period of time, the membrane should
     42     begin to "weep".
     50     If the membrane does not "weep", repace the electrolye. [N]   01-3458-25
```

FIGURE 16.4   An example of a preventive maintenance order.

tory would include one less membrane and O-ring (Figure 16.5). When the inventory of membranes or O-rings drops to the reorder point, the CMMS would generate a purchase requisition for the default order quantity. The system also allows the same type of part to be stored in multiple locations, allows users to decide which location to pull the part from, and allows for multiple suppliers of each type of part.

```
Inventory

Item Number:    01-3458-25
  Description:  DO Sensor Membrane
  Order Point:      5         Default Order Qty:   10      Unit of Measure:   Each
Single Source [Y/N]      N

Current Stock

   Location      Qty on hand      Unit Cost      Extended Cost
 Row 3 Bin 32        7             $10.50           $73.50

Supplier Information (x = primary vendor)

Vendor          Item No.                         Unit Cost     Lead Time (D)
Capital       x SM7863-2                          $10.50            21
ABC Inc         M32F23                            $10.75            15

Extended Specifications
```

FIGURE 16.5 An example of an inventory record.

To help determine breakdown patterns or unusual expenditures, the CMMS maintains a log of purchases that can be rapidly organized and analyzed. For example, the log can be organized by vendor, showing every item that was purchased from each vendor. It also can be grouped by item number, showing each number and its related supplier.

Likewise, the CMMS maintains a log of work orders that users can analyze (Figure 16.6). So, to analyze one piece of equipment, staff would enter the equipment number in the "equipment #" field. All work done on the dissolved oxygen sensor, for example, could be retrieved by querying the work-order history for "dissolved oxygen sensor 131.1". Users also could request all data on a particular type of equipment (e.g., dissolved oxygen meter) and compare the work done on each item.

Work Order History Comprehensive Report		
Starting Date: **1/1/2002**	Ending Date:	**12/31/2002**

Select Work Order History finformation by
  Work Order Number(s):
  Equipment Number(s):   Item Number(s):
  Craft Code: General Leger Number:
  Employee Code: Expense Class Code:
  Vendor Number: Cost Center:
  Work Order Type: Task Number(s):

Include the following information on the report:

  Print Employee labor [Y/N]   [Y]   Print inventory items [Y/N]   [N]
  Print Vendor Labor [Y/N]   [N]   Print task Code: [Y/N]   [Y]
  Print Comments [Y/N]   [N]   Print search Criteria [Y/N]   [N]

  Read History from [H/A/B]   [H] (H-HISTORY, A-Archived, B-Both)
  Archived file(s) on disk drive [C/D/E]   [C]

FIGURE 16.6 An example of a work-order-history query form.

EQNOCOST								
Period								
Start	1/1/2002							
Finish	########							
Equipment Numbers:			131.1	131.2	131.3	131.4		
Equip No.	WO Number	Task Code	Status Date	Workorder Status	Labor Hours	Labor Cost	Material Cost	Total Cost
131.1	P380023	0001	1/15/2003	Comp	0.50	$10.93	$10.50	$21.43
	P380183	0001	4/18/2003	Comp	0.50	$10.93	$10.50	$21.43
	P380183	0002	4/18/2003	Comp	4.50	$98.33	$158.00	$256.33
	C380197	0004	4/19/2003	Comp	2.50	$54.63	$56.75	$111.38
	C380213	0004	4/22/2003	Comp	2.00	$43.70	$56.75	$100.45
	P380456	0001	7/16/2003	Comp	0.50	$10.93	$10.50	$21.43
	P380876	0001	10/16/2003	Comp	0.50	$10.93	$10.50	$21.43
								$553.85
131.2	P380023	0001	1/15/2003	Comp	0.50	$10.93	$10.50	$21.43
	P380183	0002	4/18/2003	Comp	4.50	$98.33	$158.00	$256.33
	P380183	0001	4/18/2003	Comp	0.50	$10.93	$10.50	$21.43
	C380382	0003	6/12/2003	Comp	8.00	$174.80	$1,527.50	$1,702.30
	C380382	0001	6/12/2003	Comp	0.50	$10.93	$10.50	$21.43
	P380798	0001	9/12/2003	Comp	0.50	$10.93	$10.50	$21.43
	P381031	0001	12/12/2003	Comp	0.50	$10.93	$10.50	$21.43
								$2,065.75
131.3	P380023	0001	1/15/2003	Comp	0.50	$10.93	$10.50	$21.43
	P380183	0002	4/18/2003	Comp	4.50	$98.33	$158.00	$256.33
	P380183	0001	4/18/2003	Comp	0.50	$10.93	$10.50	$21.43
	P380456	0001	7/16/2003	Comp	0.50	$10.93	$10.50	$21.43
	P380876	0001	10/16/2003	Comp	0.50	$10.93	$10.50	$21.43
								$342.03
131.4	P380023	0001	1/15/2003	Comp	0.50	$10.93	$10.50	$21.43
	P380183	0002	4/18/2003	Comp	4.50	$98.33	$158.00	$256.33
	P380183	0001	4/18/2003	Comp	0.50	$10.93	$10.50	$21.43
	P380456	0001	7/16/2003	Comp	0.50	$10.93	$10.50	$21.43
	P380876	0001	10/16/2003	Comp	0.50	$10.93	$10.50	$21.43
								$342.03

FIGURE 16.7 An example of a maintenance cost report by equipment number.

Users can also choose which components of the work order history they wish to view (Figure 16.7). Employee labor, vendor labor, work-order comments, inventory, and task descriptions are all considered components of the work-order log.

To simplify data analysis, the CMMS allows users to display data in many types of graphs (Figure 16.8) and quickly change the formats in which the data are presented. Users also can define the time period and variables or data of interest. In a Pareto diagram (Figure 16.9), for example, the variables, problems, or problem areas are listed in descending order, beginning on the left side of the horizontal axis. Keeping in mind the 80/20 rule (i.e., 80% of the problems are caused by 20% of the activities), users would note that Basins 1 and 2 account for 80% of total labor costs associated with dissolved oxygen-related basin maintenance.

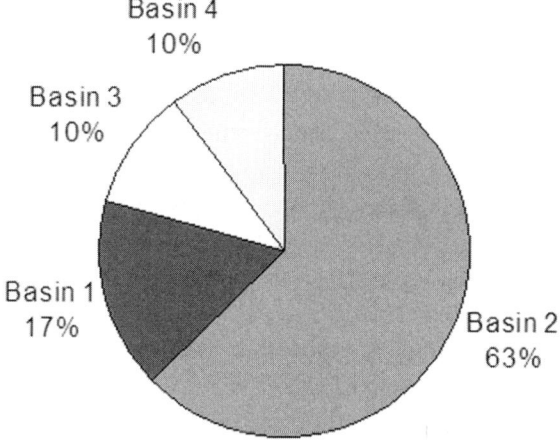

FIGURE 16.8 An example of a pie chart.

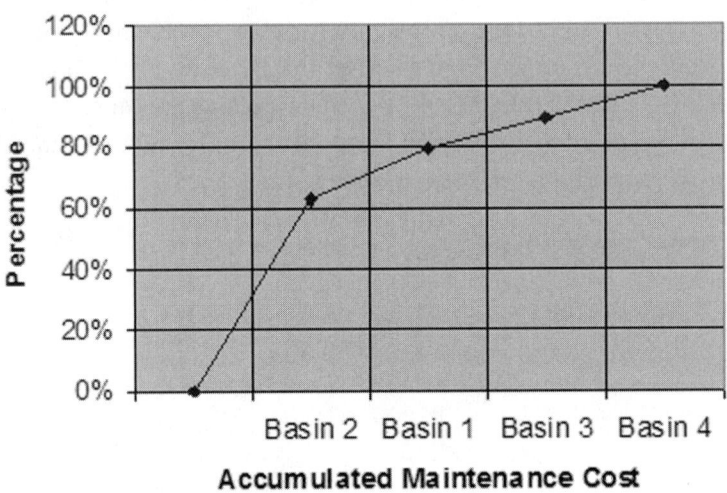

FIGURE 16.9 An example of a Pareto diagram.

VERIFICATION OF INSTRUMENT PERFORMANCE

Ideally, an I&C system should provide process data to help O&M staff run the treatment plant effectively. To ensure that the system is reliable, each instrument must be calibrated (or standardized against a known value) and tested regularly to verify that it performs as expected. Such calibrations and tests are based on statistical QC techniques for determining performance exceptions. A percent-accuracy QC chart should be used to determine when an instrument needs to be recalibrated.

QUALITY CONTROL PROCEDURES. The quality control evaluation of each instrument is based on the average (mean) and range of its data output when measuring a known standard multiple times. In this procedure, the *average* is the sum of all the test results divided by the number of tests. The *range* is the absolute value of the difference between the results of two tests.

The range and average are influenced significantly by concentration. Unfortunately, the concentrations of treatment plant samples tend to vary widely. So, wastewater professionals use modified range and average statistics to compensate (Analytical Quality, 1979).

CONTROL CHARTS. Control charts are plots of an instrument's precision (Shewhart control chart) or accuracy (percent-recovery control chart) over time. They also involve control limits for acceptable data. Basically, both plots should resemble uniformly distributed ("bell-shaped") curves, and approximately 68% of the data should be within one standard deviation of the mean—the midpoint of the curve. The *standard deviation* is a measure of the dispersion of data, equal to the square root of the mean of the squares of the deviations from the mean of the data. If less than 50% of the data are within one standard deviation, the instrument probably needs recalibrating.

The control limits for both precision and percent recovery are instrument-specific. The upper precision control limit is calculated as follows:

$$D4 \times \text{the mean of the precision data} \tag{16.1}$$

where $D4$ is the Shewhart upper-control-limit multiplier (for duplicates, D4 = 3.27). When the database consists of duplicates, "lower warning limits and lower control limits" do not exist (Analytical Quality, 1979).

The percent-recovery warning and control limits are calculated as follows:

$$\text{Upper percent-recovery control limit} = \text{mean} + 3 \text{ standard deviations} \tag{16.2}$$

$$\text{Upper percent-recovery warning limit} = \text{mean} + 2 \text{ standard deviations} \tag{16.3}$$

$$\text{Lower percent-recovery warning limit} = \text{mean} - 2 \text{ standard deviations} \tag{16.4}$$

$$\text{Lower percent-recovery control limit} = \text{mean} - 3 \text{ standard deviations} \tag{16.5}$$

Generally, 99.99% of the data in a percent-recovery chart are within three standard deviations of the mean, and 95.45% of the data are within two standard deviations of the mean. So, data outside the control or warning limits would be unusual.

For both the precision and percent-recovery charts, an *out-of-control situation* is any point or points outside the control limits. Percent-recovery charts also have an out-of-control situation when seven consecutive points are on the same side of the mean.

EXAMPLE. In this example, maintenance employees evaluated a submerged dissolved oxygen electrolytic probe (with temperature correction) that measured dissolved oxygen in an aeration basin. First, they took duplicate basin readings, using a laboratory reference probe that was calibrated via Winkler tests. Then, they used the data from both probes to determine the original probe's precision and accuracy (Table 16.1).

They calculated the precision and accuracy of each probe reading as follows:

$$\text{Precision} = \text{ABS} | (\text{field DO} - \text{lab DO}) | (\text{field DO} + \text{lab DO}) \quad (16.6)$$

On Jun 1, 2002, the probe's precision = ABS | (2.15 − 2.68) | (2.15 + 2.68)
$$= 0.53/4.83$$
$$= 0.110$$

$$\text{Accuracy} = \text{lab probe DO of standard} \times 100/\text{Winkler DO standard value.} \quad (16.7)$$

On Jun 1, 2002, the probe's percent recovery = 7.35 × 100/7.61
$$= 96.6\%$$

The control limits for precision and accuracy were calculated as follows:

$$\text{Shewhart upper control limit} =$$
$$\text{Shewhart factor for duplicates}$$
$$\times \text{ average of precision data} \quad (16.8)$$
$$= 3.27 \times 0.073$$
$$= 0.239$$

$$\text{Standard deviation} = \{[(\text{SUM R}^2) - (\text{SUM R})^2/n]/(n-1)\}^{0.5}$$
$$= [(111670.7) - (1156.3^2)/12)/11]^{0.5}$$
$$= [(111670.7) - (111410.11)/11]^{0.5} \quad (16.9)$$
$$= (23.69)^{0.5}$$
$$= 4.87$$

Upper percent-recovery control limit = (3 × standard deviation) + average
$$= (3 \times 4.87) + 96.4$$
$$= 111.01$$

Upper percent-recovery warning limit = (2 × standard deviation) + average
$$= (2 \times 4.87) + 96.4$$
$$= 106.14$$

TABLE 16.1 Original data with calculated precision and accuracy values.

Date (Month/Day)	Basin probe (mg/L)	Reference Probe (mg/L)	Precision	Ref. DO Probe Std. (mg/L)	Winkler DO Std. (mg/L)	Recovery (R; %)	Recovery (R**2)
6/1	2.15	2.68	0.110	7.35	7.61	96.6	9328.4
6/2	2.91	2.44	0.088	-			
6/5	4.62	4.48	0.015	7.52	7.90	95.2	9061.1
6/6	4.13	3.77	0.046	-			
6/7	3.43	3.40	0.004	-			
6/8	2.86	2.72	0.025	7.20	7.84	91.8	8434.0
6/9	1.98	2.34	0.083	-			
6/12	5.15	5.41	0.025	7.48	7.35	101.8	10356.9
6/13	4.80	4.68	0.013	-			
6/14	3.46	3.71	0.035	-			
6/15	2.64	2.75	0.020	7.43	7.63	97.4	9482.6
6/16	1.55	1.80	0.075	-			
6/19	4.37	4.96	0.063	7.77	7.60	102.2	10452.4
6/20	3.21	3.10	0.017	-			
6/21	2.23	2.48	0.053	-			
6/22	1.89	1.63	0.074	7.81	7.42	105.3	11078.8
6/23	1.80	1.45	0.108	-			
6/26	4.27	5.32	0.109	7.40	7.98	92.7	8599.2
6/27	3.90	4.66	0.089	-			
6/28	3.75	3.55	0.027	-			
6/29	3.54	2.04	0.269	7.28	7.93	91.8	8427.8
6/30	3.38	1.59	0.360	7.45	7.52	99.1	9814.7
7/3	5.04	5.21	0.017	4.30	4.70	91.5	8370.3
7/4	4.25	4.01	0.029	1.00	1.10	90.9	8264.5
Average			0.073			96.4	
Sum						1156.3	111670.7

$$\text{Lower percent-recovery warning limit} = (-2 \times \text{standard deviation}) + \text{average}$$
$$= (-2 \times 4.87) + 96.4$$
$$= 86.66$$

$$\text{Lower percent-recovery control limit} = (-3 \times \text{standard deviation}) + \text{average}$$
$$= (-3 \times 4.87) + 96.4$$
$$= 81.69$$

Maintenance employees then compared the reference probe's readings (Figure 16.10) to Winkler standard results. They found that the relationship is linear and in a 1:1 ratio, indicating that the reference probe can be read directly as milligrams per liter of dissolved oxygen.

When maintenance employees compared the two probes' measurements of dissolved oxygen in the aeration basin (Figure 16.11), they found that the instruments' readings began to diverge after June 23, indicating that more investigation was needed.

When maintenance employees graphed the reference probe's data to confirm its accuracy (Figure 16.12), they found that the data remained within control limits. In other words, this probe was functioning as expected.

However, when maintenance employees graphed the original probe's precision (Figure 16.13), the Shewhart chart for Jun 1 through Jul 13 revealed an out-of-control situation. The probe's precision data exceeded the upper control limit on Jun 29. It should have been removed, serviced, and calibrated then.

IMPLEMENTATION METHODS

Currently, there are four methods for developing a maintenance management program: textbook, modular, build as you go, and turnkey. Each method may or may not include a formal training program.

TEXTBOOK. The textbook method proceeds as follows:

- Identify maintenance department personnel, special skills, and crafts.
- Set up an implementation team, train its members, and assign responsibilities to each member.
- Identify, locate, and define the purpose of all instruments.

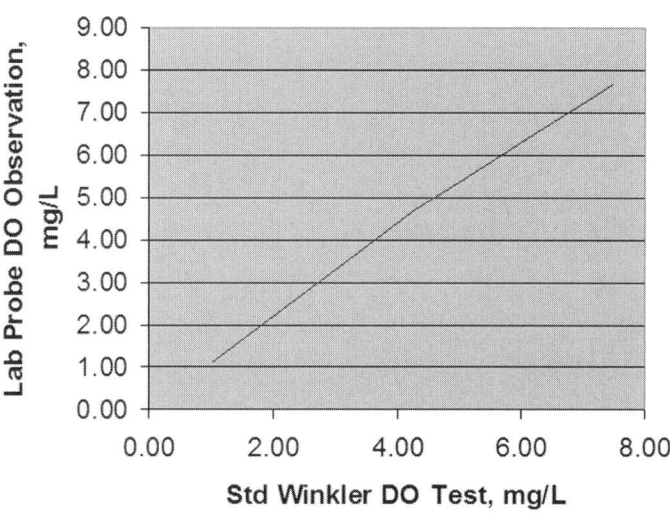

FIGURE 16.10 Calibration curve for laboratory reference DO probe.

- Establish an equipment- or instrument-numbering system.
- Inventory the parts on hand.
- Establish parts-storage and -numbering systems.
- Determine the accounting information (e.g., general ledger, cost center, and expense code) required.
- Define maintenance tasks and their frequency (e.g., based on time lapsed or meter reading).
- Determine the tools, instructions, and safety requirements for each maintenance task.

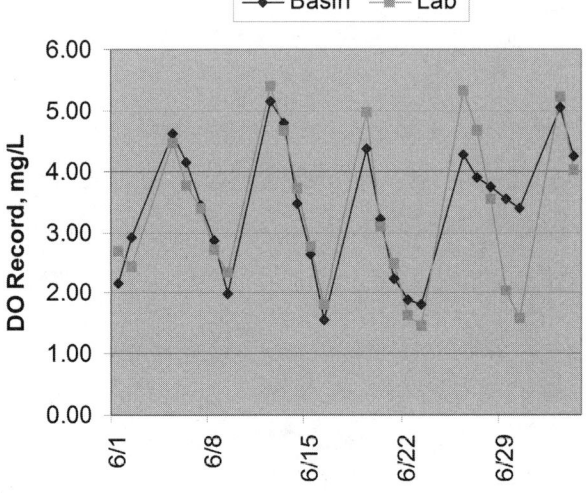

FIGURE 16.11 Graph comparing field and reference DO probes.

- Identify parts and equipment manufacturers and vendors available.
- Convert the equipment and task data into work orders.
- Issue work orders, evaluate process, and modify as needed.

This method is organized and comprehensive. As a result of being involved in this program, employees tend to be committed and well-trained and have better morale. However, this method is used less often than others because it requires maximum resources and the related data-collection and -entry effort is time-consuming.

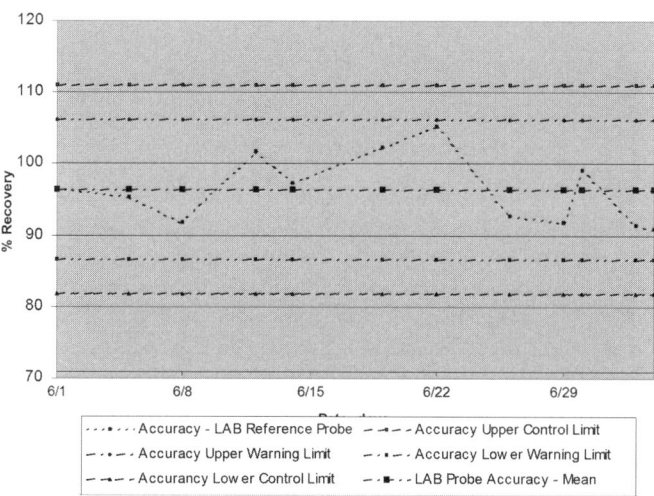

FIGURE 16.12   Percent-recovery control chart for laboratory reference probe.

MODULAR. Traditionally, the modular method refers to two options: the "task–work order" option and the "functional module" option. The "task–work order" option focuses on identifying maintenance tasks and issuing related work orders. Only after the "task–work order" system is established do employees turn their attention to parts inventories, vendors, accounting codes, personnel entries, and other maintenance program components. These components tend to grow organically until all functions are in place.

The "functional module" option starts by establishing a complete maintenance management program for a small, easily isolated area, such as the aeration-blower instrumentation and associated sensors. Then, as feedback demonstrates success, a similar program is developed for another small, easily isolated area, and another,

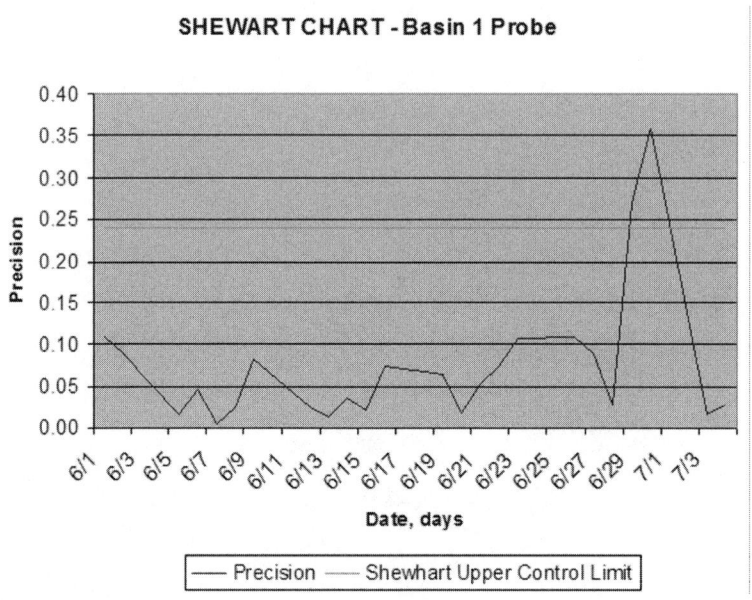

FIGURE 16.13 Shewhart control chart for aeration basin #1 DO probe.

until the entire plant is addressed and the small programs are merged into one comprehensive plant maintenance program.

BUILD AS YOU GO. The "build as you go" method basically originates with the paperwork of any maintenance activity or work order that has been filed under the appropriate equipment. It typically begins with corrective maintenance, and as the maintenance history grows and task frequency requirements are better understood, staff begin converting the corrective-maintenance activities into a preventive-maintenance program.

TURNKEY. In the turnkey method, the utility hires another organization for maintenance program setup, startup, and training. This method initially makes few demands on staff, but it is expensive and the transition period can be challenging for staff.

REFERENCE

Plant Maintenance...A Direct, Financial Impact on Profitability and Competitiveness (1987) *Practical Lubrication and Maintenance*, Sep 13.

SUGGESTED READINGS

Blackwell, L. G. (1988) Application of Computerized QC Charts and Reports for Wastewater Facilities. Paper presented at Water Pollution Control Federation Specialty Conference on Analytical Techniques and Residuals Management in Water Pollution Control, Atlanta, Georgia.

Bock, E. (2003) Advanced Options for Asset Management Software Enhance Maintenance Practices Further Than Ever Before. Presented at Instrumentation, Systems and Automation Society Expo 2003, Houston, Texas, Oct 21–23.

Cable, M. (2004) *Calibration: A Technicians Guide*. Instrumentation, Systems, and Automation Society: Research Triangle Park, North Carolina.

Dewey, A. R. (2003) Maintenance Foundation Fieldbus Networks—What Tools Will I Need? Paper presented at Instrumentation, Systems, and Automation Society Expo 2003, Houston, Texas, Oct 21–23.

Goettshe, L. D. (1995) *Maintenance of Instruments and Systems*. Instrumentation, Systems, and Automation Society: Research Triangle Park, North Carolina.

Jordan, J. K. (2000) *Maintenance Management for Water Utilities*. American Water Works Association: Denver, Colorado.

Patton, J. D., Jr. (2004) *Preventative Maintenance*, 3rd Ed.; Instrumentation, Systems, and Automation Society: Research Triangle Park, North Carolina.

Chapter 17

Instrumentation Troubleshooting

The Importance of Details	510	*Develop a Theory*	516
Useful References	511	*Test the Theory*	516
Troubleshooting Strategy	511	*Implement Changes*	516
Define the Problem	512	*Prepare a Report*	516
Observe the System Closely	512	Example 2	516
Isolate the Problem Area	512	*Define the Problem*	517
Develop a Theory	513	*Observe the System*	517
Test the Theory	513	*Isolate the Problem Area*	518
Implement Changes	513	*Develop a Theory*	518
Prepare a Report	513	*Test the Theory*	519
Test Cases	513	*Implement Changes*	521
Example 1	514	*Prepare a Report*	521
Define the Problem	514	Troubleshooting Equipment	521
Observe the System	514	Safety	523
Isolate the Problem Area	514	Suggested Readings	523

Troubleshooting is any systematic approach to problem solving. Troubleshooting methods follow an established line of inquiry for any problem, regardless of the discipline.

This chapter explores one approach to organizing standard troubleshooting techniques. Approaches will vary depending on the troubleshooter's background, experience, and understanding of the system under investigation.

THE IMPORTANCE OF DETAILS

Sometimes different component failures have similar symptoms, but a trained troubleshooter can detect the subtle differences and resolve the problem. A valuable troubleshooting resource is the operator on duty when the problem was first noted. This operator may be aware of events that could help a troubleshooter identify the source of the trouble.

Unfortunately, operators are not trained observers. Some may only note malfunctions and provide little detail:

- The pump did not start.
- The scum troughs did not rotate.
- The motor-operated valve controlling the centrifuge feed did not open.

These examples only note the events that did not occur. When a control system fails, however, related events that could reveal the troublemaker should also be noted:

- The pump did not start, but the grinder did.
- The scum troughs did not rotate, but the water spray system did come on.
- The motor-operated valve controlling the centrifuge feed did not open, but the bypass valve opened.

The additional information in these statements enables the troubleshooter to define the problem more clearly.

Sometimes, the operator knows these details but does not consider them important. So, the troubleshooter should ask leading questions that would redirect the operator's attention to other elements of the control system.

Sometimes the problem is not one component failing, but a chain reaction of component failures, and repairing one part might not necessarily resolve the problem or prevent it from happening again. In these situations, troubleshooters should investigate a cause-and-effect relationship. If they find that the failure of component "A" causes the destruction of component "B", for example, staff could try to modify the system to sever this relationship.

USEFUL REFERENCES

Contract design drawings, shop drawings (working "as-built" drawings), and operations and maintenance (O&M) manuals are valuable tools for troubleshooters. Contract drawings show how the designer intended the instrumentation and control (I&C) system to be installed, while shop drawings—if consistently updated—will show the changes that were made during and after construction and reflect the working system more accurately. Troubleshooters should verify actual conditions, though, when investigating I&C problems.

Operations and maintenance manuals provide troubleshooters with a concise collection of shop drawings, product literature, and, at times, brief descriptions of I&C system operations. These manuals typically are the best available reference materials.

The vendor's literature on an instrument or system is another important reference. This literature will answer questions about input/output specifications, proper placement, mounting, and component identification. Calling the vendor and talking to an applications engineer also can be helpful. Sometimes more or updated information is available.

When newly installed or calibrated instrumentation does not provide the anticipated measurements, troubleshooters should consult the historical log (e.g., daily site or process inspections or laboratory reports). The information may not be accurate enough to calibrate an instrument, but approximate figures and trend data can help track an instrument's performance over time.

TROUBLESHOOTING STRATEGY

The following systematic approach to troubleshooting is intended to direct the troubleshooter's attention to the sources of problems in simple I&C systems. For large, complicated I&C systems, these techniques may help the troubleshooter focus on a smaller, manageable subsystem that could cause general system failure.

The purpose of a strategy is to sort through the collected information, suggested explanations, and proposed solutions and solve the problem. Successful troubleshooters use various problem-solving strategies. One practical method is

- Define the problem,
- Observe the system,
- Isolate the problem area,
- Develop a theory,
- Test the theory,
- Implement changes, and
- Prepare a report.

DEFINE THE PROBLEM. Defining the problem is often more difficult than it sounds because problems are frequently diagnosed before the instrument or system is inspected. Available documentation will show how the I&C system was designed or what it was intended to do. Shop drawings from O&M manuals or construction files and records can help determine the scope of the system. Specific installation details, unless immediately available, are not important yet.

OBSERVE THE SYSTEM CLOSELY. Up to this point, all input comes from other sources, with varying degrees of reliability. Now, the troubleshooter should carefully observe what the system is or is not doing.

ISOLATE THE PROBLEM AREA. All relevant information has been found and reviewed. The monitored process is understood. The equipment (e.g., sensors, transmitters, transducers, indicators, and controllers) and its interconnections (wire runs and terminations) have been studied. By this point, the troubleshooter is familiar enough with the system and its components to start eliminating possible causes of the problem and narrowing the area where the potential source could be situated. Sometimes this step involves more trips into the field, depending on the troubleshooter's experience, skill, and luck. The process of elimination continues until the subsystem or device at fault has been determined.

DEVELOP A THEORY. The troubleshooter now develops a theory based on system expectations, actual output, and the corrective action(s) needed to return the I&C system to full operation. The corrective action(s) should not be done, however, until the troubleshooter has confirmed the theory.

TEST THE THEORY. With knowledge of the I&C system and theories about the problem, the troubleshooter can pinpoint the trouble by devising tests of each potential problem area and seeing whether the system reacts as predicted. Usually, each theory can be tested and the results will either confirm the theory or indicate that the problem may be elsewhere.

If the test fails, the troubleshooter should re-evaluate all the information and develop new theories and tests. A failed test provides valuable information for potential solutions.

When working with discrete devices, the only practical way to test the device may be to compare it with a similar unit (i.e., exchange it for one known to work). This method is effective, but troubleshooters should avoid damaging the "tester" device in the process. Also, all wires that are being disconnected from the instruments should first be labeled to avoid confusion.

IMPLEMENT CHANGES. If a test confirms a proposed theory, the troubleshooter should make the appropriate changes and recommendations.

PREPARE A REPORT. Each solved problem should be documented to help the troubleshooter and his or her colleagues solve similar problems in the future. The report can be based on the troubleshooter's field notes.

Accurate notes are invaluable. They serve as refreshers if work is interrupted before completion, organizers when working on concurrent projects, and historical records of past problems and solutions. Personal records can also be used as accurate tools for estimating breakdown (failure) trends, confirming whether established preventive-maintenance practices should be modified or replaced, or re-evaluating existing preventive-maintenance schedules so time can be spent more efficiently.

TEST CASES

The following two troubleshooting examples were derived from actual field problems that have been solved.

EXAMPLE 1. An I&C system has been installed at a large wastewater treatment facility to monitor the temperature in the plant's digestion tanks (Figure 17.1). The following situation occurs:

- The temperature transmitter produces a 0-mA signal.

- When the transmitter is replaced with an identical unit (known to be working) from another digestion tank's circuit, the signal remains 0 mA.

- When the remote thermocouple device (RTD) is replaced with an identical RTD (known to be working) from another digestion tank's circuit, the transmitter signal remains 0 mA.

- Operators use an ohmmeter to check the continuity of the wires connecting the RTD to the temperature transmitter, and resistance checks verify that the wiring in the problem tank is in good condition.

- When the problem tank's temperature transmitter is connected to another digestion tank's RTD, an accurate signal is produced.

The instrument technician assigned to this problem has examined it from all possible sides and cannot identify its source. Using the strategy outlined above, a troubleshooter finally solved the problem:

- *Define the Problem.* The temperature transmitter emits a 0-mA signal. Available documentation on the resistance-to-temperature curve for the RTD indicates that the temperature transmitter is a generic resistance-to-current signal converter. Its output ranges from 0 to 50 mA, but it has been calibrated to output between 4 and 20 mA for resistance within the system's operational range.

- *Observe the System.* The temperature transmitter is outputting 0 mA. The RTD's measured resistance is 87.5 ohms. According to the resistance-to-temperature curve, the actual temperature is 17 °C (62 °F). The temperature transmitter's calibrated range is 32 to 49 °C (90 to 120 °F).

- *Isolate the Problem Area.* The transmitter, probe, and wiring seem to be functional. To isolate the problem area, the troubleshooter asked: "What should the system produce?" The digester's actual temperature is 17 °C (62 °F), but the transmitter is not programmed to register temperatures below

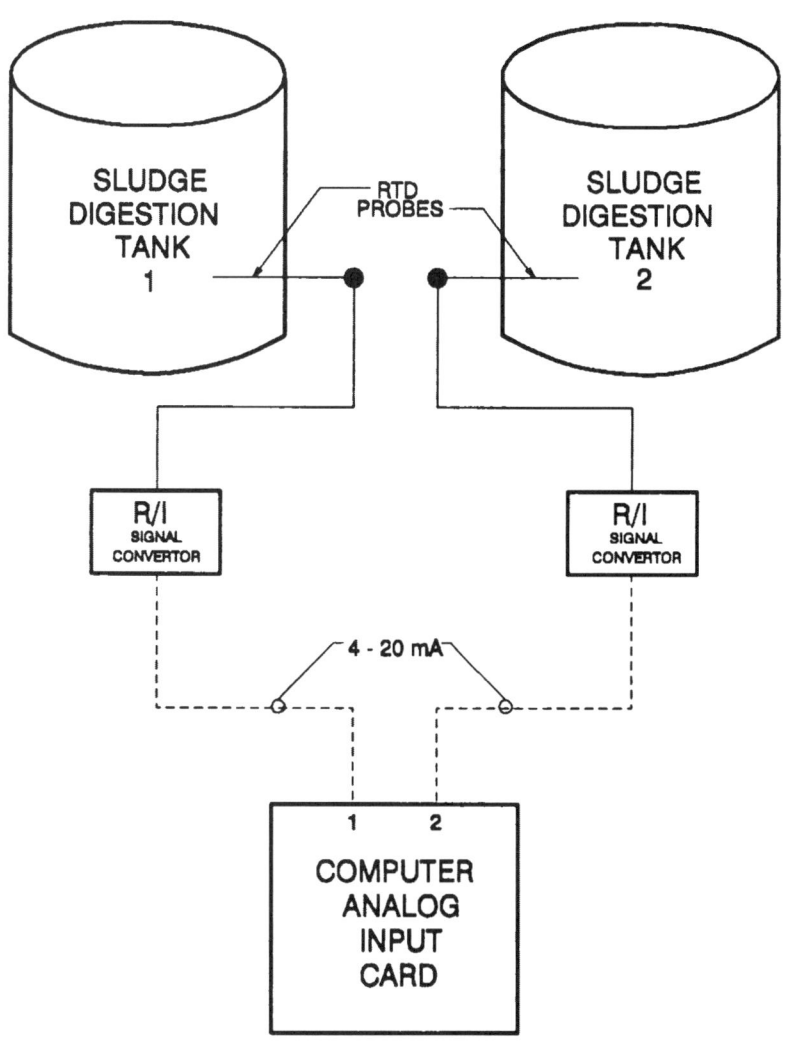

FIGURE 17.1 Schematic wiring diagram for the digestion tank temperature circuit.

32 °C (90 °F). So, the problem is the I&C system's expectation: It assumes that the transmitter will not go below 4 mA, so a 0-mA signal is interpreted as an equipment failure.

- ***Develop a Theory.*** The digester's temperature is below normal.
- ***Test the Theory.*** Human body temperature [approximately 22.8 °C (98.6 °F)] is within the RTD's range, so the troubleshooter removes the RTD from its thermowell and grasps it with a bare hand. The transmitter's output current rises above 4 mA.
- ***Implement Changes.*** The control system functions correctly and does not require any modifications. The actual problem is a clogged heat exchanger that is failing to keep the sludge warm. The only work necessary is to flush the heat exchanger and restart the digestion process, because the treatment bacteria are dead.
- ***Prepare a Report.*** The report is prepared to provide information for future troubleshooting.

Though this example seems simple, an experienced technician spent more than seven days working on the problem without finding a solution.

EXAMPLE 2. The sludge-withdrawal system for four final sedimentation tanks at a large wastewater treatment facility is not performing adequately. Operators have been unable to control the volume of sludge withdrawn from each tank. During the initial investigation, troubleshooters found that

- Sludge withdrawal is controlled by the elevation of the sedimentation tanks' telescopic valves.
- Withdrawn sludge flows over the telescopic valve into a sludge overflow box and then flows by gravity through a common header to the suction side of the return sludge pumps.
- The tank farthest from the return sludge pumps has three of its four inlet gates closed because the sludge-withdrawal system cannot remove enough sludge.
- Other problems associated with the return sludge pumping, waste sludge pumping, and aeration-tank return distribution systems are apparent.

(This example is limited to the sludge-withdrawal system. The other problem systems are listed to show the magnitude of the problem that employees face in this process area. With larger problems, it is difficult to restrict attention to one manageable area.)

Again, using the strategy outlined above, troubleshooters solved the problem:

- *Define the Problem.* All of the final sedimentation tanks influence the amount of sludge withdrawn from any of them; the rate of sludge withdrawal from each tank cannot be controlled.

- *Observe the System.* More sludge is withdrawn from the tanks closer to the return sludge pumps than from those farther away. Troubleshooters also noted that each telescopic valve is submerged (i.e., the level of sludge in the overflow box is higher than that of the telescopic valve; Figures 17.2 and 17.3).

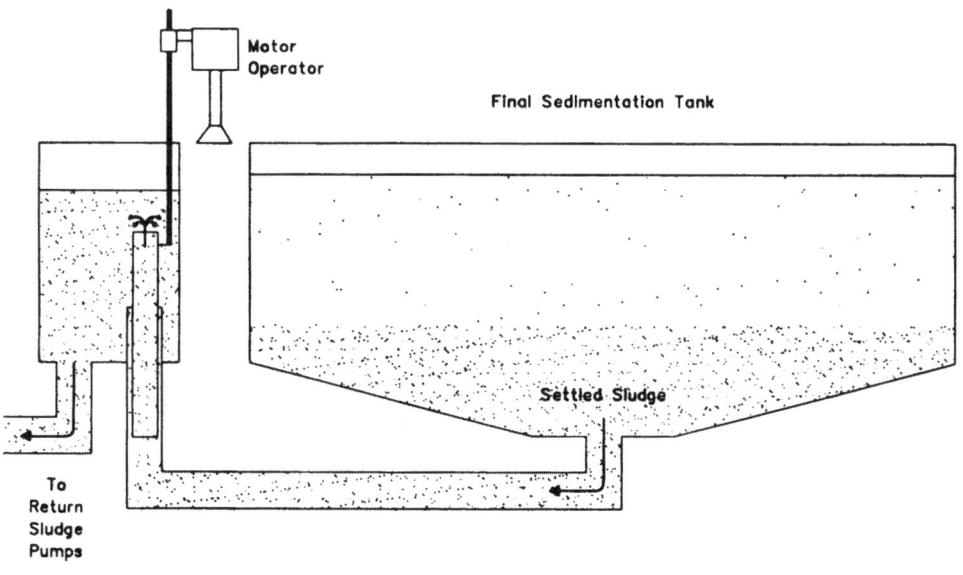

FIGURE 17.2 Final sedimentation tank sludge-withdrawal system (with submerged telescopic valve).

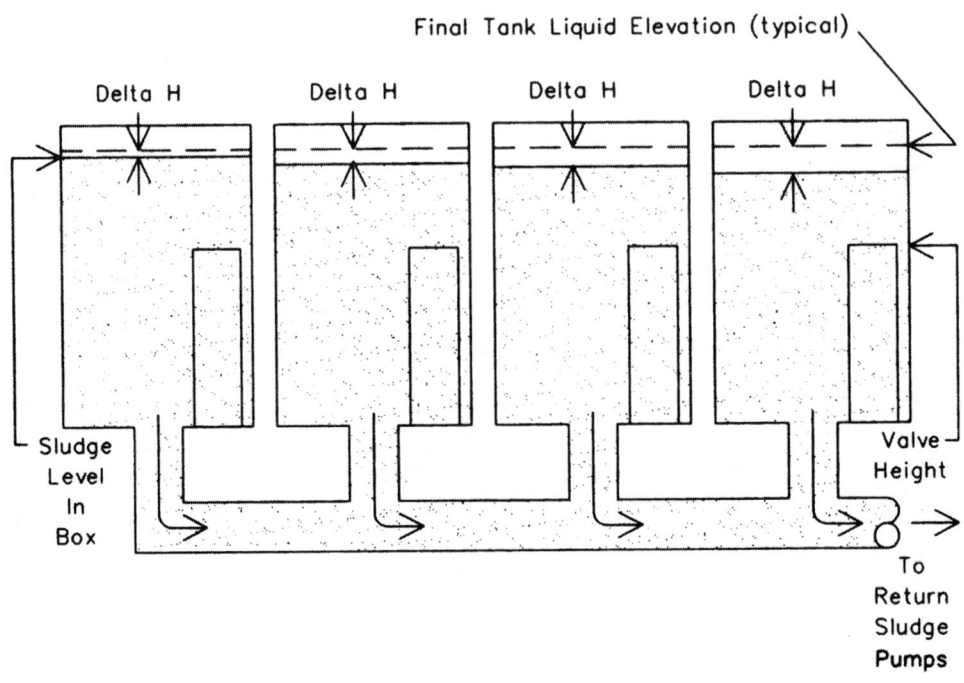

FIGURE 17.3 Comparison of liquid levels in sludge-withdrawal boxes (for submerged telescopic valve).

- *Isolate the Problem Area.* The problem is the telescopic valves.
- *Develop a Theory.* Troubleshooters discovered that lowering the telescopic valves allows more sludge to be withdrawn. They also discovered that system hydraulics limit the amount of sludge withdrawn from each tank and this amount is related to the tank's distance from the return sludge pumps. They developed the theory that the only way lowering the telescopic valve can increase sludge flow is if the sludge level in the overflow box is below that of the telescopic valve (Figure 17.4).

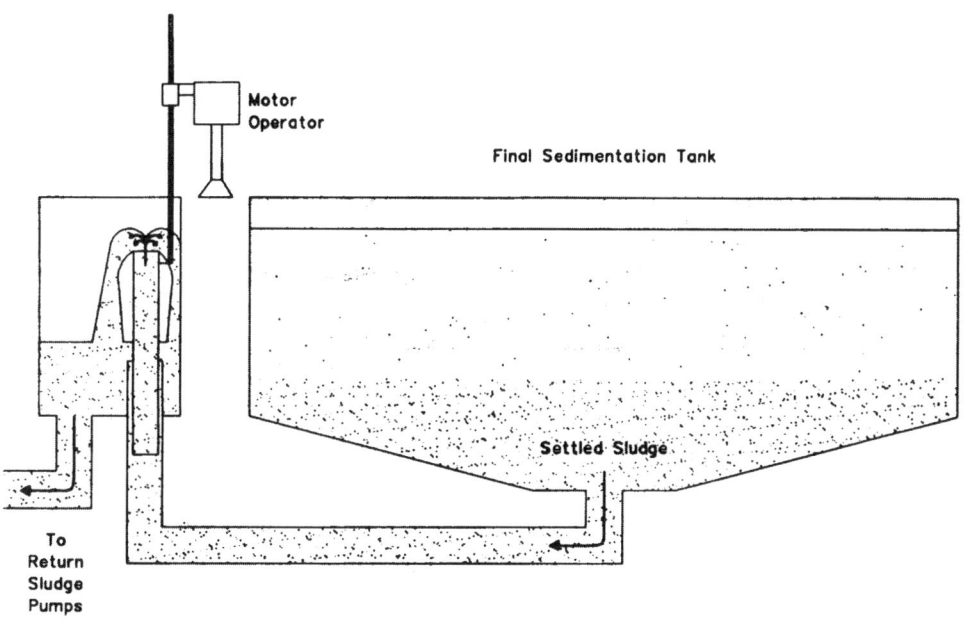

FIGURE 17.4 Final sedimentation tank sludge-withdrawal system (with normal telescopic valve).

- *Test the Theory.* Because testing this theory would upset the system's normal operations, the troubleshooters sought verification from a sedimentation system expert. The expert explained that the force withdrawing sludge from the final sedimentation tank is the difference in head pressure (H) at the telescopic valve and the sludge overflow box. When the valve is above the liquid level in the overflow box, the difference in head is that between the elevations of the valve and the liquid in the tank, and the valve's position will regulate flow (Figure 17.5). When the valve is submerged, however, the difference in H is that between the elevations of the liquid in the tank and that in the overflow box (ignoring liquid densities). Once the telescopic valve is below the liquid

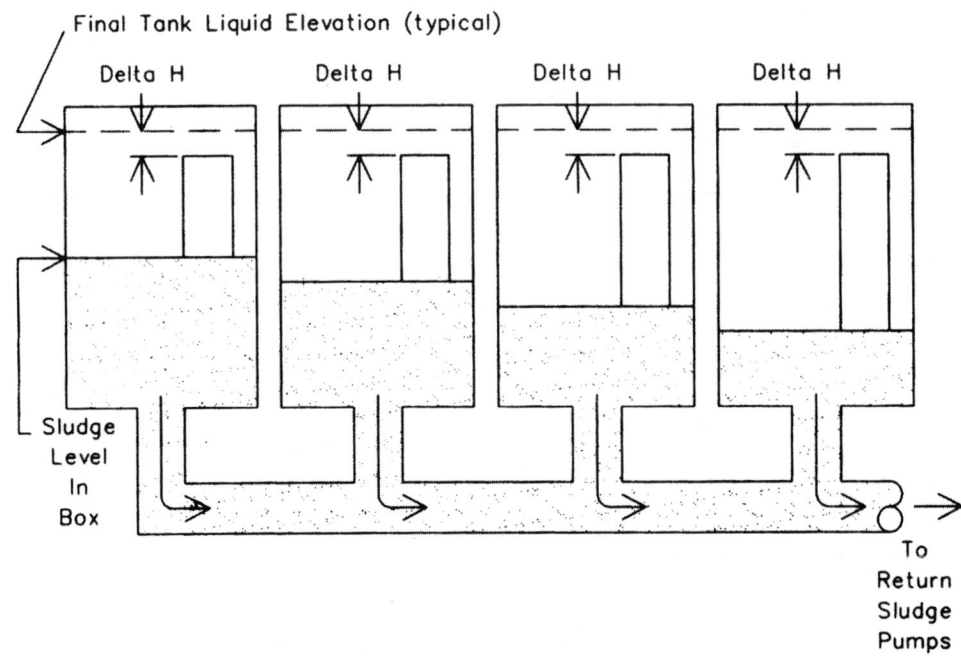

FIGURE 17.5 Comparison of liquid levels in sludge-withdrawal boxes (for normal telescopic valve).

level in the overflow box, lowering it further is meaningless (Figure 17.3). So, to solve the problem, the troubleshooters recommend the following steps:

1. Stop the return sludge pumps;
2. Raise the telescopic valves to their highest (off) positions;
3. Start the return sludge pumps;
4. Nearly empty the sludge overflow boxes;
5. Lower each telescopic valve until flow starts;

6. Adjust the telescopic valves so the flow from each tank is equal (if an equal distribution is desired) and enough sludge is being pumped; and

7. Tune the automatic controls (not mentioned previously) to keep the system operating properly.

- ***Implement Changes.*** The troubleshooters' recommended procedure is implemented, and the system changes prove the theory correct. The sludge blanket level in the final sedimentation tank farthest from the return sludge pump is now controllable; all influent gates are opened.

- ***Prepare a Report.*** The actual report covers the entire system, not just this problem. Although well-designed, the system has been operated incorrectly because the construction and operations staffs were unaware of how the final sedimentation process was designed. The control system includes two ultrasonic level detectors installed at the closest and farthest sludge overflow boxes. If the level in these boxes is not important, why is the return sludge pump's speed controlled by one of these level detectors?

This example illustrates the importance of asking questions. Sometimes it is more important to know how to ask questions and find answers than to possess actual knowledge, which may be limited by experience.

TROUBLESHOOTING EQUIPMENT

All field engineers and technicians need an assortment of tools and test equipment. A properly equipped toolbox saves time and money because projects can be completed faster and more efficiently.

A general-purpose field toolkit should include the following items:

- An industrial-quality multimeter whose test leads have a variety of replaceable tips;

- An adjustable analog current–voltage loop signal simulator;

- Several jumper leads, with shrouded alligator clips;

- Assorted crimp terminals, with crimping tool;

- Assorted fuses, with fuse puller;

- Several precision (1%) resistors;

- A small industrial-quality flashlight, with spare batteries;
- Three types of screwdrivers: flat-blade, Phillips™, and jeweler's;
- Two types of pliers: lineman's and needle-nose, with insulated handles;
- Two or three plastic hemostats;
- A 0.15-m (6-in.) crescent wrench;
- A miniature ratchet socket set, with extensions;
- Diagonal wire cutters, with insulated handles;
- A hex key set; and
- A metal inspection mirror on a swiveling handle.

Depending on the tasks, projects, and other responsibilities assigned, field engineers and technicians also may need

- A dual-trace portable oscilloscope;
- A manually operated pressure–vacuum tester;
- A clamp-on current transformer, suitable for use with the multimeter;
- Soldering/de-soldering equipment; and
- Portable radios.

The above list of tools may be augmented based on each individual's needs, ideas, and preferences.

When assembling a general-purpose field tool kit, field engineers and technicians should remember that it must be carried around, sometimes in awkward places. One possible carrying case is a molded plastic "attaché case" that can be fitted with trays for tool storage; has space for drawings, notebooks, and other papers; is lighter and easier to carry than a conventional metal box; and tolerates considerable abuse.

Field engineers and technicians also should remember the following three rules for properly maintaining tools:

- Keep tools clean. A wastewater treatment plant's corrosive environment can destroy equipment quickly.
- Put tools back where they belong so they can be found the next time they are needed.

- Use the appropriate tool for the job.

SAFETY

Because troubleshooters sometimes must enter areas outside the guard railings and doors, they should always wear hard hats and safety shoes or boots in a treatment plant, even if they are not required by regulations. Safety goggles, earplugs or muffs, or dust masks may also be needed. If the work involves entering a manhole or meter vault, troubleshooters should first check the area for hazardous gasses and oxygen deficiency. Then, they should wear a safety harness and, if necessary, a respirator while in the confined space.

The importance of following established safety rules cannot be overemphasized. Following these four basic rules can greatly reduce the risk of injury:

- Work at an established pace. Rushing reduces the ability to anticipate safety problems.

- When possible, work in pairs.

- Inform the shift supervisor and operator on duty of what work is being done, where, and which equipment is involved.

- Tag equipment out of service when necessary, especially at switchgears, motor control centers, operating control stations, and the apparatus itself. If formal warning tags are not available, makeshift substitutes can be used. When power equipment is involved, a lockout should be used. The combination padlock is best, because keys can be lost or misplaced.

SUGGESTED READINGS

Huth, D. A.; DiMartino, A. P. (2003) Remote Trouble Shooting the DCS with Real-Time Data. Paper presented at the 13th Annual Joint ISA POWID/EPRI Controls and Instrumentation Conference, Jun 15–19.

Mostia, W. L. (2000) *Troubleshooting: A Technician's Guide*. Instrumentation, Systems, and Automation Society: Research Triangle Park, North Carolina.

Shinskey, F. G. (2002) *Process Control Diagnostics*. Instrumentation, Systems, and Automation Society: Research Triangle Park, North Carolina.

Chapter 18

Instrumentation Training

Training Is a Necessity	526	Onsite Classrooms	528
Training Program Elements	526	Offsite Classrooms	528
Training-Course Classifications	527	Conferences and Web Sites	529
Operations	527	Sources of Onsite Instructors	529
Maintenance	527	Sources of Onsite Training Materials	530
Configuration	527		
Programming	527	Training Materials	530
Management Overview	528	Training Program Implementation	531
Class Types	528		
On the Job	528	Training Continuity	531
Home Study	528	Example	532
		Suggested Readings	535

525

TRAINING IS A NECESSITY

A well-trained operations and maintenance (O&M) staff is the key to a successful instrumentation and control (I&C) system, regardless of its magnitude or complexity. An appropriately planned, progressive training program is necessary for all plant personnel.

Training helps raise employee morale; the better employees understand the equipment, the more effectively they can use it. Training also should explain the need for an I&C system, its significance in plant operations, its functions, and its maintenance procedures so everyone understands the importance of having the system function properly.

The training program should be tailored to plant personnel's abilities, prior experience, levels of education, and aptitude. Also, to help personnel better understand and retain the information, trainers should use multiple types of media (e.g., written manuals, videotapes, movies, and spoken presentations) and hands-on activities whenever possible.

Training is a continual process. Staff turnover, new equipment, and new technology mean there is always something new to learn and someone new to learn it. This is especially true of the I&C industry, where technology changes are rapid.

TRAINING PROGRAM ELEMENTS

A successful training program starts with clearly defined goals. These goals should be based on the staff, funds, and specific I&C system involved. One goal should be improving employee morale. Other goals may be as basic as reducing operating costs or as specific as optimizing a single unit operation through tighter control of variables.

The second step is to determine who will direct the I&C training program. The director could be the plant superintendent, the control system engineer, a senior operator, or the plant engineer, so long as that person understands the need for training, the plant's goals, and the budget available.

The training director's first task is to understand the scope and purpose of the plant's I&C system and determine who will be involved with it. Then the director should survey plant personnel to determine each person's expertise and related coursework history. The director then should classify all onsite I&C equipment, compare existing personnel skills with I&C needs, and determine how much training will be needed for the system to be operated and maintained properly.

Next, the training director should survey the training courses available from equipment manufacturers, consultants, professional organizations, and local educational institutions and note the related costs. Once needs, budget, and course availability are compared, the director can establish a plan, with priorities to meet the training program's goals.

After the first pass through this process, training directors often feel that the needs outstrip the time and money available. Experience has shown, however, that perseverance can lead to success. Further investigation will reveal many sources, including colleagues, local educators, equipment vendors, and neighboring-plant personnel, who can provide training or training materials for little or no cost.

Then, directors should set up an onsite training room, collect the requisite training materials, schedule instructors, and notify personnel of the training schedules. Training directors should take care to schedule classes so everyone can attend and the plant continues to operate properly. Sometimes a class may need to be repeated to meet this goal.

The final step before beginning the program is to reconfirm management support. Successful I&C training programs depend on both proper planning and management support. Training directors should only do the research necessary to put together a workable plan if managers provide the funds and time to support it. Training programs that are properly funded and implemented produce good plant performance and lower operating costs.

TRAINING-COURSE CLASSIFICATIONS

Training courses are classified according to the student or students to be trained. Such classifications may include the following:

- **OPERATIONS.** These classes are designed to teach operators how to operate I&C equipment and use it in plant operations,

- **MAINTENANCE.** These classes are designed to teach maintenance personnel how to repair and maintain I&C system components,

- **CONFIGURATION.** These classes are designed to teach control-system and plant engineers how to modify a digital control system,

- **PROGRAMMING.** These classes are designed to teach the appropriate personnel how to modify programs in a digital I&C system, and

- **MANAGEMENT OVERVIEW.** Useful when dealing with large systems, these classes are designed to show managers what the I&C system can do without getting into O&M details.

Directors should not try to "short cut" the process by combining courses; this method is rarely effective. Instructors generally plan their classes according to abilities, backgrounds, and interests of their students. Faced with a broad range of students, an instructor would have to aim for the "average", so some participants would be bored and others would be confused. Training is more effective if courses match the student's needs and abilities—even if more classes must be scheduled.

CLASS TYPES

Directors have several training options, including onsite training, offsite training, and home study, to consider.

ON THE JOB. On-the-job training includes reviewing O&M manuals, equipment-instruction manuals, system-control diagrams [e.g., ladder or process and instrumentation diagrams (P&ID)], and design specifications. This training is effective but costly, because personnel learn from their mistakes, and mistakes cost money.

HOME STUDY. Home study includes any number of self-taught study programs available through local educational institutions, professional and operator societies [e.g., the Water Environment Federation (WEF)], and state or federal agencies.

ONSITE CLASSROOMS. Whether during or after working hours, in-plant classroom training can quickly and efficiently educate groups of employees. It can be used to address a variety of subjects.

OFFSITE CLASSROOMS. Educational institutions, professional organizations, and government agencies offer useful training, including seminars and short courses. However, this is a more expensive option because the courses often involve fees as well as employee travel expenses.

Most major I&C equipment manufacturers offer training on various aspects of using and maintaining their equipment. These courses, held at the factory and selected locations, are typically effective but product-specific. The classes can be held onsite if a plant has a sufficient number of students to be trained. This option also involves a fee and expenses.

CONFERENCES AND WEB SITES. Conferences sponsored by professional societies [e.g., WEF, the American Water Works Association (AWWA), the Instrumentation Testing Association (ITA), and the Instrumentation, Systems, and Automation (ISA) Society's Wastewater Division] often include courses, seminars, and workshops on wastewater-related I&C equipment. Again, this option involves fees and expenses.

Professional societies' Web sites also are a good source of textbooks, training tapes, online classes, and conference or seminar calendars.

SOURCES OF ONSITE INSTRUCTORS

There are many sources of instructors for onsite classes. Keeping in mind that employees frequently learn more from outside instructors than from colleagues, training directors should tap several sources to broaden the scope of offerings and obtain new ideas. Below is a partial list of sources.

- Staff who have become experts on various aspects of the plant I&C system can be the backbone of any training program. This includes training directors who have become knowledgeable as a result of their activities.

- Equipment manufacturers' sales representatives and technicians may provide basic operations training, and their service personnel may provide maintenance training. Factory-based personnel are also available for more specific and extensive needs.

- Design and consulting engineers sometimes offer O&M training programs as part of their services. They can teach classes on system operations and on using the I&C system to optimize and fine tune treatment processes.

- Nearby plants with similar equipment may have instructors available. An instructor-exchange program with nearby facilities can promote the sharing of different ideas and approaches.

- Instructors from nearby schools may teach courses on instrumentation or instrumentation maintenance.

- Federal and state agencies may have staff available for specific training classes.

SOURCES OF ONSITE TRAINING MATERIALS

Training materials are available from many sources, and good materials make each class more valuable to students. Below is a partial list of sources.

- The plant library houses O&M manuals containing the plant's P&IDs, ladder diagrams, and equipment-instruction manuals, all of which are primary source materials for any training program.
- Original construction specifications describe the designer's intended system and how it is supposed to work.
- The designer often has details about the concepts and considerations involved in the system design.
- Manufacturers have instructions on using and maintaining their equipment that can be downloaded from their Web sites or ordered at minimal cost (if the plant library does not already have copies). Many offer other training materials (e.g., videotapes), particularly if they have a training department.
- Neighboring treatment plants may have useful training materials to share or be aware of other useful sources.
- Professional organizations (e.g., WEF, AWWA, and ISA) offer a number of useful training materials (e.g., videotapes, manuals, and self-help courses).
- Government agencies (e.g., the U.S. Environmental Protection Agency) have manuals and material available for use in training programs.

TRAINING MATERIALS

Training materials come in a variety of forms, and directors should consider and use as many as possible to hold the students' interest and maximize retention. Below is a partial list of materials to consider.

Self-help courses are especially useful for basic training. Students should be monitored to ensure that they complete the course without outside help. These materials can be reused but should be reviewed periodically to make sure they have not become outdated.

Audio-visual materials (e.g., movies, slide shows, and videotapes) are available from professional organizations, government agencies, and equipment manufacturers. Training directors should make sure they have suitable facilities to use this material effectively. The necessary audio-visual equipment can be borrowed, rented, or purchased, depending on the plant's budget and anticipated needs.

Written materials (e.g., instruction bulletins, O&M manuals, P&IDs, ladder diagrams, specifications, self-help course materials, and factory training materials) become part of every training director's library. Properly organized, catalogued, and copied, they can be reused for years at minimal cost.

TRAINING PROGRAM IMPLEMENTATION

Training should begin immediately—even before the plant is started up. The sooner training begins, the sooner the utility will reap the benefits (e.g., better operations, lower costs, and high employee morale).

Once implemented, an I&C training program continues forever. It can be toned down, spread out, or shortened, but never stopped because there is always something new to learn. Course repetition helps employees retain information and reiterates the I&C system's purpose and needs. It also often generates more detailed questions and answers that can be incorporated to the training curriculum. In addition, new employees can be fully trained with the existing, updated program.

Again, training directors should avoid creating one course intended to address everything and everybody at once. It is more effective to separate classes into various disciplines (e.g., operations, maintenance, configuration, and programming).

Managers only need overview training if they will be actively involved with equipment. Managers who attend classes unnecessarily may inadvertently stifle questions from those who need the training, greatly reducing its benefits. Also, managers may ask questions that are beyond the scope of the class, causing the instructor to deal with the dilemma of losing the class' focus or convincing the questioner to wait after class. Keeping the disciplines separate can make courses more effective and pertinent to students' needs.

TRAINING CONTINUITY

To ensure that the program remains successful, training directors should do the following:

- Keep records of who attended which classes to track employee skill levels.

- Obtain attendee critiques of all classes and keep a composite record of them for use in determining which courses and instructors are most effective.
- Request that instructors prepare and submit course outlines before classes begin.
- Maintain a file of all course outlines, manuals, and other materials for future use or reference.
- Videotape classes whenever possible for subsequent refresher courses, as well as training new or transferred employees (close-ups of assembly and disassembly procedures are especially helpful);
- Periodically evaluate training program materials and techniques to ensure that the program is current and meets both employee and facility needs.
- Meet with employees and supervisors to assess the training's effectiveness.
- Offer training reviews every few months so plant personnel can ask questions based on their growing familiarity with the equipment and process needs.

EXAMPLE

Managers at an existing plant set goals to improve plant performance and reduce operating costs. The training director has studied the scope of the I&C system, classified the instruments involved, reviewed the employees' skill levels, and surveyed the availability of training courses, as well as their content and instructors.

Nine plant personnel need training: one superintendent, one lab technician/computer operator, one chief operator, two maintenance technicians (one senior and one junior), and four operators. The plant's I&C equipment consists of

- 20 control-panel instruments (15 recorders, four dissolved-oxygen controllers, and one residual-chlorine controller);
- 16 flow meters [one ultrasonic effluent flume; four orifice plates with differential pressure sensors (airflow); and 11 magnetic flow meters (one influent, two primary sludge, four return activated sludge (RAS), and four primary effluent)];
- 18 level sensors/transmitters [two ultrasonic; three capacitance; 10 level switches; and three pump-sequence controllers (influent, primary sludge, and RAS)];

TABLE 18.1 Training program courses.

Course	Duration	Instructor	Location
I&C system overview	1 day	Consultant	Plant
Panel instrument maintenance	4 hours	Instrument manufacturer	Plant
Controller theory and tuning	4 hours	Local college instructor	Community college
Magnetic flow meter maintenance	2 hours	Meter manufacturer	Plant
Ultrasonic flow meter maintenance	2 hours	Meter manufacturer	Plant
Air flow meter maintenance	2 hours	DP manufacturer	Plant
Flow meter basics	1 hour	Local instructor	Plant
Capacitance level maintenance	1 hour	Local representative	Plant
Pump sequence control maintenance	2 hours	Manufacturer's representative	Plant
Dissolved oxygen maintenance	2 hours	Manufacturer's representative	Plant
Residual chlorine analyzer maintenance	2 hours	Chlorine systems manufacturer	Plant
Combined gas detector maintenance	2 hours	Local representative	Plant
Transmitter operations training	4 hours	Training director	Plant
Dissolved oxygen probe operations	1 hour	Local representative	Plant
Computer operations	8 hours	Local college instructor/vendor	Plant
Safety	4 hours	Plant safety officer/consultant	Plant

TABLE 18.2 Training plan matrix.

Subject	Superintendent	Laboratory technician	Chief operator	Maintenance technician	Operator
Instrumentation and control system overview	X	X	X	X	X
Panel instrument maintenance				X	
Controller theory and tuning	X		X		X
Magnetic flow meter maintenance				X	
Ultrasonic flow and level meter maintenance				X	
Airflow meter maintenance				X	
Flow meter basics	X	X	X		X
Level measurement capacitance				X	
Pump-sequence controller maintenance				X	
Dissolved oxygen probe maintenance		X		X	
Residual-chlorine analyzer and leak detector maintenance		X		X	X
Combustible-gas detector maintenance		X		X	
Operator training transmitters/analyzers	X		X		X
Computer operations	X	X			
Safety	X	X	X	X	X

- nine analyzers (four dissolved oxygen probes, one residual-chlorine analyzer, three combustible-gas detectors, and one chlorine-leak detector); and
- One personal computer linked to the I&C system for basic operations data and monthly reports.

With this information in mind, the training director determines the courses to be offered to meet the treatment plant's goals (Table 18.1) and then generates an appropriate training-plan matrix (Table 18.2).

SUGGESTED READINGS

Instrumentation, Systems, and Automation Society (2004a) *Certified Automation Professional (CAP) Study Guide.* Instrumentation, Systems, and Automation Society: Research Triangle Park, North Carolina.

Instrumentation, Systems, and Automation Society (2004b) *ISA Catalog of Books, Standards, Video Tapes, Software, Proceedings and More ....* Instrumentation, Systems, and Automation Society: Research Triangle Park, North Carolina.

Jordan, J. K. (2000) *Maintenance Management for Water Utilities.* American Water Works Association: Denver, Colorado.

Index

A

Acceleration transducers, 219
Access levels, control system, 399
Acoustic wave touchscreens, 279
ActiveX drivers, 319
Actuators
    control valves, 234
    electric motor, 235
    hydraulic, 238
    pneumatic, 237
    solenoid, 234
Adapter modules, 420
Adaptive control, 343
Aerators, 243, 245
Alarm(s), 29, 259
    annunciator sequence, HMI, 357
    HMI software, 378
    process control narratives, 438
    sludge blanket height, 214
    stations, 263
    summary, HMI screens, 376
Ammonia analyzers, 192
    construction materials, 194
    installation, 194
    maintenance requirements, 194
Ammonia control, 460
Ammonium analyzers, 192
Amperometric bare-electrode analyzer, 186
Amperometric membrane-covered-electrode analyzer, 186
Analog
    displays, 257
    input/output, PLCs, 409
    modules, 421
Analyzers, on-line, 174

Animations, 438
Annunciators, 259, 357
ANSI-designated colors, HMI screens, 377
Application protocols, 301
Appurtenances, consoles, 392
Area
    control consoles, 262
    control panels, 262, 362
    control stations, 262
Artificial intelligence, 444
Artificial neural networks, 452
As-built documentation, 59
As-built drawings, 62
Asset management, 62
Audible alarms, HMI, 380
Auditory stimuli, HMI, 357
Autodialers, HMI alarms, 380
Automatic controls, remote, 437
Automatic process control, 331
Automatically generated lists, 62
Automation benefits, 8

B

Ball valves, 231
Bar graph displays, HMI, 360
Bellows, 166
Benchmark, IWA/COST simulation, 465
Bid documents, 33
Biological nutrient removal (BNR), 457
Bitmaps, HMI screens, 377
Block diagrams, control-system, 474
Blowers, 245
Breakfront panel, 354
Browser-based HMI, 368
Bubbler level-measurement devices
    accuracy and precision, 144

construction materials, 144
installation, 145
maintenance requirements, 146
operating principles, 143
Build as you go method, 506
Butterfly valves, 231

C

Cable list, 55
Calculations, HMI screens, 378
Calibration points, 94
Calibration standards, 93
Capacitance level-measurement devices
    accuracy and precision, 149
    construction materials, 149
    installation, 149
    maintenance requirements, 145
    operating principles, 146
Capacitive touchscreens, 278
Capacity, controller I/O, 423
Capital improvements, energy management, 443
Case-based reasoning systems, 449
Cathode-ray tube (CRT) monitors, 367
Central control room, HMI, 385
Chairs, control room, 393
Change orders, 36
Chlorine residual analyzers, 182
    accuracy and precision, 188
    installation, 188
    maintenance requirements, 189
Clark polarographic cell, 177
Class types, 528
    conferences and Web sites, 529
    home study, 528
    offsite classrooms, 528
    on the job, 528
    onsite classrooms, 528
Client, definition, 297
Closed-loop control, 9
Codes, control panel, 265
Collection systems, SCADA system, 404
Color standards, HMI screens, 377
Color, 107
Colorimetric analyzers, ammonia, 192
Colorimetric analyzers, chlorine residual, 183
Combined feedback–feedforward control, 343
Communications
    definition of terms, 297
    modules, 421
    network components, 300
    redundancy, 323
    schematics, 32
    software redundancy, 325
    tools, 57
Component-management tools, 61
Compressors, 245

Computer-aided design software, 53
Computer-based control, 345
Computerized displays, HMI, 365
Computerized maintenance management systems (CMMSs), 489
Conduit list, 55
Configuration schematics, 32
Connectivity, 297, 326
Connectors, 94
Conservation, energy, 442
Console panel, 354
Consoles
    area control, 262
    control room, 391
    main control, 263
Construction management, 41
Construction requirements, control panel, 266
Construction Specification Institute (CSI) specifications, 264
Construction, 36, 62
Contingency plans, 17
Continuous level-measurement devices, bubbler
    accuracy and precision, 144
    construction materials, 144
    installation, 145
    maintenance requirements, 146
    operating principles, 143
Continuous level-measurement devices, capacitance
    accuracy and precision, 149
    construction materials, 149
    installation, 149
    maintenance requirements, 150
    operating principles, 146
Contractors, 37
Contracts, 57
Control charts, instrument, 499
Control communications, 318
Control components, 436
Control consoles, 262
Control devices, HMI, 354
Control limits, instrument, 499
Control options, 337
    advanced, 343
    feedback, 338
    feedback–feedforward, 343
    feedforward, 342
    PID tuning, 340
    PID, 338
Control panels
    analog displays, 257
    annunciators, 259
    area, 262, 362
    data loggers, 289
    devices, 256
    digital displays, 257
    digital recorders, 289

distribution, HMI, 361
electrical systems, 269
environmental requirements, 267
factory testing, 272
flat-panel displays, 283
hazardous installations, 275
HMI software, 262
HMI, 354
identification, 267
installation, 274
local, 262
main, 263, 362
master, 362
monitors, 285
motor control centers, 270
operator interfaces, 258, 277
pilot lights, 260
power source, 269
pushbuttons, 261
remote annunciation, 290
safety concerns, 269
specifications, 264
surge protection, 272
switches, 261
thermal management, 268
touchscreens, 278
Underwriters Laboratory standards, 274
vendor-supplied, 262
wireless operator interface, 287
wiring, 271
Control, BNR process, 457
Control room, equipment layout, 390
Control room, HMI, 385
Control signal interface, 253
Control stations, types, 262
Control strategies,
controllability problem, 337
regulatory, 336
sequential, 336
Control system
collection system, 404
configuration, 347
connectivity, 326
security, 395
wastewater treatment plant, 407
Control theory, 332
Control valves, 227
actuators, 234
ball, 231
butterfly, 231
characteristics, 228
diaphragm, 231
gate, 231
globe, 229
plug, 231
positioners, 239

selection, 232
sizing, 233
solenoid, 230
standards, 232
types, 228
Controllability problem, 337
Controllers, 407
redundancy, 323
selection, 419
Control-panel designs, 32
Control-room alarms, HMI, 380
Controls
local, 437
remote automatic, 437
remote manual, 437
Control system
architecture diagram, 56
block diagrams, 474
documentation, 485
Conventional galvanic cell, 176
Copper twisted-pair wires, Ethernet, 307
Core sampler, sludge blanket level, 211
Coriolis flow meters
accuracy and precision, 139
construction materials, 139
installation, 140
maintenance requirements, 140
operating principles, 138
Corrosion, 83
COST simulation benchmark, 467
Cost–benefit ratio, 11
Costs, 10
controller maintenance, 423
controller program maintenance, 424
Counters, PLCs, 409
Course classifications
configuration, 527
maintenance, 527
management overview, 528
operations, 527
programming, 527
Cyber security, controller, 425

D

Damping, 70
Data
automation benefits, 9
collection intervals, 384
loggers, control panels, 289
storage, PLCs, 409
trending, 381
DDE drivers, 321
Dead band, 73, 384
Dead-time lag model, 335
Denitrification control, 462

Design
 considerations, 82
 experience, 36
 guidelines, controller, 422
 history, 62
 outline, 24
 process, 20
 services, 36
Device connectivity, software, 318
Diagnostics, RTUs, 415
Diaphragm
 actuator, 237
 seal systems, 166
 valves 231
Differential-pressure flow meters
 accuracy and precision, 130
 construction materials, 130
 installation, 131
 maintenance requirements, 131
 operating principles, 129
Differential-pressure level-measurement devices
 accuracy and precision, 152
 construction materials, 153
 installation, 153
 maintenance requirements, 155
 operating principles, 153
Differential-pressure transmitters, 166
Digital displays, 257
Digital-light processing (DLP) displays, 368
Digital-light processing (DLP) monitors, 285
Digital recorders, control panels, 289
Direct digital control, 343
Discrete modules, 421
Dispersive signal technology touchscreens, 282
Displacement pumps, 243
Displacement transducers, 220
Displays
 color and shapes, 377
 computerized, 365
 data trending, 381
 digital-light processing, 368
 flat-panel, 368
 graphic, 364
 internal electronics, 94
 PC-based, 367
 plasma, 368
 screen navigation, 371
 text, 365
 video projector, 367
Dissolved oxygen sensors
 accuracy and precision, 181
 DO measurement, 175
 installation, 181
 maintenance requirements, 181
 membrane design, 175
Distributed control system, 346

Distributed control units, 419
Distributed networked control system, 347
Diurnal characteristics, 107
Documentation, 98
DPD colorimetric analyzer, 183
Drawings, physical, 474
Drift, 74
Drivers
 ActiveX, 319
 DDE, 321
 definition, 298
 OPC standards, 321
 software, 318
Drives, electric motors, 250
Dry bulk conveyers, 254
Dynamic characteristics, 77

E

Electric motors
 actuators, 235
 alternating current, 248
 drives, 250
 starters, 250
Electrical concerns, 94
Electrical schematics, 56
Electrical systems, control panel, 269
Electricity, control room, 391
Electronic ice point compensation, 170
Electronic standards, 84
Elevation, 70
Embedded controllers, 420
Enclosed spaces, 86
Energy conservation, 9
Energy demand reduction, 445
Energy management, 444
Environmental concerns, 82, 84
Environmental constraints, controller, 424
Environmental requirements, control panel, 267
Equipment
 control room layout, 390
 descriptions, 435
 location drawings, 56
 SCADA security, 397
 tag numbers, 426
 troubleshooting, 521
Ergonomics, HMI, 390
Ethernet
 application protocols, 313
 encapsulation, 309
 modules, 421
 network, 306
 proprietary network bridges, 312
Event communications, HMI, 378
Expert systems, 446
External data sources, 61

F

Factory acceptance test (FAT), 37, 484
Factory testing, control panels, 272
Fans, 245
Feedback control, 338
Feedback–feedforward control, 343
Feedforward control, 342
Fiber optic cable, Ethernet, 307
Fiber optic cable, not Ethernet, 308
Fiber-optic-patch panels, 264
Fibers, 106
Field tests, 41
Fieldbus networks, 305
Field-instrument specifications, 55
Final acceptance, 42
Final design, 24
Fire ants, 87
Fire protection, control room, 390
Firewalls, 397
First-order lag model, 333
Flat-panel displays, 283, 368
Flow meters, 107
    differential-pressure, 129
    flumes, 121
    magnetic, 108
    mass, 137
    mass, coriolis, 138
    mass, thermal-dispersion, 140
    mechanical, rotary-element, 132
    mechanical, variable-area, 136
    ultrasonic, 115
    weirs, 121
Flumes
    accuracy and precision, 127
    construction materials, 127
    installation, 127
    maintenance requirements, 128
    operating principles, 121
Flush-mount vibration sensors, 222
Free-standing vertical panel, 354
Furniture, control room, 391
Fuzzy logic, 451

G

Galvanic sensor, 180
Gas-selective electrodes, ammonia, 193
Gate valves, 231
Globe valves, 229
Goals, 18
Graphic displays, object-oriented programming, 371
Graphic displays, proprietary, 365
Graphic panels, 263
Graphic panels, HMI, 364
Graphic terminals, 258
Graphics area, HMI screens, 376

Grease, 106
Grounding, 96

H

Hair, 106
Hardware testing, 38
Hardware, RTUs, 412
Hardwire compensation, 168
Hazardous installation, control panels, 275
Heaters, 254
Historical trending, 381, 384
HMI software redundancy, 326
Home study, 528
Horns, 259, 357
Human engineering, HMI, 385
Human static anthropometric data, 387, 388
Human–machine interface (HMI), 349
HVAC systems, 62
HVAC, control room, 391
Hydraulic actuators, 238
Hydrogen sulfide, 107
Hysteretic error, 73

I

I/O server, definition, 298
Identification, control panel, 267
Impedance level-measurement devices
    accuracy and precision, 149
    construction materials, 149
    installation, 149
    maintenance requirements, 150
    operating principles, 146
Incorporating technologies, 35
Indophenol blue method, 193
Inference engine, expert systems, 448
Infrared touchscreens, 278
In-plant PCS, 404
Input relays/coils, PLCs, 408
Input/output
    capacity controller, 423
    list, 54
    modules, 420
Installation, 33
Installation, control panels, 274
Instructors, 529
Instrument
    characteristics, 69
    classifications, 67
    list, 54, 474
    loop diagrams, 32, 56
    performance, 498
    redundancy, 323
Instrumentation
    diagrams, 25, 49, 57, 59
    maintenance, 487
    training, 526

Integrated circuit sensors, 172
Integrated I/O modules, 420
Integration, 61
Intelligent decision-support systems, 453
Intelligent process diagrams, 59
Interactive process models, 437
Interface/sludge blanket level (ISBL) analyzers, 211
    accuracy and precision, 216
    construction materials, 215
    installation, 216
    maintenance requirements, 217
    operating principles, 212
Intermittent aeration, 464
Internal electronics, 93
Internal utility relays/contacts, PLCs, 408
Interoperability, 61
Ion-selective electrodes, ammonia, 193
Ion-selective electrodes, nitrate, 195
Ion-sensitive field effect transistor (IsFET), 199
IWA/COST benchmark simulation, 465

K

Keys to success, 13
Knowledge base, expert systems, 446

L

Labor, 9
Layout
    control room equipment, 390
    HMI, 354
    HMI screens, 375
Level-measurement devices, 141
    continuous, 141
        capacitance, 146
        differential-pressure, 150
        impedance, 146
        microwave, 160
        sonic, 156
        ultrasonic, 156
    point, 164
Lighting, control room, 391
Linear value conversion, 78
Linearity, 74
Liquid crystal display (LCD) monitors, 367
Local control panel, HMI, 361
Local controls, 435
Local HMI stations, 382
Logs, software, 398
Loop drawings, 474
Loop testing, 41
Loop-controller displays, HMI, 360
Luminescent DO sensor, 180

M

Magnetic flow meters, 108
    accuracy and precision, 110
    construction materials, 110
    installation, 112
    maintenance requirements, 114
    operating principles, 108
Main control panel (MCP), HMI, 362
Maintenance, 16
    controller, 423
    management program, 502
    management system, 488
Management support, 14
Manual controls, remote, 435
Manual process control, 330
Mass flow meters, 137
Master control panel, HMI, 362
Master, definition, 299
Measurement method, 69
Mechanical flow meters, 132
Media, definition, 299
Membrane sensors, dissolved oxygen, 175
Memory, controller program, 423
Microprocessor-based control, 345
Microwave analyzers, suspended solids, 207
Microwave level-measurement devices
    accuracy and precision, 161
    construction materials, 161
    installation, 161
    maintenance requirements, 162
    operating principles, 160
Mixers, 243, 245
Modeling, 333
Modular I/O modules, 420
Modular method, maintenance management program, 505
Modular system, 61
Moisture, 83
Monitored variables, 28
Monitors
    cathode-ray tube (CRT), 367
    control panel, 285
    liquid crystal display (LCD), 367
Monochloramine-F method, 193
Motor control center, 270, 435
Mounting details, 474
Multiple-loop controller, 347

N

Narrative, process control, 425
National Electric Code (NEC), control panel 265
Navigation tools, 61
Navigation, HMI screens, 375
Near field imaging touchscreens, 279
Near-infrared solids analyzers, 207
NEMA rating, control panel, 267
NEMA requirements, control panel, 265
Network
    bridges, 312

components, 300
    interface redundancy, 324
    transport protocol, 300
Neural networks, 452
New developments, 59
Nitrate analyzers
    accuracy and precision, 196
    construction materials, 196
    installation, 196
    maintenance requirements, 196
Nitrate/nitrite analyzers, 195
Nitrogen analyzers, 192
Nondisplacement pumps, 245
Nonflush-mount vibration sensors, 223
Nonlinear value conversion, 79
Nonlinear, non-zero based conversion, 80
Nonpoint sensors, 68

O

Object-based schematics, 60
Object-server architecture, 277
Offset, 70
On-the-job training, 528
On-line analyzers, 174
On-line solids analyzers, 205
    accuracy and precision, 210
    construction materials, 210
    installation, 210
    maintenance requirements, 210
    operating principles, 207
OPC standards-based drivers, 321
Operation, PLCs, 410
Operator interface
    control panels, 277
    graphics, 58
    panel devices, 258
    terminals, 40
Operator involvement, 15
Operator training, 58
Optical ISBL analyzers, 213
Optical suspended solids analyzers, 205
Orthophosphate analyzers, 204
    accuracy and precision, 204
    construction materials, 204
    installation, 204
    maintenance requirements, 204
    operating principles, 204
Output relays/contacts, PLCs, 409
Overrange, 70
Owner feedback, 58
Owner involvement, 15
Oxidation–reduction potential (ORP) probes, 187

P

Pagers, HMI alarms, 381

Panels
    devices, 256
    list, 55
    specifications, 55
    fiber-optic-patch, 264
    graphic, 263, 364
    termination, 263
PC-based displays, 284, 367
PC-based terminals, 258
Peer-to-peer, definition, 299
Percent-recovery control chart, 499
Performance tests, 484
Periodicals, 36
pH sensors, 196
    accuracy and precision, 202
    installation, 202
    operating principles, 197
Phosphorus analyzers, 204
Physical drawings, 474
Physical security, control room, 390
Physical transport media, 300
PID controller, 338
PID tuning, 340
Piezo-resistive accelerometers, 219
Pilot lights, 260, 357
Piston actuator, 238
Plasma displays, 285, 368
Plastic products, 106
PLC functions, 435
Pneumatic actuators, diaphragm, 237
Pneumatic actuators, piston, 238
Point level monitors
    construction materials, 165
    installation, 165
    maintenance requirements, 165
    operating principles, 164
Point sensors, 68
Polymer dose control, 189
Portable standards, 93
Portable terminals, 259
Positioners, valve, 239
Post-design phase, 36
Power monitoring, 442
Power source, control panel, 269
Power supply, 31, 94, 424
Predesign, 23
Pressure sensors
    accuracy and precision, 167
    construction materials, 167
    installation, 167
    maintenance requirements, 168
    operating principles, 166
Primary standards, 93
Printed circuit cards, 94
Printers, HMI alarms, 380
Problem definition, 512

Problem isolation, 512
Process alarms, HMI, 378
Process and instrumentation diagram (P&ID), 435, 474
Process control
    benefits, 332
    BNR, 457
    narrative, 25, 55, 425
    objectives, 330
    system configuration, 32
Process descriptions, 435
Process diagrams, 49, 57
Process failure, 17
Process identification, 333
Process modeling, 333, 437
Process performance, 8
Product specifications, 483
Programmable logic controllers (PLCs), 407
Programming language, controller, 424
Programming, object-oriented displays, 371
Programming, PLCs, 411
Project structure, 20
Proportional–integral–derivative (PID) controller, 338
Proprietary graphical displays, 365
Proprietary networks, communications, 303
Protocol, definition, 299
Pumps, displacement, 243
Pumps, nondisplacement, 245
Purged enclosures, control panel installation, 275
Pushbutton stations, 31
Pushbuttons, 261, 359

Q

Quality assurance, 87
Quality control, instrument evaluation, 498

R

Range, 69
Rangeability, 70
Rate of return, 13
Real-time display, HMI, 354
Real-time trending, 381
Reasoning systems, case-based, 449
Recorders, data trending, 381
Recorders, digital, 289
Redundancy, communications, 323
Reflectance analyzers, suspended solids, 207
Refresh rate, HMI screens, 378
Regulations, control panel, 265
Regulatory process control, 336
Regulatory requirements, 8
Reliability, 8
Remote annunciation systems, 290
Remote controls, 435
Remote indicators, 31
Remote terminal units (RTUs), 412
Remote-facility PCS, 404

Repeatability, 75
Report preparation, 513
Reports, 9
Reset function, 30
Resistance temperature detectors (RTDs), 171
Resistive touchscreens, 278
Resource conservation, 9
Respirometry, 463
Risk reduction, 10
Ross polarographic cell, 179
Rotary-element flow meters
    accuracy and precision, 134
    construction materials, 134
    installation, 134
    maintenance requirements, 135
    operating principles, 132
RS-232, definition, 299
RS-422/485, definition, 299

S

Safety, 523
Safety, control panel, 269, 275
Sampling intervals, 384
Sampling method, 68
Sandwich displays, 364
SCADA
    collection systems, 404
    cyber security, 395
    software redundancy, 326
Screen layout, HMI, 375
Screen navigation, HMI, 375
Screen refresh rate, HMI screens, 378
Second-order lag model, 334
Security
    control system, 393
    controller, 425
    cyber, 395
    network, 395
    personnel, 393
    physical, 390
    policies, 400
SCADA equipment, 397
    software operating system, 398
    vulnerabilities, 393
Seebeck effect, 168
Sensitivity, 75
Sensors
    characteristics, 105
    chlorine residual, 182
    dissolved oxygen, 175
    nitrogen, 192
    pH, 196
    phosphorus/orthophosphate, 204
    pressure, 166
    solids analyzers, 205
    vibration, 218

Sequencing batch reactors (SBRs), 464
Sequential process control, 336
Serial network, communications, 301
Serial, definition, 299
Server, definition, 299
Setpoints, BNR process, 460
Shewart control chart, 499
Shielding, 96
Shop drawings, 36
Short-circuits, 83
Signal I/O lists, 474
Signal types, 67
Simulation software, 39
Single-loop controller (SLC), 346, 420
Site access, HMI, 398
Site demonstration test (SDT), 484
Site visits, 37
Size, RTUs, 415
Slave, definition, 299
Sludge blanket level detector, 211
Sludge core sampler, 211
Smart process instrumentation, 97
Smart pumps/drives, 253
Smoothing, data collection, 385
Soft programmable logic controllers, 412
Software
    alarms, 379
    backup procedures, 398
    change logs, 398
    connectivity, 318
    failure recovery, 398
    HMI, 262
    operating system security, 398
    redundancy, 325
    remote terminal units, 412
    testing, 39
    virus-protection, 398
    wireless communications, 317
Softwire compensation, 168
Solenoid actuators, 234
Solenoid valves, 230
Solids, 106
Sonic level-measurement devices
    accuracy and precision, 158
    construction materials, 158
    installation, 158
    maintenance requirements, 159
    operating principles, 156
Span error, 72
Span, 70
Specialty modules, 421
Specifications
    control panel, 264
    design information, 33
    I&C system, 473
    process, 474

    RTUs, 416
Standards
    control panel, 265, 274
    control valves, 232
    HMI screen colors, 377
    HMI software configuration, 370
    OPC, 321
    process control system, 346
    RTUs, 416
    specification forms, 474
    testing and quality assurance, 91
Starters, electric motors, 250
Startup, 62
Static characteristics, 71
Static process models, 437
Status-indicating lights, HMI, 357
Steady-state model, 333
Streaming current monitor (SCM), 189
    construction materials, 191
    installation, 192
    maintenance requirements, 192
    operating principles, 190
Streamlined diagram development, 60
Strobe lights, 260, 359
Sulfides, 107
Supervisory setpoint control system, 347
Suppliers, 37
Surge protection, control panels, 272
Suspended solids analyzers, 205
Suspended solids profile, 213
Swelling, 83
Switches, 261, 359
Symbol-management tools, 61
Symbols, 51
System
    advocate, 13
    alarms, HMI, 378
    availability test (SAT), 484
    navigation bar, HMI screens, 376
    observation, 512
    performance, 38
    testing, 16, 41

T

Tag numbers, 28, 426
Technical societies, 33
Technology selection, 35
Telephone-controller unit, HMI alarms, 380
Temperature, 82
    controls, 87
    sensors, 168
        accuracy and precision, 172
        construction materials, 169
        installation, 173
Terminal strip, 94
Terminals, operator interfaces, 258

Termination panels, 263
Testing, 87
Testing, reports, 88
Text displays, 365
Textbook method, maintenance management program, 502
Thermal bulbs, 172
Thermal management, control panel, 268
Thermal-dispersion flow meters
    accuracy and precision, 141
    construction materials, 140
    installation, 141
    maintenance requirements, 141
    operating principles, 140
Thermistors, 172
Thermocouples, 168
Thin client workstations, 277
Timers, PLCs, 409
Total phosphorus analyzers
    accuracy and precision, 205
    construction materials, 205
    installation, 205
    maintenance requirements, 205
    operating principles, 205
Touchscreens, control panels, 278
Traceable standards, 93
Training, 17, 35, 487, 526–531
Transient-voltage surge suppressors, 95
Transport, definition, 300
Transportation lag model, 335
Trending, HMI, 381
Troubleshooting, 511
Turnkey method, maintenance management program, 506

U

Ultrasonic flow meters, 115
    accuracy and precision, 116
    construction materials, 116
    installation, 119
    maintenance requirements, 121
    operating principles, 116
Ultrasonic ISBL analyzers, 212
Ultrasonic level-measurement devices
    accuracy and precision, 158
    construction materials, 158
    installation, 158
    maintenance requirements, 159
    operating principles, 156
User feedback, 57
UV absorbance probes, nitrate, 195

V

Value conversions, 77
Valve positioners, 239
Valve-position switches, 29
Variable blanket height, 213
Variable-area flow meters
    accuracy and precision, 137
    construction materials, 136
    installation, 137
    maintenance requirements, 137
    operating principles, 136
Variable-frequency drives, 250
Variable-torque speed control, 252
Velocity transducers, 218
Vendors, 34
    control panel, HMI, 361
    equipment, 31
    panels, 262
Vibration sensors
    accuracy and precision, 221
    construction materials, 220
    installation, 222
    maintenance requirements, 224
    operating principles, 218
Video projector displays, 367
Virus-protection software, 398
Visual stimuli, HMI, 357
Voice annunciation, HMI alarms, 380
Voltage spikes, 94
Voltage-mode accelerometers, 219

W

Wall-mount vertical panel, 354
Wastewater characteristics, 105
Wastewater treatment plants, process control system, 407
Web connectivity, 326
Web sites, 36
Weirs
    accuracy and precision, 127
    construction materials, 127
    installation, 127
    maintenance requirements, 128
    operating principles, 121
Wireless
    communications software, 317
    connectivity, in-plant, 314
    connectivity, remote sites, 314
Ethernet, 309
HMI, 369
    modules, 422
    operator interface, control panels, 287
    reliability, 316
    security, 315
Wire-wound motor control, 251
Wiring diagrams, 25
Wiring, control panels, 271

Z

Zero error, 72
Zero suppression, 70
Zeta potential, polymer dose, 189
Zullig sensor, 180